River
plants

River
plants

The macrophytic vegetation of watercourses

S. M. HASLAM
BOTANY SCHOOL, CAMBRIDGE

ILLUSTRATED BY
P. A. WOLSELEY

CAMBRIDGE UNIVERSITY PRESS

CAMBRIDGE

LONDON · NEW YORK · MELBOURNE

Published by the Syndics of the Cambridge University Press
The Pitt Building, Trumpington Street, Cambridge CB2 1RP
Bentley House, 200 Euston Road, London NW1 2DB
32 East 57th Street, New York, NY 10022, USA
296 Beaconsfield Parade, Middle Park, Melbourne 3206, Australia

First published 1978

Printed in Great Britain by
Cox & Wyman Ltd,
London, Fakenham and Reading

Library of Congress Cataloguing in Publication Data

Haslam, Sylvia Mary, 1934–
River plants.

Bibliography: p.
Includes index.
1.   Stream flora.   I.   Wolseley, P. A.   II.   Title.
QK932.7.H37   581.5′2632   76–46857

ISBN 0 521 21493 9 hard covers
ISBN 0 521 29172 0 paperback

TO DR A. S. WATT, FRS

Contents

Contents

Contents

Plates 1–27 are between pp. 210–211

Preface

Water plants have received much less attention from botanists than have land plants, and the plants of streams have had the least attention. Arber's *Water plants* [1] was the pioneer book on the larger aquatic plants, and describes how plants' structures are adapted to life in the water. Sculthorpe drew together much information in *The biology of aquatic vascular plants* [2]. The early research on ecology, on how and why plants grow in some places and not in others, is summarised in Tansley's *The British Islands and their vegetation* [3]. Since then, the water plants of lakes have received considerable attention, e.g. by Spence [4], who pointed out the importance of nutrients and water movement in controlling plant distribution. Butcher's classic papers on river plants [5, 6] described different types of stream vegetation, and the importance of currents, soil, shading, etc. in forming plant communities. Careful surveys of river vegetation have been carried out by Whitton and his colleagues, while Westlake and his colleagues have studied the physiology and productivity of chalk stream plants (see Bibliography).

This book describes the hitherto unpublished research of the author, except where the contrary is specifically stated. The parts of Chapter 8 relating plants to the nutrient status of streams are taken from the unpublished work of Mrs M. P. Everitt. It would not have been possible to write this book without the unfailing helpfulness of the Engineering, Fisheries and Pollution Prevention Departments of the River Authorities (now Water Authorities) of England and Wales, the River Purification Boards of Scotland and the Drainage Boards of England. The River Authorities, in particular, have provided what must have seemed to them endless information on plant distribution, channel maintenance, control of excessive plant growth, water analyses, effluents and pollution. To all, I express my sincerest thanks.

To make this book more easily understood by those who are not professional botanists:

(1) The contents of each chapter (except for Chapter 1) are summarised at the start of each.

(2) Towards the end of Chapter 1 is a section listing the most important river plants, classifying them according to their characteristic habitats, and illustrating them to show their typical appearance in streams.

(3) Small symbols are used, wherever possible, to represent the commonest species, both British and, in Chapters 15–17, North

xi

America. Symbols appear beside species names in the text and in place of species names on the river maps. Symbols for British species are listed in Chapters 1, 11 and 22, and for the North American species in Chapter 15. They show some characteristic feature or habit of the plant, and should help to relate unfamiliar plant names to the relevant river plants.

(4) Figures have been provided wherever practical.

(5) A Glossary is provided at the end of the text.

Streams are complex and diverse, and a book of this type cannot include all possible variations in vegetation, so plants can be found away from the habitats which are described here as characteristic. The term 'plant' is used for the larger plants of watercourse, the flowering plants (angiosperms), horsetails and water fern (pteridophytes), mosses (bryophytes) and the two groups of large algae occurring in watercourses – the stoneworts (Characeae) and *Enteromorpha*. The mosses are not otherwise named or distinguished. The smaller algae are usually excluded from the general text, but the microscopic floating algae, the phytoplankton, are referred to as causing turbidity in water and as forming part of the living entity of stream vegetation. In the illustrations of the vegetation along whole rivers, in Chapters 11 and 22, two other categories of algae are used, namely blanket weed (for filamentous algae sufficiently long and dense to be easily seen from above the water) and benthic algae (here defined as algae growing on the channel bed in sufficient quantity to be easily seen as green areas from above the water but not trailing away from the bed as does blanket weed). Only especially relevant references are cited in the text, but a list of additional references is given in the Bibliography.

I am much indebted to Mr F. G. Charlton, Mr W. H. Palmer and Mrs P. A. Wolseley for their careful reviews of the typescript; to Mr P. F. Barrett, Dr F. H. Dawson, Dr P. J. Grubb, Dr N. T. H. Holmes, Mr T. O. Robson, Professor S. G. Smith and Dr B. A. Whitton for many improvements; and to Mr D. F. Westlake for permitting the use of an unpublished manuscript. I am very grateful for the technical assistance of Mrs M. Ellis, Mrs J. Hayes, Mrs S. M. Hornsey, Miss M. Steiner, Miss J. S. Whiteside and Mr R. W. Worland, and for the careful typing of Mrs A. Hill.

The research described in this book was financed by the Natural Environment Research Council, the Commission of the European Communities (under Contract 079-74-1 ENV UK) and the Phyllis and Eileen Gibbs Travelling Fellowship of Newnham College, Cambridge, to whom I express my appreciation.

July 1976   S. M. H.

I

Introduction

Plants and animals, large and small, live in, on and beside streams and man-made channels. This book describes the larger plants, mainly the angiosperms (flowering plants). Streams are complex habitats, ranging from mountain torrents to quiet, still lowland waters, and may be deep or shallow, large or small. These differences, and many more, affect the vegetation. The first part of this book describes the effect of different physical features of the river on the plants. These features are often associated: for instance a swiftly flowing stream will also have a stony substrate, be sited in the hills and be liable to frequent fierce storm flows from heavy rainfall. In order to understand and interpret the plant community, however, it is necessary to have a basic understanding of all these features and so the early chapters separate out single factors or simple combinations of factors, such as channel width and channel width plus slope. Inevitably, there is some overlap between these chapters, since the vegetation of, for example, a large slow-flowing river is controlled by the slow water movement, silty bed, deep water and wide channel, and so is described in relation to each of these facets of the habitat.

The most important physical variables are:

(1) the water movement, the quantity of water, or **flow**;
(2) the soil or **substrate** on the bed of the watercourse;
(3) the **width** of the channel;
(4) the **depth** of the channel;
(5) the general position of the channel in the river system, its **drainage order**;
(6) the downwards **slope** or gradient of the channel.

In Chapter 2 flow and substrate are considered (as independent factors), with types of flow and substrate being described and then linked to the plants characteristic of those types. Chapter 3 considers how flow and substrate, singly and in combination, affect plants: the effects of erosion, battering, etc. and the responses of different plant species to these. In Chapter 4 three more factors – width, depth and drainage order – are, as in Chapter 2, described and linked to plants characteristic of different degrees of these. Chapter 5 takes up the wider aspects of flow, and considers the effects of storm damage in different streams, and the different streams and different annual patterns of flow that occur in Britain. The last chapter in this group, Chapter 6, again widens the picture by discussing the integration of the flow and substrate characters in the combination of width and slope known as the **width–slope pattern**. The habit and distribution of the different common river plants are closely linked to these width–slope patterns, and much of the interpretation of plant behaviour in relation to physical factors is given here.

The next two chapters consider habitat factors not here linked to the earlier ones: Chapter 7 describes the effect of **light**, since trees, etc., may reduce the light reaching the water, and deep or turbid water may prevent light reaching the channel bed; and Chapter 8 discusses the plant **nutrients**, the substances in the soil and water which are needed for plants to grow well, including the **dissolved gases** in the water (oxygen and carbon dioxide) needed for metabolism.

Chapter 9 is independent, as it describes **plant production** in rivers, that is, how much vegetation is produced in different streams.

The emphasis then moves to the **plant communities**. First, in Chapter 10, **small-scale plant patterns** are described. These are often due to variations in the factors discussed in Chapter 2 to 7, and are linked back to these. Chapter 11 increases the scale of the plant pattern by dealing with the **downstream variation** found between the source of a stream and its mouth; i.e. the changes with increasing size and decreasing water movement.

Once all these natural reasons for plant and vegetation behaviour are understood, a picture can be given of the **general vegetation** of British water courses. Chapter 12 describes communities of streams on soft rocks, Chapter 13, those on hard rocks, and Chapter 14, those of dykes and canals.

Interpolated next are three chapters, 15 to 17, on North American vegetation, describing the vegetation in the temperate East and Mid-west, and discussing some reasons for the differences between the plant communities there and in Britain.

The final chapters consider river plants from the point of view of man. Their **uses**, and recommendations for the optimum amount of vegetation in different stream types are given in Chapter 18. The practical objection to river plants is the **flood hazards** they cause, and Chapter 19 describes both where and how these occur, and the amount of vegetation which is safe in different stream types. Abstraction, water regulation and water transfers all **alter flow patterns**; their effects are discussed in Chapter 20. Chapter 21 describes the various forms of channel **management** (dredging, cutting, herbicides, etc.) and their effects on the vegetation. Finally, Chapter 22 describes the effects of **pollution** on river plants, and how plants can be used to monitor and assess the general health of river systems.

As mentioned earlier, with such a complex and interrelated subject, some overlap between the material in different chapters is necessary. The principle adopted is that information is always included in a chapter when it is needed in order to understand the point or argument

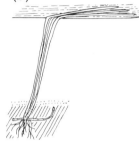

Fig. 1.1

being developed; cross-references are given to related but less directly pertinent information.

TYPES OF RIVER PLANTS

Watercourse plants must, by definition, tolerate or indeed prefer having water round some or all of their parts. Aquatic plants have been grouped in many ways (e.g. [2]), but a simple classification for river plants is as follows:

1. *Leaves and stems within, or floating on the water.* They are flexible, moving and bending with the current (Fig. 1.1*a* and *b*).

(*a*) Leaves and stems submerged under water and usually anchored to the soil by roots. The gases needed for photosynthesis and respiration come from the water, while mineral nutrients come mainly from the soil but partly from the water. These plants are particularly vulnerable to strong poisons in the water because a large proportion of the plant is submerged and the leaves are usually translucent, thin and delicate.

(*b*) Leaves floating flat on the surface of the water and stems usually rooted in the soil, though some species are free-floating. Gases can be taken up from and released into the air through the upper surfaces of the leaves, and so photosynthesis is not likely to be slowed down by a lack of carbon dioxide in the water. Water transpires (evaporates) into the air, so is pulled into the upper leaf from below. Floating leaves are usually opaque with waxy cuticles on the upper surface and are flexible enough to lie flat on the surface of the water.

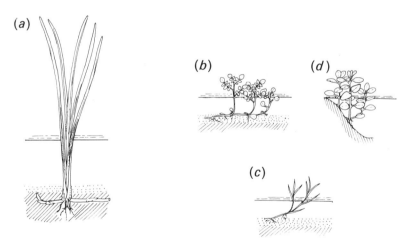

Fig. 1.2

2. *Upper leaves or shoots able to grow above the water, lower ones able to grow submerged (Fig. 1.2a, b, c and d).* The leaves and stems are stiff enough to stay erect without being supported by water. The leaves are opaque and most, but not all, can live under water. Gas exchange is mainly through the air but is possible through the water. Water transpired from the

upper leaves is replaced by water drawn up from below the stems. Plants falling into this group are:

(*a*) Tall monocotyledons – grasses, sedges and rushes.

(*b*) Short dicotyledons – the fringing herbs – typically found fringing the sides of small brooks.

(*c*) Short monocotyledons, mainly grasses, again frequently at the sides of brooks. A smaller group than (*b*).

(*d*) A very few tall dicotyledons.

3. *Bank plants growing above normal water level, flooded after heavy rain (Fig. 1.3)*. These range from emerged aquatics to true land species, and they do not necessarily have any special adaptations for aquatic life.

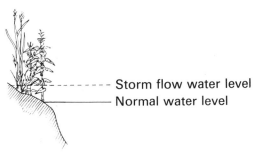

Storm flow water level
Normal water level

Fig. 1.3

Some species occur in only one of these groups, e.g. duck weed (*Lemna minor* agg.) in group 1(*b*) and nettle (*Urtica dioica*) in group 3, while others may occur in several, e.g. strapweed (*Sparganium emersum*) in groups 1(*a*), 1(*b*) and 2(*c*).

HISTORICAL CHANGES AFFECTING STREAMS

Most of Britain was wooded for several thousand years before man started felling trees, but for many centuries trees have been removed from near streams, formerly to make open land for pasture and arable farming and nowadays to allow large machines access to channels for maintenance. Streams still flow through woodland, though, and this is well illustrated in parts of North America. Plants grow well in the frequent breaks in the canopy, where patches of sunflecks or open sunlight reach the water, and there are as many species as there are in fully open British streams. Along large rivers the trees shade only the edges of the channels, but plants cannot usually grow in the centre because the depth of the water prevents enough light from reaching the river bed.

In addition to forest clearance another important change over the years is that the water level in lowland Britain has been lowered as wet woodland, swamps, water meadows, flood meadows, etc. have been largely replaced by well-drained farmland. More recently, ground water has been abstracted for domestic water supplies, and this further lowers the water table. In parts of North America though, the water table is still high, and road-side ditches, unlike those in Britain, still contain water plants. Swamps are also frequent in lower ground and many streams rise in the lowland tree swamps (Plate 1). Here the ground level is uneven, stream flow negligible and the water often brown-stained from the peat. Downstream the swamp becomes channelled and eventually turns into a slow-flowing brook. On higher, drier ground, in contrast, the brooks arise as gullies which carry flood water. This is too infrequent to allow aquatic plants to live in the channels and too swift to permit land ones to become established. Where springs rise on limestone there is perennial flow from the source, but this too may be swampy. In high-rainfall areas in both Britain and North America, rivers start as flood rills running down hills, which soon merge and form perennial streams.

In lowland Britain, channels have been managed by man for many centuries and there is a network of ditches between fields and beside roads on poorly drained soils, particularly in clay country, and, to a lesser extent, elsewhere. Because of the drop in the water table most of these ditches are now dry, though, even after storms. There is no clear boundary between these ditches and the upper part of the stream. As it goes from mouth to source, the channel first bears flood water in winter, then has intermittent puddles in summer and flow in winter and, finally, perennial flow. In the past many of these sources would have been tree swamps, and when the ditch system was made it must have contained water. Thus there has been more change in the upstream than the downstream reaches of these streams (also see [7]). Management over the centuries has much decreased flooding from the streams on to the land beside them, but such flooding hardly affects the river plants (unless it alters the amount of scouring). Flow patterns have been artificially altered by falling water tables, culverts, reservoirs, weirs, mills etc., and also for navigation purposes and to provide urban water supplies. The effects of these changes on the plants are described in later chapters.

In the mountains the course of a stream must necessarily follow the V- or U-shaped valleys. In a wide lowland valley, however, the channel could be anywhere on the alluvial plain, and indeed may have been in several different positions in the past. Its present position is likely to be firmly established by channel maintenance, and to have been fixed partly by chance, partly by the location of built-up areas, partly in order to free land for farming and partly by the boundary between properties. Some rivers are braided, having two or more channels. Braiding depends on the physical features of the river, but where there are several channels these can be used for different purposes, e.g. navigation and domestic water supply. Different flow patterns resulting from moving the position of a channel, or from using braided rivers for different purposes, may lead to differences in the river vegetation (see later chapters).

Most towns and villages were sited on streams so that they were provided with a water supply. Consequently sewage and other effluents

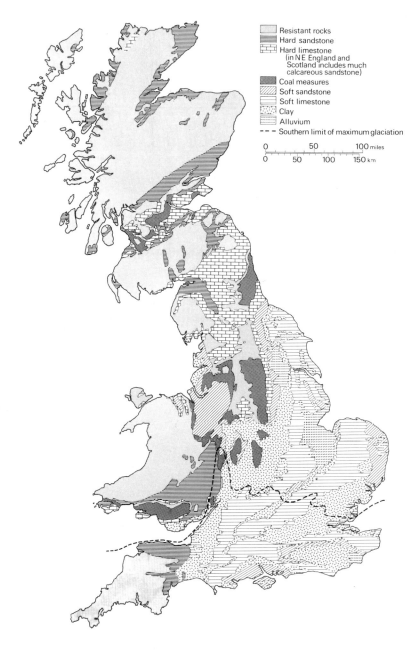

Fig. 1.4. The main rock types of Britain. British geology is complex, and this diagram has been simplified by omitting small areas of differing rock types, and where several types occur close together by amalgamating these and labelling them as the type which affects stream vegetation the most. Thick Boulder Clay, deposited in the Ice Age, covers some land north of the line marking the limit of the last glaciation. Streams on thick Boulder Clay over soft limestone or soft sandstone are, vegetationally, clay streams. A detailed geological map should be consulted for field investigations. The alluvial plains in Yorkshire are less flat than those elsewhere and have only local areas with the dyke vegetation characteristic of the more southerly alluvial plains. (Modified from a copyright map by George Philip & Son Ltd.)

6

were discharged into the rivers, and towns in lowland areas released raw sewage into rivers until very recently (e.g. until the 1960s along parts of R. Great Ouse). Treated sewage effluent is of course far less damaging than raw sewage, but, unless much diluted, still affects the river plants. Industrial effluents are very harmful in some regions too, particularly in coal-mining areas and near Birmingham. Farming also affects river plants: changing the vegetation and soil texture affects the pattern of storm flow; the ploughing regime affects the amount of silt entering the stream, and thus its substrate and turbidity; and fertilisers and pesticides may also be washed in the stream.

STREAM TYPES

The vegetation types of watercourses may be classified according to the flow pattern of their channels and the geology of their catchment area. Rock type influences topography to some extent and affects the chemical composition of the substrate and water, the amount of sediment entering the stream and the physical composition of the substrate. Fig. 1.4 shows the main rock types of Britain, grouped as they affect the river plants. The basic groups are the relatively recently formed soft rocks and alluvium, and the hard rocks which are mostly Palaeozoic or Pre-Cambrian. The soft rocks are divided into limestone, sandstone and clay, and the hard ones into limestone, sandstone, Coal Measures and the Resistant rocks. The Resistant rocks comprise those rocks (excluding Coal Measures) which are resistant to both erosion and solution, and include schists, gneisses, (hard) shales, slates, andesites, granites, felsites and basalts. The general topography of Britain is shown in Fig. 1.5 and the rainfall in Fig. 1.6. Rainfall and hills are both higher in the north and west, on the harder rocks.

Flow patterns are determined by a complex of factors. The dykes and drains of alluvial plains such as the Fenland and Romney Marsh, together with the canals, are an extreme case, for the channels are nearly flat and water movement is little without sluices and pumps, or, in the case of canals, locks. Vegetation can be abundant. In the other topographical types water is free-flowing, draining under gravity. The vegetation is affected by the force of the water flow and its variations, these in turn being determined by:

(1) height of the land, of the hills in general;
(2) height of the hills above the stream, i.e. the fall from hill to channel;
(3) downstream slope of the channel;
(4) seasonal distribution of rainfall;
(5) total annual rainfall;
(6) porosity of land surface, i.e. how much of the rainfall runs to the streams, and how much passes through the soil and rock to replenish the ground water below;
(7) distribution of springs.

7

Fig. 1.5. The general·
topography of Britain.
(Modified from [9].)

Fig. 1.6. The rainfall of
Britain. (Modified from [9].)

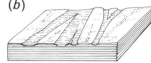

Fig. 1.7. Lowland streams. (a) Single channel; (b) braided channel.

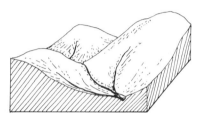

Fig. 1.8. Mountain stream.

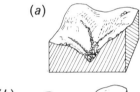

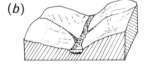

Fig. 1.9. Downstream changes in a hill stream. (a) Source, with steep slopes and shallow, small, bouldery channels. (b) Middle region, with gentler slopes and a larger, more gravelly channel. (c) Lower region, with a large, winding silted, channel in a flood plain.

The most important of these are usually the fall from hill to channel and the channel slope, while the total annual rainfall is usually the least important factor. Local conditions, however, frequently alter these priorities.

Areas classed as lowlands have hills that are not over about 800′ (250 m), the height of the hills above the stream channel is usually less than 200′ (60 m) and the slope of the channels near their sources can be as little as 1:100 (Fig. 1.7). Vegetation in the streams can be abundant, and although storm flows do some damage to plants they are not strong enough to sweep the vegetation away or to move boulders. Upland streams are found where the hills are usually 800–1200′ (250–375 m) high, there is a fall of around 400′ (125 m) from hill top to stream channel, and the slope of the upper tributaries is usually between 1:40 and 1:80. The vegetation may become as thick as in the lowlands, but upland streams are liable to the fierce storm flows termed spates, which, when particularly severe, can sweep away almost all vegetation and can move large stones and boulders. Mountain streams usually rise in hills at least 2000′ (650 m) high, with falls from hill top to stream channel of 600′ (200 m) or more, and the slope of the upper tributaries is often steeper than 1:40 (Fig. 1.8). Severe spates, swift flows or unstable substrates usually prevent plants from becoming dense. In the subgroup of extremely mountainous streams the water force is even greater – either because the fall from hill to channel is often over 1000′ (310 m), or because rainfall is greater – and plants, if not entirely absent, are very sparse.

Streams change as they flow from their sources to their mouths (Fig. 1.9a, b and c), the channels becoming wider and deeper as more water is carried, the hills typically becoming lower, and falls from hill to channel, and channel slopes, becoming less. Thus the flow pattern changes and, also, nutrient status increases. These downstream changes are further complicated when the rock type changes and when the topography is not just a simple transition from hills to lowlands. In general, when this happens, factors causing increased flow override those which decrease flow. For example, if a slow-flowing stream plunges into a gorge, flow and plants immediately become those appropriate to a torrent, while if a swift stream enters a plain it will be a considerable distance before it develops the flow and plants of a channel typical of a plain.

The pattern that the channels make across the land differs in different land forms. It has already been mentioned that in an alluvial plain many channels are man-made and sited for man's convenience, and so are typically straight and form a criss-cross pattern, with small dykes connecting the larger drains (Fig. 1.10a). Elsewhere the channels must be sited in the valleys between hills. On chalk and oolite (i.e. soft limestones) streams are sparse because the rock is very porous and so much of the rain passes through the rock to replenish the ground water below, leaving little to reach the streams (Fig. 1.10b). In contrast, lowland

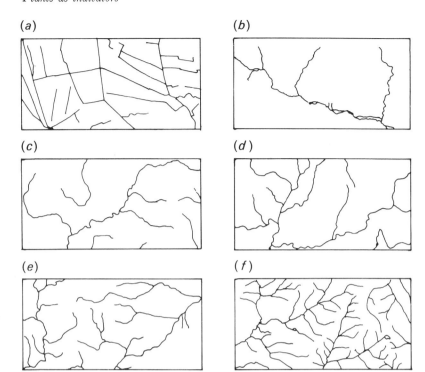

Fig. 1.10. Channel patterns in different land forms (all drawn to same scale). (*a*) The larger dykes and drains in an alluvial plain (Fenland). (*b*) Lowland chalk stream (R. Wylye). (*c*) Lowland soft sandstone stream (R. Tern). (*d*) Lowland clay stream (R. Chelmer). (*e*) Upland streams (Welsh Borders). (*f*) Mountain streams (Grampian Mountains).

clay shows a pattern of dense streams (Fig 1.10*d*), for clay is not porous and most of the rain runs off the surface into the streams. The low hills are also less regular on clay than on chalk and the tributaries vary their direction more. Lowland sandstone shows a pattern intermediate between chalk and clay (Fig. 1.10*c*). The mountain pattern in Fig. 1.10*f* shows the high density of streams typical of a high-rainfall area. The tributaries tend to be parallel because they flow from the side of large mountains. Fig. 10.10*e* is of an upland/low mountainous region with a lower rainfall; here there are fewer tributaries and they are less parallel.

PLANTS AS INDICATORS

River plants are sensitive indicators of the conditions in which they live. Because they are affected by a wide variety of factors, from land form to pollution, if the plants present at a site are known a good deal can be deduced about that site from the plants and, conversely, if the habitat is known, the expected type of vegetation can be predicted. If some but not all of the habitat factors are known then the plants can be used to monitor and assess the remaining factors: for example, when flow pattern, geology, etc. are known, pollution can be diagnosed by the plants.

IMPORTANT BRITISH RIVER PLANTS

There are, fortunately, only about fifty river plants which are both widespread and of diagnostic importance in Britain. To this basic list can be added some common bank plants, and some channel species of diagnostic importance in a more limited range of waters. These eighty or so important plants are illustrated here and their habitat preferences summarised, so that this section may be used as a guide to identification and diagnosis both in the field and for readers who are unfamiliar with some of these plants. Names are given here in both English and Latin, though later in the text they are in Latin only. *British water plants* by Haslam, Sinker & Wolseley [10], can be used to identify the plants of the channels (but not those of the banks, unless these can also grow as emerged aquatics). *British water plants* has many illustrations and most species can be identified without flowers or fruit. Its nomenclature is followed here. The standard British flora, which includes plants of the bank as well as those of the water, is *Flora of the British Isles* by Clapham, Tutin & Warburg [11].

It is, unluckily, not possible to use only a single classification for all these important species, and they are placed in five groups. They are listed according to their typical habitat, but may in some cases be found elsewhere.

Group 1 classifies species according to their general nutrient preferences, with notes about other habitat factors. The nutrient terms used are dystrophic (not-feeding), with negligible nutrients, up through oligotrophic (few-feeding) and mesotrophic (medium-feeding) to eutrophic (well-feeding), with ample nutrients. These terms, and the differences between, for example, mesotrophic limestone and mesotrophic sandstone, are discussed more fully in Chapter 8.

Group 2 lists *Ranunculus* spp.

Group 3 comprises the short semi-emerged dicotyledons or fringing herbs.

Group 4 is of species commonly found in the channel which do not fit into the preceding categories.

Group 5 comprises those species which are frequent on the banks but rare in the channel. (Some of the fourth group may be more common on the bank than in the channel.)

The scale bar to be found on most illustrations is 2 cm long.

GROUP I: ARRANGED BY NUTRIENT STATUS

A

Common cotton grass *Eriophorum angustifolium* ⎫
                                                 ⎬ Smaller
Bogbean *Menyanthes trifoliata*   ⎭ streams

Bog pondweed *Potamogeton polygonifolius*

Lesser spearwort *Ranunculus flammula*

Dystro-
phic
streams
of acid
peat
bogs.

Shore weed *Littorella uniflora*

Larger
streams

B

Needle spike-rush *Eleocharis acicularis*

Floating clubrush *Eleogiton fluitans*

Small to medium oligotrophic
streams, often with little spate
Substrate both
mineral and acid peat. On
Resistant rocks or soft
acid sandstone.

Bulbous rush *Juncus bulbosus*

Jointed rush *Juncus articulatus*

Swifter oligotrophic
streams.

13

C

   Intermediate water-starwort *Callitriche hamulata*

 Alternate-flowered water-mil-foil *Myriophyllum alterniflorum*

 Hemlock water-dropwort *Oenanthe crocata*

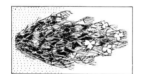 Water crowfoot *Ranunculus aquatilis/peltatus*

 *Ranunculus penicillatus*

Oligotrophic upstream reaches on Resistant rocks or acid sandstone streams. *Myriophyllum alterniflorum* tolerates much spate, often occurring upstream of *Myriophyllum spicatum*; and *Callitriche hamulata* upstream of other *Callitriche* spp.

D

 Monkey flower *Mimulus guttatus*

 Red pondweed *Potamogeton alpinus*

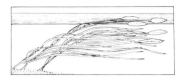

Potamogeton × sparganifolius

Semi-oligotrophic to semi-meso-
trophic usually swift waters. *Ranunculus fluitans* also into richer waters.
Potamogeton alpinus northern, *Potamogeton × sparganifolius* in Scotland.

 Water crowfoot *Ranunculus
fluitans*

E

Water-violet *Hottonia palustris*

Frogbit *Hydrocharis morsus-
ranae*

 Ivy-leaved duckweed *Lemna
trisulca*

Water crowfoot *Ranunculus
aquatilis/peltatus*

Mesotrophic unpolluted dykes
(canals). *Lemna trisulca* also in unpolluted chalk brooks.

 Ranunculus circinatus

Ranunculus trichophyllus

F

 Water celery *Berula erecta*

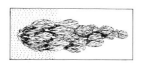

 Water crowfoot *Ranunculus aquatilis/peltatus*

Mesotrophic Calcareous streams

 Ranunculus calcareus

 Ranunculus penicillatus

G

  Blunt-fruited water starwort *Callitriche obtusangula*

  Common water-starwort *Callitriche stagnalis*

 Long-styled water-starwort *Callitriche platycarpa* (less frequent)

(*a*) Common in shallow streams, particularly on sandstone, less on limestone, least on clay streams with scours and in spatey streams. Regenerate well from fragments. (Less *Callitriche platycarpa*.) (*b*) Dykes, canals etc., least in eutrophic or polluted ones. (More *Callitriche platycarpa*.)

16

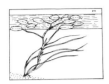

Canadian pondweed *Elodea canadensis* In a wide range of habitats, but rare without some eutrophic influence, e.g. clay in catchment, alluvial silt in channel. Absent from dystrophic and very oligotrophic streams, and upstream on chalk. Usually on silt, so in swift waters confined to sheltered banks, etc. Frequently luxuriant in still and slow-moving clear waters. Regenerates well from fragments.

Broad-leaved pondweed *Potamogeton natans* Frequent in medium sandstone streams with moderate flow and in dykes, etc. on clay. Occasionally in lower reaches of essentially oligotrophic streams and in upper reaches of essentially eutrophic ones.

Water crowfoot *Ranunculus aquatilis/peltatus*, etc.

H

Rigid hornwort *Ceratophyllum demersum* (*a*) Semi-eutrophic dykes, canals, etc. (*b*) Sheltered parts of slow streams, usually on clay. Is non-rooted and easily washed away.

Spiked water-milfoil *Myriophyllum spicatum* When luxuriant, usually on deep eutrophic (little-polluted) silt, or in dykes, canals, etc. on silt or clay, but occurs sparsely into almost mesotrophic habitats.

River water-dropwort *Oenanthe fluviatilis* On soft limestone with some eutrophic influence such as clay in the catchment, or in chalky clay streams, etc. Usually in a fairly large volume of water.

Amphibious bistort *Polygonum amphibium* Often in mesotrophic (or not very eutrophic) non-chalk, and usually large streams or canals. Tolerates some spate, but not a swift normal flow. Absent from eutrophic streams with eutrophic pollution. Shoots usually grow out from the bank and float.

 Perfoliate pondweed *Potamogeton perfoliatus* Usually in semi-eutrophic streams or canals with a medium or large water volume. In flow, usually on a firm substrate with sediment above. Absent from eutrophic streams with eutrophic pollution.

Water crowfoot *Ranunculus fluitans*

I

Water plantain *Alisma plantago-aquatica* Shallow semi-eutrophic to eutrophic places on fine substrate.

Opposite-leaved pondweed *Groenlandia densa* In semi-eutrophic streams, usually with a small water volume, particularly in chalk–clay streams, and the middle of clay ones. Less in hard limestone with some spate, and in semi-eutrophic unpolluted dykes.

 Curled pondweed *Potamogeton crispus* Usually on semi-eutrophic or eutrophic shallow silt over a hard bed, sometimes on mesotrophic non-limestone silt. Most often in sandstone, chalk–clay and the middle of clay streams and in lower reaches on Resistant rocks. Commonly in streams with medium water volume.

Water crowfoot *Ranunculus penicillatus*

 *Ranunculus trichophyllus*

Horned pondweed *Zannichellia palustris* In semi-eutrophic streams, often on shallow silt, usually those with rather small water volumes. Most often on chalk–clay streams and in the middle of clay ones. Also in unpolluted dykes, etc.

J

Yellow water-lily *Nuphar lutea* On eutrophic soft substrates, mostly in streams or canals, etc. with large or moderate water volumes. Characteristic of lower clay streams, eutrophic canals, drains, etc. Less frequent on sandstone. Intolerate of spate.

Great yellow-cress *Rorippa amphibia*

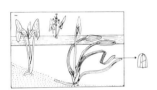

Arrowhead *Sagittaria sagittifolia* On eutrophic soft substrates, particularly in canals, slow clay streams, drains, etc. Rare on sandstone. Intolerant of much non-eutrophic pollution.

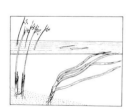

Greater rush or bulrush *Schoenoplectus lacustris* In eutrophic and semi-eutrophic streams of variable substrate and flow, though little spate. Usually in moderate to large water volumes. Most common in clay streams, frequent in lower chalk streams, rare on sandstone.

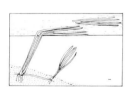

Strapweed *Sparganium emersum* In eutrophic and semi-eutrophic streams on soft, preferably silty soils, and at least moderate water volumes. Characteristic of clay streams; frequent in sandstone ones, both lowland and upland; less common on limestone and elsewhere.

GROUP 2: BATRACHIAN 'RANUNCULUS' SPP. (WATER CROWFOOTS)

 Ranunculus aquatilis/peltatus (a) Very short-leaved forms. In small swift streams without severe spates, mesotrophic to oligotrophic, on soft limestone or Resistant rocks. Tolerate summer drought on chalk. Can develop terrestrial habit. Also in unpolluted semi-eutrophic or mesotrophic dykes, canals, etc.

(*b*) Medium-leaved forms. In medium-sized, fairly swift lowland streams or small hill ones, usually mesotrophic to somewhat oligotrophic, rarely on (eutrophic) clay. Often found upstream of *Ranunculus fluitans* in hill streams and, where both occur together, *Ranunculus fluitans* grows in the deeper water.

Ranunculus aquatilis is the more frequent on hard rocks and *Ranunculus peltatus* on soft ones.

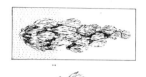

Ranunculus calcareus (Strictly, *Ranunculus penicillatus* var. *calcareus*) In medium to large chalk streams, less often on other basic rocks.

Ranunculus circinatus Uncommon. In unpolluted mesotrophic to semi-eutrophic dykes, canals, etc., sometimes in slow mesotrophic peaty streams.

Ranunculus fluitans Usually in mesotrophic to semi-eutrophic unpolluted to slightly polluted waters in: (*a*) large not-too-spatey hill rivers; and (*b*) smaller lowland streams of at least moderate flow.

Ranunculus penicillatus (Strictly, *Ranunculus penicillatus* var. *penicillatus*)
(*a*) Short- to medium-leaved forms. In swift streams on Resistant rocks, particularly in the West and the Pennines.
(*b*) Medium- to long-leaved forms. In medium-sized mesotrophic to eutrophic streams of at least moderate flow, particularly on limestone–clay streams, less on chalk or upland clay ones, rare elsewhere. In less eutrophic sites it tolerates some pollution.

 Ranunculus trichophyllus   Infrequent. Occurs (*a*) in medium-sized clay or part-clay streams; (*b*) in canals, dykes, etc; (*c*) in streams on Resistant rocks.

GROUP 3: FRINGING HERBS

(Short perennial dicotyledons, usually semi-emerged, also emerged or submerged)

 Water celery   *Apium nodiflorum*

 Water celery   *Berula erecta*

 Water mint   *Mentha aquatica*

 Monkey flower   *Mimulus guttatus*

 Water forget-me-not   *Myosotis scorpioides*

 Water cress   *Rorippa nasturtium-aquaticum* agg. (*Rorippa nasturtium-aquaticum, Rorippa microphylla* and their hybrid; the former is commoner, especially in the south)

21

Water speedwell *Veronica anagallis-aquatica* agg. (*Veronica anagallis-aquatica* and *Veronica catenato*; the former is the commoner)

Brooklime *Veronica beccabunga*

(*a*) Small, usually mesotrophic streams on chalk, hard sandstone, Resistant rocks with silting and little water force. Here there are typically at least three species present, often in wide or long fringes at the sides. Often submerged, on chalk frequently right across the channel. Other small streams have less.

(*b*) Small clay streams often have patches of large, emerged plants, typically one to three species per site.

(*c*) Banks of larger streams. Small, scattered, often temporary patches may occur in muddy places where the water force is usually low, and the banks have not recently been swept by swift storm waters.

(*d*) Small- and medium-sized streams on soft limestone frequently have, in addition to (*a*) above, one or more of: *Berula erecta* present (often as a submerged carpet); *Apium nodiflorum* as a submerged carpet; *Rorippa nasturtium-aquaticum* agg. dominant over a substantial part of the site.

(*e*) *Veronica beccabunga* is the most frequent on the Resistant rocks (except in the south-west, where it is replaced by *Apium nodiflorum*).

(*f*) *Myosotis scorpioides* is the most frequent in clay and part-clay streams.

(*g*) *Veronica anagallis-aquatica* agg. may occur submerged after clay and part-clay streams have been dredged.

(*h*) *Mimulus guttatus* Usually on hard rocks, more often emerged than semi-emerged. It tolerates considerable pollution.

Great yellow-cress *Rorippa amphibia* In eutrophic streams with little scour, most often clay or part-clay. Also in eutrophic to nearly mesotrophic dykes, etc. Ecogically separate from the others and usually excluded from references to 'fringing herbs'.

GROUP 4: CHANNEL SPECIES NOT INCLUDED IN GROUPS 1 TO 3

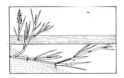

Fiorin *Agrostis stolonifera* Frequent on edges, somewhat less so in the centre, where submerged plants (in shallow water) are often temporary. Often occurs soon after dredging.

Lesser pond sedge *Carex acutiformis* Infrequent on the edges of dykes and lowland streams, particularly chalk and lower clay ones.

Water whorl-grass *Catabrosa aquatica* Frequent at the sides of smallish chalk streams, uncommon elsewhere.

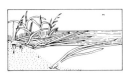 Flote-grass *Glyceria fluitans, Glyceria plicata* Infrequent on edges of smallish streams, rarely in the centre.

 Reed grass or reed sweet-grass *Glyceria maxima* In channels of shallow dykes, at sides of canals and slow large streams, particularly those with alluvial banks; less often on clay, chalk and flatter Resistant rocks; sparse elsewhere.

 Common duckweed *Lemna minor* agg. (*Lemna gibba* cannot be distinguished from *Lemna minor* when it is not swollen, and when swollen may grow with *Lemna minor*.)
   Often abundant on still and very slow waters, but restricted to sheltered places in swifter streams.

 Reed grass, reed canary-grass *Phalaris arundinacae* Frequent in swift hill streams, on the banks and on intermittently flooded gravel spurs, etc. Also in intermittently flooded ditches and dykes, and on the banks of silt and clay dykes.

 Reed *Phragmites communis* Commonly dominant in shallow eutrophic and mesotrophic dykes, and on the banks of deeper dykes and drains.

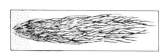 Fennel pondweed *Potamogeton pectinatus* In unpolluted places, mainly lower mesotrophic (or semi-eutrophic) hill streams without very severe spates, particularly in those on sandstone, infrequently in chalk–clay streams. Also in dykes with some sea water pollution or recent dredging. Elsewhere, it is the best indicator of pollution.

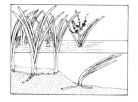 Bur reed *Sparganium erectum* The most widespread watercourse species. Luxuriant stands are usually on thick silt. In larger streams patches are usually marginal. It avoids severe scours.

23

Reed-maces, bulrushes *Typha angustifolia* and *Typha latifolia* Occasionally dominant in shallow dykes, or at sides of lowland, particularly clay, streams.

GROUP 5: BANK SPECIES NOT INCLUDED IN GROUPS 1 TO 4

Great hairy willow-herb *Epilobium hirsutum* Common except in the higher hills and mountains.

Meadow-sweet *Filipendula ulmaria* Frequent, especially on chalk and in the hills.

Soft rush *Juncus effusus* Frequent, especially in the hills and after dredging.

Bramble *Rubus fruticosus* agg. Common in the lowlands, sparse in the hills.

Woody nightshade *Solanum dulcamara* Rare except on chalk, where it may grow into the channel.

Nettle *Urtica dioica* Very common except in the mountains. (Characteristic of disturbed ground.)

Species listed in Groups 1 to 4 which typically also occur on banks out of the water are: *Eriophorum angustifolium, Polygonum amphibium,* all the fringing herbs of Group 3, *Agrostis stolonifera, Catabrosa aquatic, Glyceria* spp., *Juncus articulatus, Phalaris arundicacea, Phragmites communis, Sparganium erectum* and *Typha* spp.

INTRODUCED PLANTS

River plants may be transferred from one stream to another by animals (e.g. on birds' feet) or by man. Introductions by man may be made accidentally, when restocking with fish, or on boots, machinery, etc., or species may be planted deliberately. Deliberate introductions tend to be of species suitable for the site, of suitable sizes and parts of the plant for growth, and may also be anchored in place until their own roots have developed. They are, therefore, more likely to grow well than are accidental introductions.

Plants may be introduced to protect banks from erosion. These are likely to be the tall monocotyledon (e.g. *Phragnites communis*) which have a deep network of roots and rhizomes to bind the soil of the bank and dense thick shoots to protect its surface from scour. More commonly, though, plants are introduced to improve fisheries, most often trout fisheries, and *Ranunculus* is usually preferred by the anglers. However, *Ranunculus* will only grow under certain conditions, and introductions into slow clay streams, or mountain streams with excessively severe spates, for example, are likely to prove unsuccessful. Introducing plants is valuable when a stream is recovering from damage but the plants have not yet recolonised by natural dispersal. When R. Mimram was recovering from sewage pollution, small weirs were built to increase water flow locally, wash away accumulated silt and produce fast gravel reaches. In these, *Apium nodiflorum, Callitriche* and *Ranunculus* were introduced and grew well [12]. When R. Wear was recovering from pollution from coal mines, *Ranunculus* was introduced in 1959, and spread successfully [13].

Before introductions are contemplated, however, the likelihood of explosive growth should be considered, for excessive vegetation can cause flood hazards and hinder angling. *Ranunculus*, for example, causes more flood hazards than any other plant of flowing waters, and *Phragmites communis* more than any other plant of shallow still waters, so these should be introduced only with caution. It is probable that introductions within Britain have had little effect on the general composition of British streams, because of the restricted habitat requirements of each species and the likelihood of natural dispersal by animals. One important, and several minor river plants have, however, been introduced from abroad: *Elodea canadensis* is now an integral part of British river vegetation and *Elodea callitrichiodes, Elodea nuttallii, Sagittaria rigida* and *Vallisneria spiralis,* are also found here.

2

Flow, substrate and plant distribution

Water is not only present in rivers but moving, and the flow of the water is crucial in determining river vegetation. Moving water influences plants directly, and also because it controls the soil; larger particles being found in faster flows. Some plant species grow best in still water, others in swiftly flowing streams. This means that, if habitats are equal in other respects, similar vegetation occurs in streams of similar flow types.

Boulders, stones, gravel, sand, silt and mud, and peat all occur on stream beds. Some plants grow better in the fine-grained silt, mud or peat, while others do better on coarser beds. Plant species may prefer a certain soil type, or may grow there because they prefer the flow type associated with that soil, or because of the particular combination of flow and substrate. Vegetation is related to the substrate on which it grows.

Water movement is one of the most important factors influencing river plants. Flow affects plants directly, and indirectly through its effects on the channel bed. One direct effect of flow is that currents move plant parts, and faster currents may batter and tangle leaves, stems and flowers in the water, scour soil from around their roots, and pull plants up. Flow also brings to the plant the gases (dissolved oxygen and carbon dioxide) necessary for respiration and photosynthesis, as well as other dissolved substances such as nutrients, e.g. phosphates and harmful or neutral compounds.

Moving water can, of course, move the soil particles of the substrate, the water velocity needed to move particles on the channel bed depending on the particle size, grading and specific gravity of the soil. Above a critical velocity the soil particles are initially moved along the bed by rolling and saltation, while at higher velocities some of the material is carried away suspended in the water. If water velocity decreases, these rolling particles stop moving and the suspended ones are deposited again. Naturally the larger the particle the faster the flow needed to move it, so that lowland brooks may carry a lot of silt and fine sand but little coarse sand and less gravel, while fast mountain streams can move quite large stones. Most river plants are anchored in the soil and can remain so only when the soil remains in place also. Plants also get most of their nutrients from the soil (though some from the water; see Chapter 8), these soil nutrients being mainly in the fine particles, the silt and mud. Swift flows erode, removing smaller particles, while slow ones do the reverse, accumulating sediment on the bed.

26

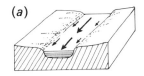

(a)

(b)

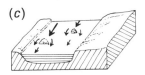

(c)

Fig. 2.1

FLOW TYPES

In a straight channel without obstructions, the water moves fastest near its centre and more slowly near the banks (Fig. 2.1a). When the river bends, the faster flows occur near the outside of the bend. (Fig. 2.1b) Obstructions such as stones, boulders, plants and bridge piers have a dual effect; they increase turbulence and slow down both the average speed of flow and the speed near the obstructions (Fig. 2.1c). Each plant is affected by the conditions of the flow around it and by the water passing through the plant clump itself, but it is, unfortunately, not practicable to measure the flow conditions near many plants over a long period of time. This means that the situation must be simplified and the average flow conditions over a stretch of channel used to describe the hydraulic parameters affecting the plant.

Discharge is defined as the total volume of water per unit time flowing through the channel, and this is one method of describing the flow. However, in a deep channel with smooth sides and a gentle slope to the bed, flow will be slower and less turbulent than in a wide, shallow, steeply sloping channel with a rough bed of boulders, even if the discharge is the same. Thus other measures of flow must be used as well. The speed of the water is another possibility, but since a swift, smooth flow damages plants less than a slower, turbulent one, it is important to consider turbulence as well as velocity.

When the water particles move in smooth paths and one layer of water slides over another, flow is said to be laminar. When the water particles move in irregular paths the flow is said to be turbulent. Flow is laminar when the viscous forces (the forces holding the water particles together) are strong compared with the inertial forces (the forces preventing movement) and turbulent when the viscous forces are weak compared with the inertial forces. The ratio of the inertial forces to the viscous force is called the Reynolds number, so turbulent flow occurs at high Reynolds numbers, and when the channel bed is relatively rough.

The water surface in a channel may be rough, and the flow then appears to be very turbulent. This happens when the ratio of inertial to gravitational forces is high, or when velocities are high, and the water is shallow. It appears, therefore, that it may be difficult to define criteria for turbulent flow in natural channels, as these would depend on both the Reynolds number and the roughness of the channel bed.

Useful visual categories of the flow conditions affecting river plants (which can be correlated with physical parameters) are: (1) negligible flow, as in canals, fen dykes, etc. (Plate 11); (2) slow flow, in which trailing plants hardly move (Plate 9); (3) moderate flow, in which trailing plants clearly move and the water surface is slightly disturbed (Plate 18); and (4) fast flow, in which trailing plants move vigorously and the water surface is markedly disturbed (Plate 10). The flow type at any one place often changes very little over long periods of time, though flow will vary during and after storms, or if the water is controlled and the discharge deliberately altered by dredging, river draining works or water storage schemes.

A fifth category of flow must be taken into consideration when assessing the effect of flow on plants: liability to spate. In lowland brooks storm flows do some damage but normally leave most vegetation unharmed; in the highlands, however, the channels and hill slopes are steeper and the rapid spate flows subject the plants to greater forces. Liability to spate can occur when normal flow is fast, moderate or even slow.

As regards the distribution in Britain of channels with these different flow types, channels with negligible flow (flat gradients and low velocities) occur in alluvial plains like the Fenland, Broadland and Romney Marsh, though they also occur in small pockets of drained land throughout Britain; canals are more frequent in England than Scotland or Wales; streams with slow and moderate flow are widely distributed; and fast-flow sites are more frequent in Highland Britain (Scotland, N. England, Wales and the West Country), liability to spate of course being confined to these regions (see Chapter 5).

VEGETATION CORRELATED WITH FLOW

Watercourses with similar flows have similar vegetation, other factors being equal, so that plant distribution is clearly correlated with flow. Dykes, for example, bear a generally similar vegetation whether in the alluvial plains of S. Wales, Sussex, Norfolk or Lincolnshire, and the plants of upland streams of Herefordshire and Lanark are likewise similar, as are those of mountain streams in Snowdonia, the Lake District and the Scottish Highlands. The river habitat is a rigorous one, difficult for plants to live in, and the flow pattern is an overall controlling factor, comparable to the action of drought in the desert, or cold and dark in the Arctic. Thus flow type affects, and may entirely determine, the species present in a habitat, their performance and the relations between them, other factors acting only insofar as they are permitted to by the flow. In very mountainous districts the natural flow is so swift and fierce that it (and the substrate determined by it) are almost the sole controlling factors for the vegetation, but as flows become less fierce in less hilly regions, other factors such as rock type, pollution and management become increasingly influential. Competition between plants, which is such a noticeable feature of land vegetation, is of no importance in the swiftest streams and only of minor or local importance in slower-flowing waters, but may be very important in still waters, where it can determine the general type of the vegetation (see Chapters 3, 5, 6, 11 and 17).

CORRELATION OF FLOW TYPE WITH INDIVIDUAL SPECIES

Species can be grouped according to their response to flow, but these groups are arbitrary in that (as is the case with most other habitat

Fig. 2.2. Species most closely associated with negligible flow. Histograms I, on the left, show percentage occurrence in each flow type; histograms II, on the right, show number of occurrences in each flow type. Flow types are as defined on p. 27 n, negligible; s, slow; m, moderate; f, fast; sp, liable to spate (this category is independent of the others).

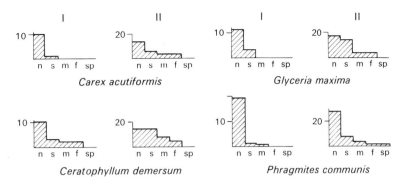

Carex acutiformis *Glyceria maxima*

Ceratophyllum demersum *Phragmites communis*

factors) the species show a wide and nearly continuous range of variation. The strength of their preferences varies also. Some distributions are shown in Figs. 2.2–2.7. There are two histograms given for each species, the first (on the left) showing the flow type with which the species is best correlated and the second flow type in which the species is most abundant. If a species actually prefers a certain flow type then this will be the one with which it is best correlated, but a plant may be abundant in a particular habitat because the habitat itself is extremely frequent and not necessarily because it is particularly suitable for the plant.

The species most closely correlated with negligible flow are shown in Fig. 2.2. One of these (*Ceratophyllum demersum*) does not have roots or rhizomes and its submerged shoots are easily washed away by fast or storm flows. The other three are tall emerged monocotyledons. These will also grow on the banks above the average water level, and up on the banks they may more often be found by streams with moving water.

Fig. 2.3 shows those species that are best correlated with negligible flow but which also occur frequently in other flow types. One (*Lemna*

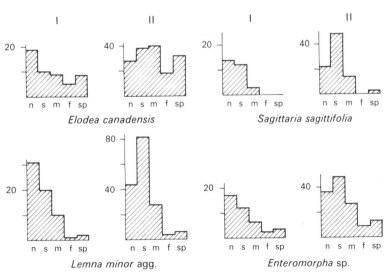

Elodea canadensis *Sagittaria sagittifolia*

Fig. 2.3. Species associated with negligible flow to a lesser extent than those of Fig. 2.2. Details as for Fig. 2.2.

Lemna minor agg. *Enteromorpha* sp.

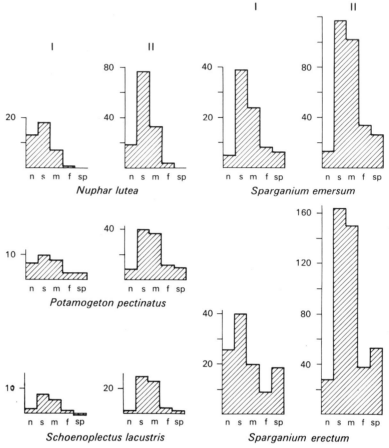

Fig. 2.4. Species most closely associated with slow flow. Details as for Fig. 2.2.

minor agg.) has very small plants which are free-floating on the water surface and are therefore easily removed by currents. The others are submerged but are susceptible to scouring, etc. (see Chapter 3).

The species whose distribution is best correlated with slow flow are shown in Fig. 2.4. These include *Sparganium erectum*, the commonest and most ubiquitous watercourse plant in Britain. However, at any one site a plant may be sparse or dense, and the habitat a plant needs for good growth may not correspond to the range of habitats where the species can be found. For example, when plants of this group grow abundantly these dense stands may be best correlated with other flow types, viz:

Negligible flow: *Sparganium erectum*
Negligible to slow flow: *Nuphar lutea*
Slow to moderate flow: *Sparganium emersum*
Moderate flow: *Potamogeton pectinatus*

The species best correlated with moderate flow (Fig. 2.5) include those characteristic of chalk streams. Compared with *Ranunculus* spp., *Apium nodiflorum* and *Berula erecta* occur more in fast flow and less in sites liable to spate. Abundant populations of *Ranunculus* are, compared to sparse ones, more frequent in fast flow. The different *Ranunculus* species have different habitat preferences, but conclusions about the

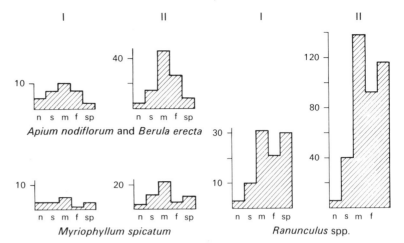

Fig. 2.5. Species most closely associated with moderate flow. Details as for Fig. 2.2.

separate species are tentative because hybridisation and intermediates are common, so identification is often difficult. Some general conclusions are:

Sometimes occurring in negligible flow: *Ranunculus aquatilis, Ranunculus circinatus, Ranunculus peltatus, Ranunculus trichophyllus*

Primary correlation with moderate flow, secondary with fast flow: *Ranunculus calcareus*

Primary correlation with moderate flow, secondary with spate liability: *Ranunculus peltatus, Ranunculus penicillatus*

Primary correlation with fast flow, secondary with spate liability: *Ranunculus aquatilis*

Primary correlation with spate liability, secondary with fast flow: *Ranunculus fluitans*

Also see [5, 6, 14, 15]

The few plants best correlated with fast flow include the mosses, as an aggregate group (Fig. 2.6), *Myriophyllum alterniflorum, Ranunculus*

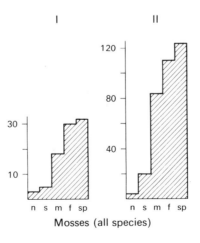

Fig. 2.6. Species most closely associated with fast flow. Details as for Fig. 2.2.

31

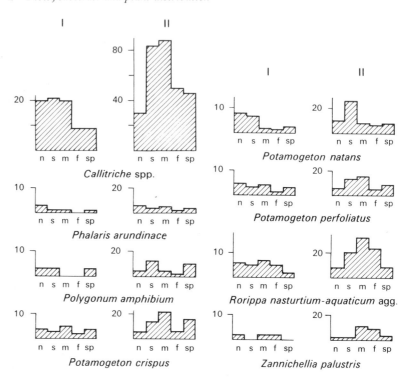

Fig. 2.7. Species poorly associated with flow type. Details as for Fig. 2.2.

aquatilis and *Ranunculus fluitans*. Fig. 2.7 shows some channel species which do not fit into the other groups.

Sometimes, on these histograms, a plant is recorded frequently in one habitat and rarely in another. Such rare records are usually due to some local conditions not shown on the histograms, such as a stream normally kept deep and slow by flood gates being recorded on an unusual occasion when the gates were open.

THE FAST FLOW AND THE SPATEY HABITATS

Habitats with fast flow have continuous scour which prevents particles accumulating in the path of the main current, but since the flow is steady, the bed is stable. Spatey reaches in times of spate are subjected to an extremely turbulent flow of great force which disturbs and erodes the bed, but if the normal flow is slow or moderate then fine sediment is deposited between spates.

Fast flow and spates usually occur together, as in many mountain streams, but can be found independently, for spates occur without fast flow in flat places in the hills and near the flood plains of hill rivers, while there is fast flow without spates in the steeper sections of lowland streams. When these two components of swift flow do occur independently they result in different vegetation types, so it can be seen which features of habit and morphology enable certain plants to tolerate which aspect of swift flow.

Plants tolerating fast flow better than spates are more easily uprooted by extra scour and water force, though are more tolerant of being battered and tangled in the water, e.g.

🌿 *Apium nodiflorum*   🌿 *Ranunculus aquatilis*

🌿 *Berula erecta*   🌼 *Ranunculus calcareus* (usually)

🌱 *Nuphar lutea* (perhaps)   🌿 *Rorippa nasturtium-aquaticum* agg.

🌿 *Oenanthe fluviatilis*   🌾 *Schoenoplectus lacustris* (perhaps)

🌿 *Zannichellia palustris*

Species tolerating spates better than fast flow may do so because they need fine substrates for rooting and anchorage (see Chapter 3), or are tougher plants, or grow very quickly so that damaged parts are rapidly replaced. They are more tolerant of battering than the last group and are not usually easily uprooted (or if they are can quickly regrow from fragments).

🌿 *Elodea canadensis*   🌿 *Potamogeton perfoliatus* (perhaps)

🌿 *Myriophyllum spicatum*   🌿 *Ranunculus fluitans*

🌿 *Polygonum amphibium*   🌿 *Sparganium erectum*

🌿 *Potamogeton crispus*

Some species are particularly intolerant of both fast flow and spate. For example:

🌿 *Carex acutiformis*   🌿 *Phragmites communis*

🌿 *Ceratophyllum demersum*   🌿 *Potamogeton pectinatus* (often)

🌱 *Nuphar lutea*   🌿 *Sagittaria sagittifolia*

🌾 *Schoenoplectus lacustris* (usually)

No plants grow where there is both very fast flow and severe spates, but a few can sometimes tolerate a considerable degree of both. For example:

🌿 *Callitriche* spp.   *Potamogeton* ×*sparganifolius*

🌿 *Myriophyllum alterniflorum*   🌿 *Sparganium emersum*

Three of these are firmly rooted, two regrow quickly from fragments, three tolerate much battering and one avoids much turbulence.

SUBSTRATE TYPES

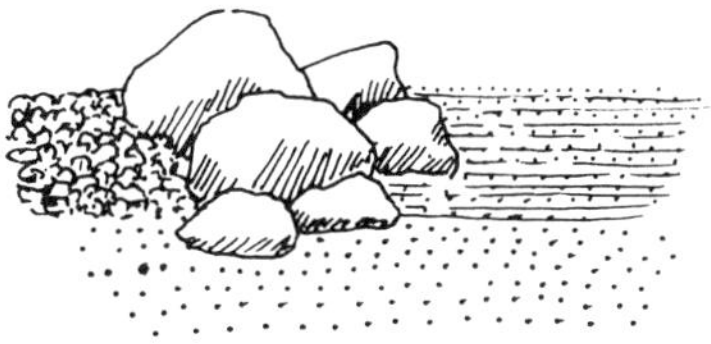

Stream beds may be made up of many different kinds of substrate (Fig. 2.8). In mountain streams the solid bedrock is commonly exposed, and this is sometimes the case elsewhere (e.g. on clay), though more

Fig. 2.8

usually the substrate is composed of particles derived from the rock. These may come directly from the rock in the channel, as with boulders, stones and gravel from hard rocks, sand from sandstone and fine particles from clay, or alternatively the particles may come from the soil or subsoil of the catchment area, being washed in from the stream banks with the run-off water, or coming down from upstream through the scouring of the channel. As well as the inorganic component, organic particles will be present, from the decay of aquatic plants and animals, and from humus, peat and undecayed material washed in from the catchment. Substrates can also be man-made, either intentionally, as in concrete culverts, or inadvertently, as with rubbish dumps.

Natural channels may be incised or alluvial. The incised channel typical of fairly steep catchments scours its bed, unless it meets material it cannot erode, and so the bed is made of material older than the river itself. On this bed river-borne material is deposited when the flow falls and is picked up again and carried on when the flow increases. The alluvial river, found in flatter areas, gradually builds up its own plain composed of deposited sediment. The channel moves within this sediment, and the bed and banks can be eroded more easily. Velocities in the steep channels are usually higher than those in channels with more gentle slopes, and their flows vary more rapidly with the rainfall. These variations in flow cause, in turn, variations in erosion and deposition of sediment, the overall pattern being that of little deposition upstream and much downstream, while in the middle reaches sedimentation occurs in areas of least flow (which include plant clumps).

Most channels have a consolidated bottom (i.e. a hard bed) which is commonly in the subsoil, though it can be man-made as in clay-bottomed canals, etc. Above this there may be sediment deposited which is not consolidated and can thus be moved easily in storm flows. There is usually little fine sediment in the smaller highland streams because of the swift flow, or in chalk streams because the weathering of chalk leads to little sediment. Clay streams usually have the most inorganic sediment, then sandstone, limestone and finally the Resistant rocks.

The substrate types discussed here and elsewhere in this book were classified on size as boulders or hard solid rock, stones, gravel, sand and silt. Silt here includes the finer particles of mud, since these cannot be distinguished easily from silt. The additional category used was peat, which forms a very organic soil. In this chapter only fen peat, which is alkaline and moderately rich in nutrients, is considered; bog peat forms a much more specialised habitat which is described in Chapter 13. In the histograms of Figs. 2.9–2.13 the substrate types are those which are prominent on the exposed part of the bed at a site, not necessarily those in which the relevant species are actually rooted. Silt or sand which is almost confined to within plant clumps is not included.

34

VEGETATION CORRELATED WITH SUBSTRATE

Similar types of soil, if they are equally stable, bear similar vegetation, other factors being equal. There are also similarities between the histograms relating plant distribution and flow and those showing distribution against substrate type (see next section). This is because the particle sizes and the stability of the substrate are largely controlled by flow. For example, faster flows have coarser substrates and streams with widely fluctuating flows have a wide range of scour and deposition during a season, and so have unstable substrates. Thus although flow exercises an overall control, the plants may be responding to the flow, the substrate, or both. Mosses are a good example. Being without roots they are easily washed away by basal scour and being short they can be smothered by accumulating silt or sand. Consequently, mosses occur in places which remain stable and are not silted. These are usually boulders, etc. (but may be the peat of fen dykes, where flows are too slow to move the fine soil, or be stray bricks or bridge piers [16]) and so the correlation of mosses with spates and fast flow is due to a causal association between mosses and the boulders etc. of hill streams.

CORRELATION OF SUBSTRATE TYPE WITH INDIVIDUAL SPECIES

Some plants grow best in fine soil, some among coarse particles. This is partly a genuine preference of a species for a particular soil, partly a preference for a particular flow type (which is necessarily correlated with substrate) and partly a joint effect of flow and substrate. This last

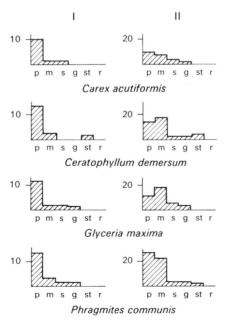

Fig. 2.9. Species most closely associated with fen peat substrates. Histograms I, on the left, show percentage occurrence in each substrate type; histograms II, on the right, show number of occurrences in each substrate type. Abbreviations for substrate: p, peaty; m, mud and silt; s, sand; g, gravel; st, stone; r, rock or boulder.

35

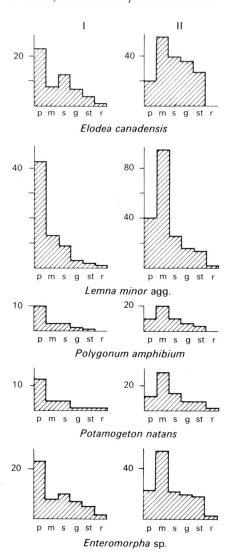

Fig. 2.10. Species associated with peaty substrate to a lesser extent than those of Fig. 2.9. Details as for Fig. 2.9.

occurs when, for instance, a plant can grow in its optimum flow type only when a particular substrate allows it to anchor firmly.

Figs. 2.9–2.13 are histograms of plant distribution in relation to substrate, arranged in the same way as the flow histograms. As with flow types, species can be arranged in groups with similar responses to types. The plants show wide range of variation and once again the grouping is arbitrary. Where several substrate types occur in one site, each species present is recorded as though it occurred in each substrate type. This shows correctly that habitats of particular particle sizes can support particular species, but it also wrongly implies that each species present grows on each substrate type, which may not be the case (e.g. a particular species may grow only on the silt banks in a gravel stream).

No angiosperm actually grows on boulders or large stones, though many grow in streams containing these as well as smaller particles. Where a species is recorded with a certain substrate type on only a few occasions, it probably does not grow on this type, but only on other substrates of the site.

The species most closely correlated with fen peat or organic alluvium are shown in Fig. 2.9. These are the same large, deep-rooted emerged plants and the non-rooted submerged ones which are most closely correlated with negligible flow (Fig. 2.2), and will therefore be found in fen dykes, which have both peat beds and still water.

Fig. 2.10 shows the species best correlated with fen peat which also occur frequently in other soil types, and Fig. 2.11 those best correlated with silt and mud. Most species of fine substrates have long deep roots,

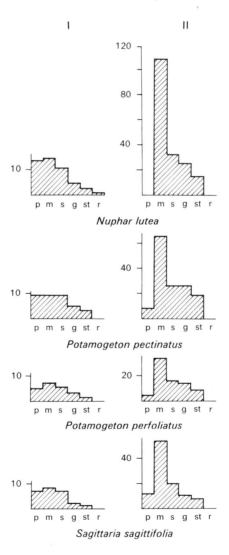

Fig. 2.11. Species most closely associated with silty substrates. Details as for Fig. 2.9.

Nuphar lutea

Potamogeton pectinatus

Potamogeton perfoliatus

Sagittaria sagittifolia

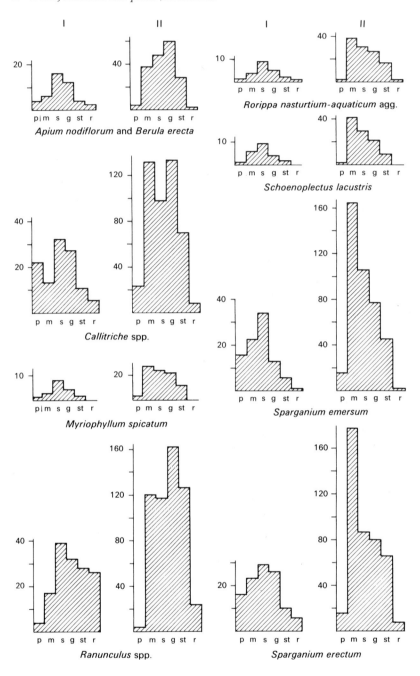

Fig. 2.12. Species most closely associated with medium-grained substrates. Details as for Fig. 2.9.

though some (e.g. *Potamogeton* spp.) may grow on shallow soil. Species often occur in a wider range of soils than those in which they grow well (e.g. *Nuphar lutea*).

Medium-sized particles (sand or gravel) can be found in substrates that are mainly fine, medium or coarse, and the species of Fig. 2.12

38

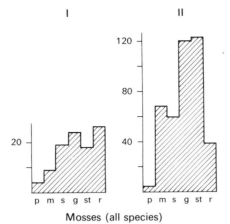

Fig. 2.13. Species most closely associated with coarse substrates. Details as for Fig. 2.9.

comprise both deep-rooted members (e.g. *Sparganium* spp.) and shallow-rooted ones (e.g. *Callitriche* spp., *Ranunculus* spp.). If tentative identifications are included for *Ranunculus* spp., then these have the following habitat preferences:

Primary correlation with sand, secondary with gravel:

Ranunculus aquatilis

Ranunculus calcareus

Ranunculus penicillatus

Primary correlation with sand and gravel, secondary with stone:

Ranunculus peltatus

Relatively high correlation with boulders and stones:

Ranunculus fluitans

Mosses can anchor to boulders and rock, and these are the only macrophyte group to be best correlated with large particles (Fig. 2.13).

3

Flow, substrate and how they affect individual plants

Water movement affects plants in several ways. Swift flows pull plants away from the ground, the force needed to do this depending on the species and size of the plant and on the time of year. If a plant is broken above ground then the underground parts are left in place and can regrow, but if the plant is pulled up completely it is washed away and lost from the community. Turbulent flow tangles and batters leaves and stems in the water. In storm flows, silt and other material carried by the water can cause abrasion, and so damage plants.

In general, plants of slow flows and fine substrates have long deep roots which anchor the plants firmly even if the upper silt is disturbed in storms, while plants of faster flows and consolidated gravel substrates usually have tangled roots curling around the gravel particles near the surface, which also remain anchored in this swifter-flowing habitat. When flows are swift enough to move the soil plants may be eroded, the ease of erosion depending on the plant's species and size and on the type of substrate. Where the current is checked silt is deposited and some plants can grow upwards into the loose sediment. When it is washed away then the plants will be washed away also.

Vegetation and flow are normally in equilibrium, the plant species present in any one reach being those which can, in the long run, tolerate both the normal flows and the storm flows of that reach. This is either because there is little damage to individual plants or because any damage is quickly made good.

Individual plant shoots are affected by the flow actually around each of them, and the roots, similarly, by the substrate around each of them. As flow and substrate vary, their effects on shoots and roots vary as well, and so in one river site there may be many microhabitats each with different flows and substrate and thus different vegetation, or the site may be fairly uniform in flow and substrate and have only small variations in vegetation.

The flow pattern within and around a plant varies with both time and space. A plant will usually (though not always) grown from quieter water near the stream bed into swifter water nearer the surface, so the upper parts are likely to be moved more and battered more than the lower ones (Fig. 3.1). Large plants tend to impede flow, so, as thick plants grow, patches of quiet water tend to develop in them, while (if discharge remains stable) flow increases between the plants (Fig. 3.2). Plants therefore alter the flow pattern of a stream as well as being altered by it. Altering the flow pattern may also alter the substrate pattern, since sediment is more likely to be deposited in the quiet water within the plant than between plants, and conversely if the flow is fast

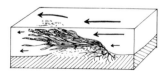

Fig. 3.1

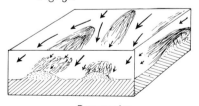

Ranunculus

Fig. 3.2

40

enough to cause erosion, this will be greater between the plants (Fig. 3.2; Plate 5). When the discharge changes the water level changes, and thus turbulence and flow pattern both change. When the water level rises above the plants, turbulence due to the plants usually decreases, and vice versa.

Storm flows have greater force than normal flows, they raise the water level, usually increase turbulence and are more likely to have eroding velocities. They are thus capable of causing major damage. In general, though, plant communities are adjusted to the storm flows liable to occur in a habitat. This may be because the individual plants are not damaged much in storms; or damage may be great to above-ground parts but there is quick regeneration from rhizomes; or some stands may suffer major damage but recolonisation from fragments or fruits is rapid; or although there may be major damage and slow recolonisation the community has a repeating pattern over several decades.

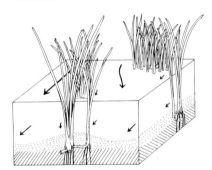

Fig. 3.3

Sparganium erectum

Obstructions, including plant clumps, check flow and increase turbulence. Water velocity is greater beside plant clumps, and least downstream of them (Figs. 3.2, 3.3). This means that different habitats suitable for other plants occur around the larger clumps. Downstream of a clump of a trailing plant there will be long waving shoots or leaves, which would hit other plants growing there, so species wishing to grow in shelter do not find this a favourable habitat. Upstream of a clump in shallow streams, however, velocity, water force and shoot movement are low, yet turbulence, which aids aeration and nutrient supply, is high, and some short species in streams of moderate or fast flow often grow upstream of a clump (e.g. *Berula erecta, Potamogeton crispus* when short, and *Zannichellia palustric*). Thus a clump of one species can allow another plant to develop upstream of it in a site which would otherwise have too fast a flow.

Water velocity is very low on and inside a plant clump, dropping sharply from close to (e.g. 3 cm from) the edge. Thick vegetation, whether as clumps or as carpets, has negligible flow inside it unless the flow outside is very turbulent indeed, and in fact as few as four shoots of

41

Ranunculus in a path 10 cm wide (the shoots being only *c.* 2 mm wide) can halve the water velocity. Thus an important effect of varying flow patterns and storm turbulence is to bring flow – and therefore the gases essential for metabolism – inside clumps. Some species, including *Ranunculus* spp. [17] (and probably *Ranunculus fluitans* especially), require considerable water movement for good growth. The outside shoots of a *Ranunculus* clump will grow well if the outside current is suitable even though losses from battering may be great, though the inside shoots are weaker, both because the habitat is less favourable and because they are more liable to be eaten by aquatic snails, insect larvae, etc. The following diagram links the physical effects of flow. Each of these is described separately below.

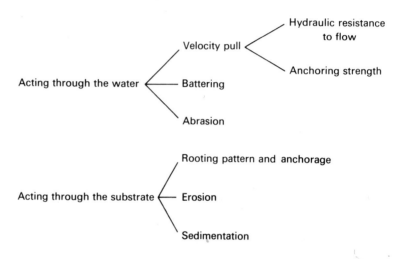

HYDRAULIC RESISTANCE TO FLOW

Any stationary object in moving water resists the movement of that water, so all aquatics offer some hydraulic resistance to flow, though the amount depends on the plant's structure, its size (length, breadth, depth) and shape (branching pattern, leaf and stem shape). The greater the plant's resistance, the greater the 'pull' on it exerted by the water and the more likely it is to be damaged. In general a 'bushy' plant has a high resistance, a streamlined plant a low one, while the bigger a plant is, the greater the impact of the water on it (Figs. 3.4, 3.5; Table 3.1). Small plants, if they are long enough, are curled by turbulence and battered (see below). Large plants are less vulnerable, both because their shoots shelter one another and because they usually have longer roots and so are less likely to be uprooted by erosion. They are, however, much more likely to be pulled away from the substrate, because their large size makes them vulnerable to flow forces. Damage depends on three factors: the force of the water, the resistance of the plant (its 'drag', to use an aerodynamic term), and the force needed to

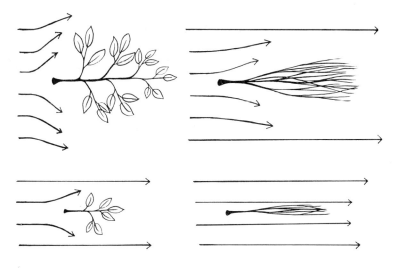

Fig. 3.4

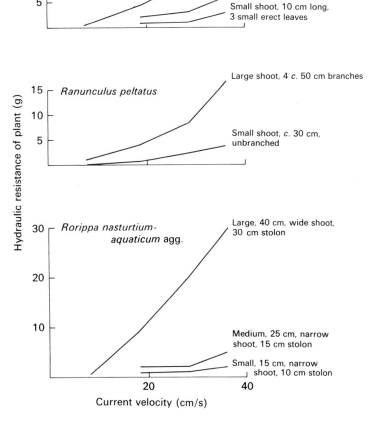

Fig. 3.5. The effect of current velocity on the hydraulic resistance of plants.

TABLE 3.1   *Hydraulic resistance to flow of different river plants*

Tests were done in a flume tank, in water 6.5 cm deep with a surface velocity of 43 cm/s. The plant parts were attached to a spring balance and the pull of plant on the balance was recorded

(a) Hydraulic resistance of typical shoots (branched if this is typical) c. 30 cm long

1   High resistance

Apium nodiflorum	Bushy, semi-emergent
Berula erecta	Bushy, semi-emergent
Epilobium hirsutum	Bushy, emergent
Elodea canadensis (when much branched)	Semi-streamlined submergent
Myriophyllum spicatum (when much branched)	Semi-streamlined submergent
Nuphar lutea	Very large leaves, floating and submergent
Ranunculus peltatus (when much branched)	Streamlined submergent
Rorippa nasturtium-aquaticum agg.	Bushy, semi-emergent

2   Medium resistance

Elodea canadensis (when less branched)	Semi-streamlined submergent
Mentha aquatica	Not bushy, semi-emergent
Myosotis scorpioides	Not very bushy, semi-emergent
Myriophyllum spicatum (when less branched)	Semi-streamlined submergent
Polygonum amphibium	Large leaves, floating shoots
Potamogeton pectinatus (when much branched)	Streamlined submergent
Potamogeton perfoliatus (when branched)	Large-leaved submergent
Ranunculus peltatus (when less branched)	Streamlined submergent
Sparganium erectum	Streamlined emergent

TABLE 3.1 (*cont.*)

3   Low resistance

Callitriche spp. — Semi-streamlined but delicate submergent

Potamogeton crispus — Semi-streamlined submergent

Potamogeton pectinatus — Streamlined submergent

Potamogeton perfoliatus — Large-leaved submergent

Ranunculus penicillatus — Streamlined submergent

Schoenoplectus lacustris — Streamlined emergent and submergent

Sparganium emersum — Streamlined submergent

Zannichellia palustris — Delicate submergent

(*b*) *Hydraulic resistance of different shoots of species, to show the effects of shoot size and branching*

1   Semi-emerged fringing herbs

	Length (cm)	Habit	Resistance (g)
Apium nodiflorum	55	12 large leaves	40
	35	5 leaves	8
	13	3 leaves	5
Myosotis scorpioides	60	15 side shoots	40
	60	10 side shoots (same plant as above)	28
	60	5 side shoots (same plant as above)	10
	60	0 side shoots (same plant as above)	4
Rorippa nasturtium-aquaticum agg.	100	Well-grown	65
	50	Well-grown	20
	6	Delicate	< 0.5

2   Submerged or floating species

	Length (cm)	Habit	Resistance (g)
Callitriche obtusangula	30	Tuft of *c*. 50 shoots	7
	30	Tuft of *c*. 20 shoots	< 0.5
Polygonum amphibium	115	11 leaves	15
	25	8 leaves (same plant as above)	4
Potamogeton pectinatus	190	Large bushy shoot	15
	190	Side shoots over 90 cm removed (same shoot as above)	0.5
Potamogeton perfoliatus	110	3 shoots	15
	110	1 shoot	< 0.5
Ranunculus peltatus	85	5 side shoots	15
	85	Side shoots removed (same plant as above)	3
Ranunculus penicillatus	100	No side shoots, many leaves	8
	200	No side shoots, few leaves	5

break or dislodge the plant. Plant resistances increase rapidly with water speeds, particularly in bushy specimens (Fig. 3.5).

Emerged and semi-emerged dicotyledons are bushy and thus have high resistances. Bushy species at the channel edge may be removed by resistance to flow and by erosion (e.g. *Epilobium hirsutum, Mentha aquatica, Myosotis scorpioides*). Some species growing both submerged and emerged (like *Apium nodiflorum* and *Berula erecta*) are usually short when submerged, and taller and bushier when emerged. The short submerged shoots are close to the ground and so are in the zone of the stream which receives little flow anyway (see Chapter 2), being short, they have little hydraulic resistance, and because they are often overlapping, protect each other from flow (Fig. 3.6). The emerged shoots

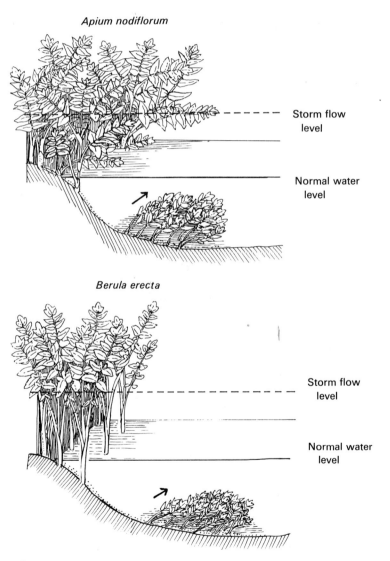

Fig. 3.6

46

are generally larger and set farther apart, and so even though they are flooded only after rain they are more often washed away than are the submerged ones because they offer more resistance to flow. Of the two examples cited, *Apium nodiflorum* has the greater resistance to flow, with more spread-out emerged leaves and sprawling stolons with large terminal shoots. It occurs submerged less often than *Berula erecta*. A different habit is illustrated by *Polygonum amphibium*, which is anchored firmly in the watercourse bank with shoots bearing large tough leaves floating in the water nearby. This plant is associated with slow flow and liability to spate. The high resistance of the floating shoots means they avoid swift normal flows, but because the plants are anchored in the bank, shoots destroyed in storms can be replaced easily. Submerged species are generally more streamlined than emerged ones and thus have lower resistances, though there are exceptions. A submerged species which is exceptionally bushy and so offers a large resistance to flow is *Myriophyllum spicatum* (Fig. 3.7). It often grows in rather soft soils and is easily washed away in storms. At the other extreme are the *Ranunculus* spp., with *Ranunculus fluitans* the most so (Fig. 3.8). This species is the most characteristic of large highland rivers, liable to frequent spates. *Ranunculus penicillatus* is only a little streamlined and in the hills may occur upstream of *Ranunculus fluitans* [18].

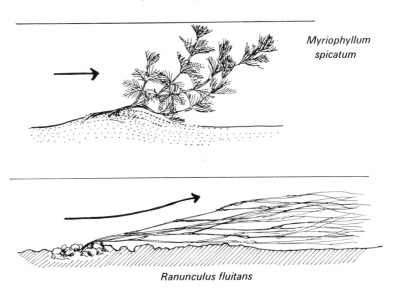

Myriophyllum spicatum

Fig. 3.7

Fig. 3.8

Ranunculus fluitans

Long narrow leaves offer the least hydraulic resistance to flow and thick wide leaves the most. Resistance is also less in flexible leaves than stiff ones. A plant that has both thick, fleshy floating leaves and thin, frail submerged ones is *Nuphar lutea*. When positioned so as to give the least resistance, floating and submerged leaves of the same size offer a similar resistance to flow, but if the floating leaves are twisted they can

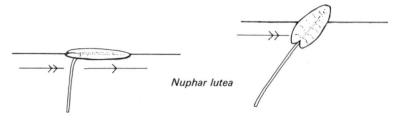

Fig. 3.9

Nuphar lutea

more than double their resistance, even in a flow of very low velocity (Fig. 3.9), and this is one cause of the greater storm damage to floating leaves.

It is not always the case that the streamlined plant has the least resistance. For example, size as well as structure affects resistance and although *Ranunculus fluitans* is streamlined and *Apium nodiflorum* bushy, *Ranunculus fluitans* clumps are often 4 m long and 2 m wide while *Apium nodiflorum* often grows as a carpet about 20 cm high. In this case the *Ranunculus fluitans* clump would offer the greater resistance.

Flow pattern also varies across the stream channel and so a plant at the sides, or low on the bed in the middle, offers less resistance than a trailing plant in the centre.

If shoots are broken or eroded from the bed but not washed away from their clump, then these shoots add extra resistance to the rest of the plant, as does debris lodged in plant parts. After rain the increased flow rate and greater amount of debris in the water both increase the likelihood of damage to the plant by increasing its resistance.

ANCHORING STRENGTH

Anchoring strength is the force holding the plant to the ground. If the pull of the water at ground level exceeds the plant's anchoring strength then, depending on the relative strengths of the above-ground and below-ground parts, the plant will either break off at the stem or be completely uprooted. If the below-ground parts are stronger and are left in place when the stem breaks then the shoots can regrow (Fig. 3.10), but if the below-ground parts are structually weaker then the plant will be uprooted and washed away, and so be lost to the community. Table 3.2 shows the results of testing the anchoring strength of different species. The plants used were of a size typical of each of the shallow-water habitats considered. Larger plants have a greater anchoring strength (i.e. more force is required to break them or dislodge them from the soil) than smaller ones, and small or bitten parts can be torn with less force than is shown in Table 3.2. Some species are strongly anchored (e.g. *Sparganium erectum*) while others are easily broken or dislodged (e.g. *Elodea canadensis*).

The anchoring strength of some species varies with season. For example, *Berula erecta* is anchored more firmly in spring and summer than it is in autumn and winter. This is because its rhizomes, which

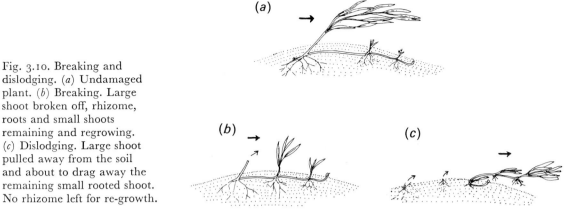

Fig. 3.10. Breaking and dislodging. (*a*) Undamaged plant. (*b*) Breaking. Large shoot broken off, rhizome, roots and small shoots remaining and regrowing. (*c*) Dislodging. Large shoot pulled away from the soil and about to drag away the remaining small rooted shoot. No rhizome left for re-growth.

grow mainly in spring, are short-lived, so that in autumn the older shoots are no longer linked by rhizomes and are more easily uprooted. In other species, e.g. *Ranunculus peltatus*, anchorage varies little with season. Substrate also affects anchorage and species which can grow in various substrates are usually more firmly anchored in coarser soils, where their roots can curl round particles that are too large to be moved easily.

TABLE 3.2   *Anchoring strength*

A pull applied to a plant just above ground (in the direction of the flow) may break the stem, leaving roots and rhizomes in the soil, or it may dislodge the whole plant. In these experiments the pull was applied on a spring balance, to shoots of average size for the site. 'Low force' is defined as up to 200 g, 'moderate' as 250–750 g, and 'high force' as greater than this.

Species capable of rooting well in substrates of different textures are usually more firmly held in coarser ones. The terms 'fine', 'medium' and 'coarse' in brackets after species names refer to substrate texture.

(*a*) *Species usually breaking when pulled*

1   Breaking with low force
     Potamogeton crispus (fine, medium and coarse)

     Ranunculus penicillatus (medium)

2   Breaking with moderate force

     Myriophyllum alterniflorum (coarse)   *Ranunculus peltatus* (fine and coarse)

     Oenanthe fluviatilis (medium)   *Ranunculus penicillatus* (fine)

     Potamogeton crispus (medium and coarse)   *Ranunculus trichophyllus* (medium)

49

TABLE 3.2 (*cont.*)

Ranunculus aquatilis (coarse)

Sparganium emersum (medium and coarse)

3   Breaking with high force

Ranunculus fluitans (coarse)

Ranunculus peltatus (coarse)

Sparganium emersum (medium)

(*b*) *Species usually dislodging when pulled*

1   Dislodging with low force

Apium nodiflorum (fine)

Berula erecta (fine)

Epilobium hirsutum (medium)

Myriophyllum spicatum (medium)

Phalaris arundinacea (coarse)

Rorippa nasturtium-aquaticum agg. (fine and medium)

Sagittaria sagittifolia (fine)

Zannichellia palustris (medium)

2   Dislodging with moderate force

Apium nodiflorum (medium)

Berula erecta (medium)

Epilobium hirsutum (coarse)

Myriophyllum spicatum (medium)

Phalaris arundinacea (medium and coarse)

Rorippa nasturtium-aquaticum agg. (fine and medium)

3   Dislodging with high force

Apium nodiflorum (coarse)

Berula erecta (coarse)

Epilobium hirsutum (fine and (medium)

Nuphar lutca (coarse)

Phalaris arundinacea (coarse)

Rorippa nasturtium-aquaticum agg. (fine and medium)

Sagittaria sagittifolia (coarse)

(*c*) *Species which break or dislodge equally when pulled*

1   Damaged with low force

Agrostis stolonifera (fine)

Myriophyllum verticillatum (fine)

TABLE 3.2 (*cont.*)

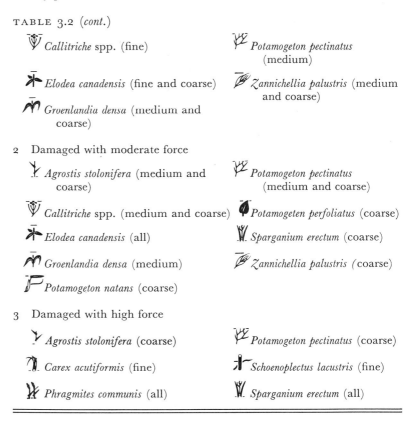

    Callitriche spp. (fine)   *Potamogeton pectinatus* (medium)

    Elodea canadensis (fine and coarse)   *Zannichellia palustris* (medium and coarse)

    Groenlandia densa (medium and coarse)

2   Damaged with moderate force

    Agrostis stolonifera (medium and coarse)   *Potamogeton pectinatus* (medium and coarse)

    Callitriche spp. (medium and coarse)   *Potamogeten perfoliatus* (coarse)

    Elodea canadensis (all)   *Sparganium erectum* (coarse)

    Groenlandia densa (medium)   *Zannichellia palustris* (coarse)

    Potamogeton natans (coarse)

3   Damaged with high force

    Agrostis stolonifera (coarse)   *Potamogeton pectinatus* (coarse)

    Carex acutiformis (fine)   *Schoenoplectus lacustris* (fine)

    Phragmites communis (all)   *Sparganium erectum* (all)

VELOCITY PULL

When a plant is uprooted by flow, this is the result of the interaction of a number of factors, which have been individually discussed above: the force of the current pulling on the plant, the hydraulic resistance to flow of the plant, and the anchoring strength. Together these can conveniently be called velocity pull (see diagram on p. 42). The species most susceptible to velocity pull are those with high hydraulic resistance and low anchoring strength (compare Tables 3.1 and 3.2). For example:

Very
   susceptible: *Apium nodiflorum*   *Berula erecta*

Susceptible: *Epilobium hirsutum*   *Potogameton pectinatus*

     Myriophyllum spicatum   *Rorippa nasturtium-aquaticum* agg. (extremely susceptible when poorly rooted)

The hydraulic resistances in Table 3.1. are less than they would be in faster flows. Even so, where the measured resistances are below 1 g, healthy shoots would be unaffected by the usual pull in their normal habitat. As shoots are frequently grouped into clumps or carpets, they shelter each other and lessen the likelihood of damage. A plant with a low anchoring strength but also offering little hydraulic resistance to flow, e.g. *Callitriche*, is less easily removed than one with a very high resistance to flow, such as *Rorippa nasturtium-aquaticum* agg., which has such a high resistance that its relative anchoring strength is low. Two *Sparganium* spp., *Sparganium erectum* and *Sparganium emersum*, form an interesting contrast because although both have rhizomes and long narrow leaves, other morphological differences cause them to respond differently to flow. *Sparganium erectum* has leaves which are strong in summer and rhizomes near the surface of the stream bed, so that when storm flows pull at the leaves the rhizomes can be dragged away and the plant uprooted. *Sparganium emersum*, in contrast, is much frailer, with very thin rhizomes. This is seldom pulled up, mainly because the leaves tear off before any great pull is placed on the rhizomes, though partly because the rhizomes are lower in the soil.

Shallow-rooted short carpets are formed by *Berula erecta*. If a gap is made in the carpet, either from natural causes (see below) or disturbance, then some shoots and roots become separated from the carpet and the combination of the increased velocity pull on the exposed parts and the erosion of exposed soil can result in a large carpet being destroyed. It is possible, therefore, for an intact carpet to continue spreading in a flow swift enough to cause large-scale erosion if a gap is made in that carpet.

BATTERING AND TANGLING

While one component of flow tends to pull plants from the ground, another component – turbulence – tangles and batters submerged parts. Species may respond differently to the two components since different structural features are involved. For example, a tough bushy shoot is more susceptible to velocity pull (e.g. *Epilobium hirsutum*) while a frail one is more susceptible to battering (e.g. *Sparganium emersum*) (Table 3.3; Fig. 3.11).

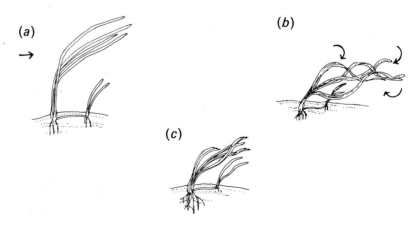

Fig. 3.11. Damage by battering. (*a*) Undamaged plant. (*b*) With turbulence. (*c*) With torn leaves.

TABLE 3.3 *Susceptibility to turbulence*

Tests were performed in flume tanks, typical shoots being placed in different turbulences for different periods.

1 Very susceptible

   Potamogeton obtusifolius[a]

2 Susceptible

   Potamogeton compressus[a]   *Potamogeton pusillus*[a]

   Potamogeton friesii[a]   *Sparganium emersum*[b]

   Potamogeton perfoliatus   *Sparganium erectum* (submerged leaves)

3 Intermediate

   Agrostis stolonifera   *Potamogeton pectinatus*

   Callitriche hermaphroditica[a]   *Sagittaria sagittifolia*[b]

   Myriophyllum verticillatum[a]   *Schoenoplectus lacustris* (submerged leaves)[b]

   Potamogeton alpinus

   Potamogeton crispus   *Sparganium minimum*[a]

4 Tolerant

   Apium nodiflorum   *Nuphar lutea*[b]

   Berula erecta   *Rorippa nasturtium-aquaticum* agg.

5 Very tolerant

   Callitriche spp.   *Phalaris arundinacea*

   Ceratophyllum demersum[a, b]   *Potamogeton natans*

   Elodea canadensis   *Ranunculus aquatilis*

   Lagarosiphon major[a]   *Ranunculus peltatus*

   Mentha aquatica   *Ranunculus penicillatus*

   Myriophyllum spicatum   *Sparganium erectum* (emerged leaves)[b]

   Oenanthe fluviatilis

[a] Characteristic of still water; [b] characteristic of slow waters; unmarked species characteristic of moderate or swift water.

Small parts are more easily moved by turbulence than large ones, and isolated shoots more than ones in clumps. The possible damage is greatest when shoots are separate and in the zone of maximum turbulence, so that each is individually moved with the force of the whole current. Turbulence varies with channel roughness, configuration, etc. (see Chapter 2), short plants on the bed at the sides of the channel receiving least turbulence (Figs. 2.1, 3.2) and long plants in the centre usually the most. When a plant has more than one shoot or large leaf, these protect each other, so plants with susceptible leaves can survive in swift flow provided they are not tangled.

Streams with a continuous high turbulence can support only those species which are tolerant of battering. Where high turbulence is intermittent, occurring only during storms, there may be plant damage during the storm but quick-growing species remaining in place can recover quickly.

The field distribution of various species in relation to flow is shown in Table 3.3 (see also Figs. 2.2–2.7). This distribution is usually linked to their tolerance of turbulence, e.g.

Intolerant of battering. Negligible flow, *Potamogeton obtusifolius*
Intolerant of much battering. Fairly slow flows, *Sparganium emersum*
Tolerate battering. Swift flow, *Ranunculus* spp.

However, more species can tolerate battering than would normally occur in fast flow, since distribution is also related to velocity pull, erosion, etc. Some species without anchoring roots, for instance, are confined to slow or still waters although they tolerate battering (e.g. *Ceratophyllum demersum, Hydrocharis morsus-ranae*).

Turbulence battering is independent of the size and shape of the leaves. Thick leaves are normally resistant, but thin ones may be either resistant (e.g. *Callitriche* spp., *Elodea canadensis*) or susceptible (e.g. *Potamogeton compressus, Potamogeton obtusifolius*). Some species have two or more types of leaves with different susceptibilities. For example, most leaves of *Sparganium erectum* are emerged and tough, but there are submerged leaves which occur in spring and on isolated shoots in deeper water (over *c.* 0.5 m) in summer. These are more susceptible and this may reduce the species in faster deeper channels. The early strap-shaped leaves of *Schoenoplectus lacustris* are more susceptible than the later cylindrical shoots. In contrast the early strap-shaped leaves of *Sagittaria sagittifolia* are more tolerant than the later bladed ones.

It is the damaged parts of plants that are the most vulnerable to turbulence: a partly bitten stalk breaks easily, a leaf with a hole tears easily, a stalk bent by a heavy bird breaks easily, and so on. Thus damage initiated during a storm can be continued in the normal flows, or in the next storm, so that losses due to a storm can actually happen some time later. Debris and silt, which are washed down particularly in storms, much accentuate the damage. The tough leaves or *Nuphar lutea*, for instance, tear easily in storms, but the tears only slightly increase in a further 6 hours of high turbulence.

When whole healthy shoots are damaged, the type of damage depends on the habit of the plant. Long narrow stalks or leaves are bent, and the tissue at the bends is harmed. On flat leaves patches of epidermis are first lost, and then with further turbulence the patches get wider and also deeper as mesophyll is lost, and so large holes develop until finally the whole leaf is lost (Fig. 3.12). Young and adult parts are usually equally tolerant, though the young parts are often protected by older ones. However, if dead or dying parts are present in a clump, these are usually harmed first.

The length of time the turbulence lasts is important, several short periods of high turbulence being less damaging than one long period because the plants can recover and grow again between the storms. *Apium nodiflorum*, for instance, is viable, though damaged, after 6 hours of an extremely high turbulence, but is non-viable after 15 hours.

Many plants that live in flow are damaged by turbulence, and so the relative speeds of growth and loss are important:

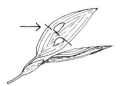

Fig. 3.12

Potamogeton pectinatus	Fast flow: plants small, loss equals growth
	Moderate flow: plants often rather small in main current, large out of it
	Slow flow: plants large, growth exceeding loss
Potamogeton perfoliatus	Largest plants in slower flow than *Potamogeton pectinatus*
Myriophyllum spicatum	Largest plants in both slow flow and faster flow than *Potamogeton pectinatus*
Ranunculus fluitans	Swift large highland rivers: plants often 4 m × 2 m
	Torrential flow: plants to about 75 cm long, with e.g. three shoots
	Very turbulent flow: plants to about 75 cm × 20 cm, but in this site, when turbulence and battering were decreased, the plants reached 3 m × 1 m in a few weeks.

A number of land plants were also tested for their resistance to battering and were all found to be tolerant, or fairly tolerant – the more susceptible water plants were damaged much more quickly than any land plant. Tolerance to battering is not, therefore, one of the special modifications to life in water.

ABRASION

Damage to plants by abrasion is the result of silt and other particles carried in flowing water hitting exposed leaves and stems as they pass by. All storm damage tends to be increased by silt, sand, or debris in the

TABLE 3.4 *Rooting depths of river plants*

(*a*) *Shallow-rooted species*
  (Most roots in the upper 15 cm of the substrate)

1 Species characteristic of moderate or fast flow

Agrostis stolonifera   *Myriophyllum alterniflorum*

Apium nodiflorum   *Myriophyllum spicatum*

Berula erecta   *Potamogeton crispus*

Callitriche hamulata   *Ranunculus* spp.

Callitriche obtusangula   *Rorippa nasturtium-aquaticum* agg.

Callitriche stagnalis   *Veronica anagallis-aquatica* agg.

Groenlandia densa   *Veronica beccabunga*

Mentha aquatica   *Zannichellia palustris*

Myosotis scorpioides

2 Species characteristic of negligible or slow flow

Agrostis stolonifera   *Elodea canadensis*

(*Callitriche* spp.)   *Rorippa amphibia*

(*b*) *Deep-rooted species*
  (Many roots deeper than 15 cm into the substrate)

1 Species characteristic of moderate or fast flow

Sparganium erectum

2 Species characteristic of negligible or slow flow

Alisma plantago-aquatica   *Potamogeton pectinatus*

Carex acutiformis   *Potamogeton perfoliatus*

Carex riparia   *Sagittaria sagittifolia*

Glyceria maxima   *Schoenoplectus lacustris*

Nuphar lutea   *Sparganium emersum*

Phragmites communis   *Sparganium erectum*

Polygonum amphibium   *Typha angustifolia*

Potamogeton natans   *Typha latifolia*

water, both because these cause abrasion and because, by settling on and catching in plant parts, they increase the plant's hydraulic resistance to flow. Abrasion is least where the leaf surface is smooth, as in *Ranunculus* spp., and high where there are projections such as hairs on the leaf surface. In a hairy plant the upper leaves on a clump can be completely removed by a silty storm flow, leaving bare stems above and leafy ones only where they are sheltered in the lower part of the clump (e.g. *Myriophyllum spicatum*). Leaves with epiphytic algae on them are intermediate between the two extremes, because although the algae cause increased abrasion, they can themselves be washed off as a result without damaging the leaf surfaces (e.g. *Elodea canadensis* and *Potamogeton perfoliatus*).

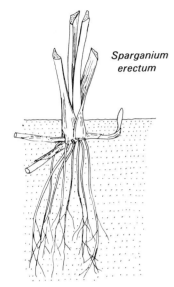

Fig. 3.13. A deep-rooted plant.

ROOTING DEPTH AND PATTERNS

Two characteristic rooting patterns occur in rivers: long deep roots, which are characteristic of plants of slow flows; and short curly roots, which are characteristic of plants of faster flows (Table 3.4). There are, of course intermediates, and the actual pattern varies with the size and the species of the plant, as well as the habitat.

In slow flows fine sediments usually accumulate, into which roots can grow deep. As the upper soil is easily disturbed in storms, long-rooted species can remain in place better than short-rooted ones. The roots usually grow straight down (Fig. 3.13). In faster flows substrates are usually coarser, and if plants are present then the beds are at least partly consolidated. The plants typically have a dense weft of shallow, often horizontal, roots which curl around gravel and stones in the upper soil. The weft is not only securely anchored, but it also helps to consolidate the soil and so increase the plant anchorage (Fig. 3.14). A few shallow-rooted species do, of course, grow in slow flow though in somewhat specialised habitats. *Callitriche* spp. and *Elodea canadensis*, which can quickly recolonise from fragments, tend to grow in fine sediment or in more stable soil at the sides of the channel, while in the larger watercourses *Rorippa nasturtium-aquaticum* agg. and small grasses are usually found at the sides when these are stable. In fairly swift shallow streams with coarse beds silt often accumulates at the sides and a temporary shallow-rooted vegetation may grow in this (e.g. *Apium nodiflorum, Mentha aquatica, Rorippa* spp., *Veronica* spp.) until storm flows remove both the substrate and the plants in it. Larger species with deeper roots (e.g. *Carex acutiformis, Sparganium erectum*) anchor below the temporary silt, and are less likely to be affected.

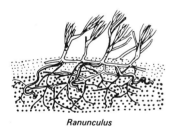

Fig. 3.14. A shallow-rooted plant.

EROSION

Erosion removes soil, and it harms plants when soil around roots is removed or disturbed sufficiently for the plant to be uprooted (Fig. 3.15; Table 3.5). Tolerance to erosion differs from anchoring strength

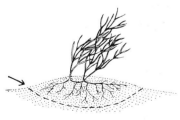

Fig. 3.15

TABLE 3.5   *Susceptibility to erosion*

A horizontal jet of water of constant velocity was directed at the plants, from upstream, to the soil at the base of the shoots, and the time taken for the plant to be eroded was noted. The classification here is based on these times.

The habitats used were typical for each species. Many species, of course, grow in several habitats, and Table 3.7 lists species according to their habitats.

1   Very easily eroded

    Agrostis stolonifera (submerged)   (*Epilobium hirsutum*, rooting fragments)

    Ceratophyllum demersum   *Rorippa amphibia*

    Elodea canadensis   *Rorippa nasturtium-aquaticum* agg.

2   Easily eroded

    Callitriche spp.   *Myriophyllum spicatum*

    Epilobium hirsutum   *Sparganium erectum*

    Myosotis scorpioides   *Zannichellia palustris*

3   Rather difficult to erode

    Apium nodiflorum   *Potamogeton perfoliatus*

    Berula erecta   *Schoenoplectus lacustris*

    Potamogeton crispus   *Sparganium emersum*

4   Difficult to erode

    Glyceria maxima   *Potamogeton pectinatus*

    Nuphar lutea   *Ranunculus calcareus*

    Oenanthe fluviatilis   *Ranunculus fluitans*

    Phalaris arundinacea   *Ranunculus penicillatus*

    Phragmites communis   *Ranunculus trichophyllus*

in that erosion is concerned with the substrate whereas anchoring strength refers to plant parts, and a plant can have a high anchoring strength but a low erosion tolerance, and vice versa (Table 3.6). When underground parts are exposed or moved by erosion, the plants are vulnerable to velocity pull and further erosion, and are likely to be washed away by moving water. Erosion may actually remove soil, but just churning the soil up *in situ* is as effective in removing the plants.

TABLE 3.6   *A comparison of erosion susceptibility and anchoring strength*

Tests were carried out as described in Tables 3.2 and 3.5, using the same plants. Categories as used in these Tables.

	Anchoring strength	*Erosion susceptibility*
Apium nodiflorum	Moderate and high force	Rather difficult to erode
Berula erecta	Moderate	Rather difficult to erode
Callitriche spp.	Low Low Low	Very easily eroded Easily eroded Difficult to erode
Eleocharis palustris	Moderate	Rather difficult to erode
Epilobium hirsutum	Moderate and high High	Easily eroded Easily eroded
Myosotis scorpioides	Moderate	Very easily eroded
Oenanthe fluviatilis	Low to moderate	Difficult to erode
Potamogeton pectinatus	Low	Rather difficult to erode
Ranunculus fluitans	Low to moderate Low to moderate	Very easily eroded Difficult to erode
Ranunculus penicillatus	Low	Rather difficult to erode
Sparganium emersum	Low	Difficult to erode
Sparganium erectum	High High	Easily eroded Difficult to erode
Zannichellia palustris	Low	Rather difficult to erode

Velocity pull and battering occur during normal and storm flows, while erosion is associated only with storm flows (or dredging, etc.).

The various aspects of erosion are grouped here as the variations at different times of year, the behaviour of different species, and the effects of different types of soil.

Many plants are easily eroded in the spring, when the shoots are growing but there is little rhizome and root. During the summer the vegetation is denser, so the soil below the plants is protected from erosion, and underground parts are larger. Isolated plants and broken plant carpets are susceptible, though. Erosion tends to be greatest in winter when the soil is less protected, as much vegetation has died back, and storm flows are more frequent and have greater discharges. Some species have more seasonal variation in plant structures than others, and therefore more differences in seasonal erosion. For example:

Ranunculus spp.	Usually winter-green, with perennial root wefts. Little variation in erosion susceptibility.
Sparganium erectum	Summer-green with deep roots, winter die-back, and slow death of roots in winter. Particularly susceptible to erosion in late winter.
Potamogeton pectinatus	Overwinters in the form of small buds and fruits which are not harmed by soil disturbance. Small plants vulnerable in spring. In summer the rhizome system is deep and complex, and difficult to remove, though easy to damage.

The erosion tolerance of various species is given in Table 3.5. Susceptible plants often have small root systems, so that only a little soil need be disturbed before the whole plant is loose. Similarly, larger plants of any species are usually more tolerant than smaller ones, as their root systems are larger. Rhizomatous species are less easily eroded if a rhizome network is present than if shoots are isolated. If other factors are equal, plants able to grow on several substrate types (e.g. *Callitriche* spp., *Rorippa* spp.) are usually eroded more easily on fine soil, as this is moved by slower water flows than coarser particles.

Differential effects of erosion are demonstrated well by *Epilobium hirsutum*, a common plant of stream banks and frequent in channels of narrow lowland brooks with little erosion. It has annual stolons about 20 cm long. When these are even partly held in the compact soil of the bank they are almost immovable, even if the soil around them is lost. However, shoots of the second year's growth, away from the bank, are likely to have no connection with the bank, and are easily eroded, particularly in spring before the roots are full-grown (Fig. 3.16). These bushy shoots have a high hydraulic resistance to flow, and of course this increases losses. Eroded shoots may become established farther downstream, but though shoots grow well under water the stolon fragments necessarily lie on the soil, above the ground, and the plants are extremely vulnerable, combining as they do a high resistance to flow, easy erosion and low anchoring strength. Most are washed away before midsummer. In the same way some fringing herbs are stable in ordinary storms if anchored in the bank, but easily washed off if they grow away from the bank.

Soil is consolidated by root wefts and dense rhizome networks and so erosion is made difficult (Fig. 3.14). *Zannichellia palustris* is unusually erosion-tolerant for a short-rooted plant in fine soil, since its root weft is extremely intricate. However, if the weft is broken, erosion can be very

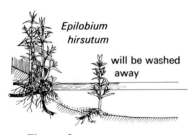

Epilobium hirsutum

will be washed away

Fig. 3.16

quick. Even in a soft substrate, the close rhizome network of *Schoeno-plectus lacustris* prevents large losses.

The very tolerant species form two groups: those that have root and rhizome wefts firmly entangled in coarse consolidated substrates (though these plants are susceptible to erosion when they happen to occur in unconsolidated soils); and those that have large networks of rhizomes and deep roots. *Nuphar lutea* is an example of the latter. It has deep rhizomes which are sometimes eroded in storms, and these are so large that their high hydraulic resistance to flow means that they are easily washed downstream. *Glyceria maxima*, another of this group, has shallow rhizomes, so that parts losing contact with a firm anchorage are vulnerable.

(a)   **(b)**

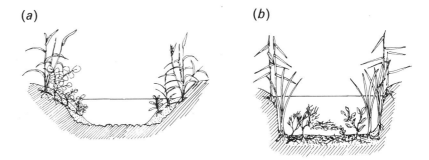

Fig. 3.17. Brook and dyke sections. (*a*) A brook: unconsolidated sediment at the sides. (*b*) A dyke: sediment consolidated by the roots of living and dead plants.

The effects of substrate type on the erosion susceptibility of various plants are shown in Table 3.7. Erosion is greatest for species growing in unconsolidated fine soils and is minimised when the plant has deep roots, a root weft, or if plants are attached to the bank. An interesting difference occurs between brooks and dykes, the latter having little flow or scouring so that their fine soil is hardly moved and can be consolidated by roots (Fig. 3.17). Once it is consolidated, erosion is difficult. Unconsolidated sand of course erodes rather more slowly than unconsolidated silt. Consolidated sand or stone is firm. When there is a hard bed with loose sediment above, rhizomes in the silt above (e.g. of *Sparganium erectum*) are more easily eroded than those in the hard bed (e.g. of *Schoenoplectus lacustris*). A hard bed of clay is eroded more easily than one of gravel.

TABLE 3.7   *Effect of substrate type on erosion susceptibility*

Erosion was tested as described in Table 3.5, in habitats of different texture and type. Erosion classification is as in Table 3.5.

1   Entirely rooted in unconsolidated silt
     Very easily eroded

    Agrostis stolonifera   *Myriophyllum spicatum*

    Callitriche spp.   *Ranunculus fluitans*

61

TABLE 3.7 *(cont.)*

ǰ *Ceratophyllum demersum*   ʇ *Rorippa amphibia*

✱ *Elodea canadensis*   ✿ *Rorippa nasturtium-aquaticum* agg.

ɣ *Myosotis scorpioides*

Easily eroded

ʃ *Epilobium hirsutum*   ψ *Sparganium erectum*

Rather difficult to erode

ψ *Glyceria maxima*   ʄ *Sparganium emersum*

2   Entirely rooted in a silt or peat which, through lack of disturbance and flow, has become consolidated by dead roots

Easily eroded

ψ *Callitriche* spp.   ʃ *Epilobium hirsutum*

Rather difficult to erode

❀ *Apium nodiflorum*   ʄ *Eleocharis palustris*

Difficult to erode

ψ *Glyceria maxima*

3   Entirely rooted in sand

Easily eroded

ψ *Callitriche stagnalis*

Rather difficult to erode

ʠ *Potamogeton perfoliatus*   ✎ *Zannichellia palustris*

✻ *Ranunculus penicillatus*

4   In a mixed, somewhat unconsolidated bed

Easily eroded

ɣ *Agrostis stolonifera*   ʃ *Epilobium hirsutum*

Rather difficult to erode

❀ *Berula erecta*   ʄ *Ranunculus fluitans*

TABLE 3.7 (*cont.*)

 *Callitriche* spp.   *Ranunculus penicillatus*

Myriophyllum spicatum   *Sparganium emersum*

Potamogeton pectinatus

Difficult to erode

Glyceria maxima   *Sparganium erectum*

Schoenoplectus lacustris

5   In a mixed, somewhat consolidated bed

Rather difficult to erode

Apium nodiflorum

Difficult to erode

Epilobium hirsutum   *Potamogeton crispus*

Oenanthe fluviatilis   *Ranunculus penicillatus*

6   In a firm gravel and stone bed, with or without some sediment above

Difficult to erode

Berula erecta   *Ranunculus penicillatus*

Callitriche stagnalis   *Ranunculus trichophyllus*

Potamogeton pectinatus   *Schoenoplectus lacustris*

Potamogeton perfoliatus   *Sparganium emersum*

Ranunculus fluitans   *Sparganium erectum*

Ranunculus peltatus

7   Wedged in havens (see text)

Difficult to erode

Callitriche spp.   *Sparganium emersum*

Myosotis scorpioides   *Sparganium erectum*

Nuphar lutea

(a)   *Veronica beccabunga*

(b)

Fig. 3.18. Effect of a secure haven on plant damage. (a) Before the storm. (b) After the storm.

The effects of erosion may depend on havens, i.e. places where plants may be wedged and secured by stones or similar objects which are not moved during normal storm flow (Fig. 3.18; Table 3.7). These havens for plants are most important for the distribution and continuity of many species on softer soils. Here, fringing herbs may be almost restricted to havens in winter, spreading to form new clumps in spring. After storms, only the plants in the havens survive to form the nucleus of new clumps, and so as the haven is in the same place year after year, the clumps also occur repeatedly in these same places. Where there are no havens, however, clumps which have become established at the sides from fragments from upstream are lost in storms, and when the next clumps grow again from new fragments they are likely to be in different places to the previous patches. So, on softer soils, non-erodible havens determine the positions of stable patches of fringing herbs. Similar havens may be important in the centre of the channel (e.g. *Callitriche* spp., *Sparganium* spp.), though more often most of the centre is either firm or soft and the combination of loose silt and local firm stones that provide havens is less common.

A few other points arise from Table 3.7. Silt banks in streams of moderate flow frequently bear *Callitriche* spp. or *Elodea canadensis* which are easily eroded in storms, and regrow from fragments. *Myriophyllum spicatum* has shallow rhizomes, and although it can grow quickly over soft substrates, there are major losses in severe storms. In sandy streams *Berula erecta* and *Callitriche* spp. are more easily eroded from sand than gravel, and then more easily recolonise sand than gravel. *Ranunculus* (and *Potamogeton perfoliatus*) rhizomes are firmly held in the hard floor below the sand, except for young plants not yet established. Root systems often develop differently in different substrates. *Sparganium emersum* roots, for instance, grow large in deep nutrient-rich silt, but are smaller in coarser, more nutrient-poor soils. This means the plant is eroded more easily from coarser soils, even though the particles there are eroded less easily.

Much information can be gathered from streams containing several types of substrates, and one such site is shown in Fig. 3.19. It was found that vegetation pattern was closely correlated with soil stability. *Potamogeton pectinatus* dominated in deep silt and slow flow (a), its density downstream depending on flow in terms of how many winter buds or fruits were washed down. In spring, young plants were fairly dense on the softer soils, but these are easily eroded, and if storms in May were severe hardly any plants survived, while if storms were absent *Potamogeton pectinatus* could be nearly co-dominant. The softer mixed substrates (b) bore *Sagittaria sagittifolia* and *Sparganium emersum*, and there were sufficient stones present to protect the plants from erosion. Unlike *Potamogeton pectinatus*, these two species have rhizomes living through the winter, so some of the plants occurred in havens. The mosaic of mixed-grained substrates (c) bore *Sparganium emersum* and *Potamogeton pectinatus* on the softer patches. These have straight roots

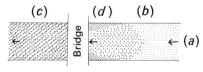

Fig. 3.19

which anchor by growing deep. On the firmer patches *Oenanthe fluviatilis* and *Potamogeton crispus* grew. These have curly root wefts, anchoring in gravel. The firmest substrate was the hard gravel (*d*), where *Ranunculus fluitans* was dominant. Shoots grew out on to the softer soils around, but if growth was poor and shoots small, the plants would be washed off in the next storm. Larger plants develop root wefts, which consolidate the soil somewhat and make erosion less easy, yet *Ranunculus fluitans* was effectively confined to the firm gravel at this site, because only there was it resistant to erosion.

SEDIMENTATION

When moving water carries suspended particles these are deposited when the flow is checked for any reason, such as the land becoming flatter or plants forming obstructions. Deposited sediment varies from nil to several metres deep, but it is normally unconsolidated and unstable. In some reaches sedimentation may be only inside plant clumps, so that when the clump is damaged or dies this sediment is lost also. Up to 15 cm can be deposited in 2 months in normal flows, though slower rates are more common.

There are two basic responses to sedimentation. Either the plants can keep their rooting level on the hard bed, while sediment builds up above, or they can vary their rooting level so that it remains at ground level, even though this ground level is changing (Fig. 3.20). All species vary somewhat between these two alternatives (although individual plants can show just one extreme), but Table 3.8 is a grouping of the common behaviour of some species.

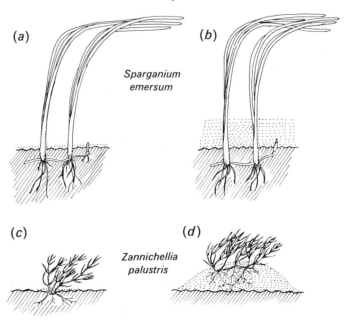

Fig. 3.20. Effect of sedimentation. (*a*) and (*b*), plants with constant rooting level; (*c*) and (*d*), plants with a variable rooting level.

TABLE 3.8 *Rooting level in relation to sedimentation*

Species vary considerably in their behaviour, and this classification is on typical behaviour only. There are no species which never alter their rooting level.

(*a*) *Plants whose rooting level varies*

1 Shallow-rooted species

    Apium nodiflorum   *Rorippa nasturtium-aquaticum* agg.

    Berula erecta   *Veronica anagallis-aquaticum*

    (*Callitriche* spp.)   *Veronica beccabunga*

    Myosotis scorpioides   *Zannichellia palustris*

2 Deep-rooted species

    Nuphar lutea   *Sparganium erectum*

    Potamogeton pectinatus

(*b*) *Plants whose rooting level remains constant*

1 Shallow-rooted species

    (*Callitriche* spp.)   *Ranunculus* spp.

    (*Potamogeton crispus*)   (*Elodea canadensis*)

2 Deep-rooted species

    Carex acutiformis?   *Schoenoplectus lacustris*

    Phragmites communis   *Sparganium emersum*

    (*Potamogeton pectinatus*)

Ranunculus and *Zannichellia palustris* are two hummock-forming species with contrasting responses to sedimentation. In *Ranunculus*, the sediment in the plant clump commonly stays above the root weft, and the thickness of the hummock varies with flow. Sediment is deposited between storms, and removed either during storms or when the hummock is tall enough to reach a zone of faster flow. So in faster currents a point is reached when scour equals accretion. But removal of the upper sediment need not mean removal of the plants. In slow flows silting can continue, and short plants be smothered as a result. In *Zannichellia palustris* the root weft stays around ground level, growing upwards as sediment accumulates. If there is much deposition the plant community is unstable, as both plant and hummock are washed away together.

Thus the plants most vulnerable to sedimentation are (like *Zanni-chellia palustris*) shallow-rooted species with a varying rooting level. In these plants, when sediment accumulates, the rhizomes move up from the hard bed, in which they were secure, and as the sediment deepens the roots lose this contact as well, so that the underground parts are entirely in the loose sediment and are easily washed away in storms.

Harm can be done to the plants when shoots are smothered by silt, though this is rare because the short species with constant rooting levels which might be susceptible (e.g. some *Ranunculus* spp.) usually grow where little sediment accumulates anyway, while the constant-level species of silting streams are large and can tolerate thick sediment above their rhizomes. Damage from sedimentation may also occur when flow is altered, for instance by sluice gates. In this case if flow is checked and much sediment deposited, not only will short species with a constant rooting level be smothered (though fragments may grow in the sediment above), but also plants with a variable rooting level will rise with the accumulating sediment to become rooted in the sediment only and, when the gates are opened and the current increased, be washed away along with the sediment. Another source of harm is when silt and mud are deposited on submerged leaves, because this cuts down the light and so decreases photosynthesis. Interestingly, strap-shaped leaves seldom become covered with silt, while wide leaves, and those divided into thread like segments, are often covered. On the other hand, silting of leaves and stems may be an important source of nutrients (see Chapter 8).

Silt banks frequently develop in sheltered places, between major storms, and are churned up during storms. Submerged banks are usually bare if very short-lived, or bear *Callitriche* spp. or *Elodea cana-densis* if lasting rather longer. Banks near the water level typically bear fringing herbs. As the unconsolidated silt of these banks offers little resistance to erosion and no firm anchorage, and the clumps have a high hydraulic resistance to flow, they are easily washed away in storms.

Gaps in plant cover allow erosion to take place. When species with a variable rooting level rise with increasing sediment, the unconsolidated sediment is easily washed away if a gap is made by animals, cutting, or changes in flow pattern, etc. Carpets of *Berula erecta* often occur on sand or gravel. In shallow streams (e.g. those 20–30 cm deep) current paths are diverted between these carpets, so while sand accumulates inside the carpets, erosion takes place between them, making the ground level lower. Erosion then proceeds sideways, removing the sand under the edge of the carpet, while the tangled rhizomes and roots of *Berula erecta* hold a 10–15 cm strip of plants to the main carpet (Fig. 3.21). As more sand is washed away under the carpet the plants at the edge are lost too. In deeper water this undercutting does not happen, and the sediment can be deeper before erosion occurs.

Once erosion has occurred, and the plants are lost, there are often other plants present which have newly grown on to the hard bed, and

Berula erecta

Fig. 3.21

so far have no accumulated sediment, which can continue the population. This leads to plant cycles (Fig. 3.22). The fastest cycle seen was in *Zannichellia palustris*, where silt hummocks about 15 cm high developed in 2 months, after which a change in flow pattern brought erosion. In *Berula erecta* the fastest cycle seen took less than a year. The length of the cycles depends upon sedimentation rates. Where this is slow, hummocks also develop more slowly and the plants remain anchored to the hard bed for longer, so that in perennial plants the cycle may take several years, while in annuals the autumn die-back occurs before the cycle is complete. Where there is negligible sedimentation, even perennial plants do not show cyclical development.

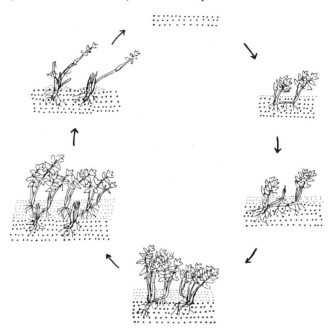

Fig. 3.22. Cyclical development of *Berula erecta* carpet in a shallow chalk stream with moderate flow and some sedimentation.

VEGETATIVE SURVIVAL

River plants can often tolerate bad conditions for a long while, though they stay very small. Many can remain alive for a year with living stems only 1–3 cm long (e.g. *Myriophyllum spicatum, Rorippa nasturtium-aquaticum* agg.). Therefore, in general, uprooting is more dangerous to river plants than battering, since a battered plant can remain alive and regrow when conditions improve. In unfavourable conditions, land plants are usually able nearly to halt their growth. Many aquatics, on the other hand, continue to grow at the apex quite noticeably, even though their photosynthesis is negligible and they do not get larger. They merely take food from the lower end of the plant, which dies, and transfer it to the tip, which grows. This means the rhizomes and the anchoring roots may die in order to allow shoot tips to grow. This can be an advantage in still water, as shoots that had been shaded can float to the surface and perhaps grow better. In moving water, however, plants like this are washed downstream, out of the community.

CONCLUSIONS

Plants in natural streams are in equilibrium with the usual flow in the stream, both storm flows and normal flows, and the plants usually recover quickly from the peak and drought flows which may happen in the river. Thus really drastic damage seldom happens. If the plants are damaged in flow the growth rate must, on average, be at least equal to the rate at which the species is lost as a result of the flow. Different plants live in different types of flow: still and slow waters with soft deep soils have mainly deep-rooted and floating plants; faster flows have species which benefit from flow, either because they grow better in it or because they can anchor better to the coarser soils; and in swifter flows plants are increasingly those which can anchor near immovable objects (e.g. *Ranunculus* spp.) or those propagating easily from fragments (e.g. *Callitriche* spp.). Many streams bear no plants at all, and two possible causes for this are that the flow is too swift and the soil too unstable (see Chapters 5, 6, 11 and 13).

The susceptibility of river plants to the physical components of flow is usually linked to their field distribution. (The correlation is incomplete, though; the physiological and chemical changes with flow variations are described in Chapters 7, 8, 12–14 and 22.) Each species is most closely correlated with the component of flow to which it is most susceptible, for what is important in determining plant distribution is which factor has a lethal effect first as conditions deteriorate. In swift waters the most usual lethal factors are, in order: erosion, velocity pull, sedimentation and battering. These are, however, often linked – for instance, plants of faster currents growing in slow waters are likely to be harmed both by erosion, and by physiological effects of the lack of dissolved gases, etc. reaching the shoots.

The potential hydraulic resistance to flow offered by a plant depends on its structure and size, but its effective resistance depends on its position in the stream, being least if it grows flat in quiet water at the sides of the channel or below an obstruction and most if it is in the fastest current. Plants with a high hydraulic resistance and a low anchoring strength are the most damaged by velocity pull. Even here though, position is important, for a susceptible plant may be protected, by more tolerant plants beside it. Anchoring strength is low when roots are not spread out, when they grow straight in a coarse substrate, when they are short in a fine substrate, or when roots and rhizomes are sparse.

Plants which are most tolerant of erosion usually have a rhizome and root weft in a hard substrate; or deep underground parts; or are protected by immovable objects which provide havens for a few plants in the community and from which susceptible areas can be recolonised after each storm. Tolerance to erosion, battering and some forms of velocity pull increases with increased vegetation cover (except when this leads to sedimentation and therefore poor anchorage), clumps and carpets being less vulnerable than single plants and larger plants providing cover and protection for small ones. However, if the plant cover is broken, damage can spread over a large area.

The amount of sedimentation that plants can tolerate depends both on the species and the sediment. Plants with short roots, which move upwards when sediment accumulates, erode easily and small plants which do not move upwards can be smothered by sediment, while large plants with deep roots are harmed the least by sedimentation.

If parts in the water are harmed the plant seldom dies, as new shoots can grow from below-ground parts. If, on the other hand, a plant is uprooted, it is then lost from the community.

4

River width, drainage order, depth and plant distribution

Width, drainage order and depth are all measures of the size of a stream, and so they usually all increase downstream. Plant distribution is related to width though this is generally because width is in turn associated with flow type, substrate type, and shading from the banks, rather than being the result of the changes in width themselves.

As streams flow to the sea tributaries join them and drainage order (Fig. 4.5) is a measure of this stream pattern. Some plant species increase as drainage order increases, while others increase as it decreases.

Plants are also related to water depth. Those which have most leaves above the water show the greatest connection with depth, usually growing in shallow places, while floating-leaved and submerged plants can be found in various depths of water. Drought is an important factor in shallow places, and summer droughts may kill some submerged plants.

WIDTH

The narrowest watercourses are some headwater streams and some drainage ditches. These channels may be heavily shaded by tall plants (herbs) on their banks, which prevents the growth of water plants in the channel itself (Plate 2). Hedges on channel banks, which are common particularly in the lowlands, can also produce the same effect (Plate 3). Wider channels cannot be shaded by herbs, or completely shaded by bushes or trees, so provided other conditions are suitable, river plants are able to grow where there are gaps in the tree canopy and in the centre of the channel (see Chapters 1, 7).

Some plants can anchor firmly to channel banks, but are only loosely attached to the channel beds (e.g. *Epilobium hirsutum* and, in some habitats, *Apium nodiflorum*). They are able to grow perhaps 0.5 m into the channel while remaining anchored to the bank, but if they grow further they are likely to lose contact with the bank and be swept away in storms (see Chapter 3). As 0.5 m is a much greater proportion of a narrow stream bed than a wide one, narrow streams are likely to support relatively more of this group of plants. If washed away, this narrow band of plants can quickly regrow from the populations that remain anchored to the bank.

Small lowland streams do not receive much storm flow. Thus although the discharge increases during storms, the rise in water level is usually small and emerged and semi-emerged plants remain above water level, with storm flow affecting only their lower parts. Velocity

71

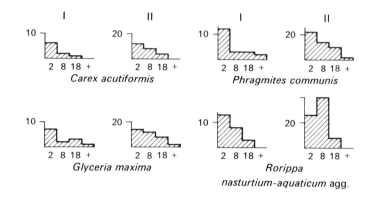

Fig. 4.1. Species most closely associated with narrow channels. Histograms I, on the left, show percentage occurrence in each channel width; histograms II, on the right, show number of occurrences in each channel width. Abbreviations for channel width: 2, up to 2 m wide; 8, 2.5–8 m wide; 18, 9–18 m wide; +, over 18 m wide.

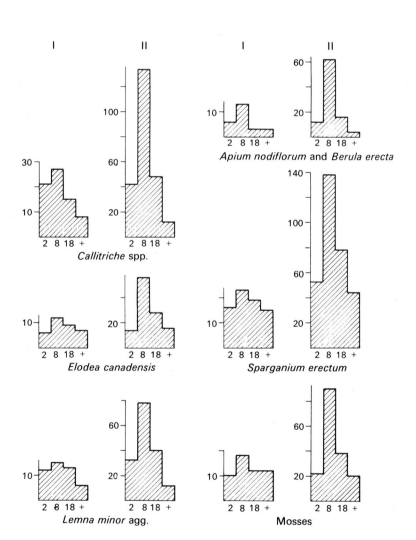

Fig. 4.2. Species most closely associated with channels of medium width. Details as for Fig. 4.1.

pull (Chapter 3) therefore remains low, and damage is also low. In wider streams, where the water level usually rises higher, storm flows are more likely to cover, and then wash away, emerged plants.

The widest watercourses are the lower reaches of large rivers. When watercourses become wider they usually become deeper also, and in wide, deep rivers, plants tend to be near the sides, where the water is shallower, the substrate is often more stable, and they are able to anchor to the banks. Some correlations between plant distribution and channel width are shown in Figs. 4.1–4.4. Plants best correlated with channels 1–2 m wide are mainly large deep-rooted species characteristic of shallow still waters, and also ones very easily washed away (Fig. 4.1). Those best correlated with channels 2.5–8 m wide form a more varied group. Two species more often grow best in narrow channels, though they may grow well in wider ones also (*Lemna minor* agg., *Sparganium erectum*), while the others, when growing abundantly, are best correlated with medium-width channels (Fig. 4.2).

Those best correlated with wide channels are also a varied group ecologically (Fig. 4.3), including shallow-rooted species of swift flows, deep-rooted ones of slow flows, and intermediates (see Chapter 3).

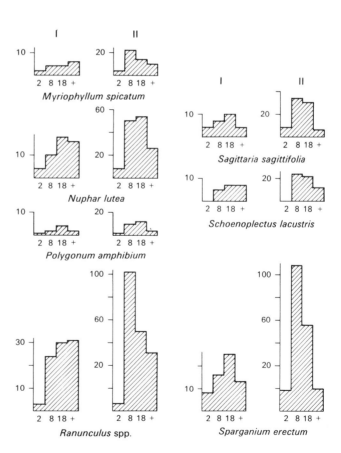

Fig. 4.3. Species most closely associated with wide channels. Details as for Fig. 4.1.

Using tentative identifications of *Ranunculus* species (see Chapter 2), the preferences of the different species are:

> Best correlated with wide channels (including abundant stands):
> *Ranunculus fluitans, Ranunculus penicillatus*
> Best correlated with medium width: *Ranunculus calcareus*
> Avoids narrow channels, correlated equally with other widths:
> *Ranunculus peltatus*

Channel plants not well correlated with width (Fig. 4.4) are also varied ecologically. The bank species are even less likely to be directly affected by channel width.

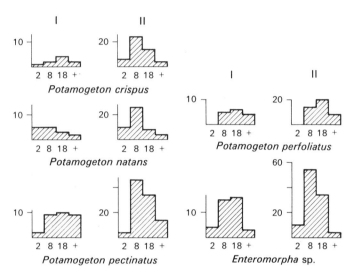

Fig. 4.4. Species poorly associated with channel width. Details as for Fig. 4.1.

DRAINAGE ORDER

As streams flow towards the sea, tributaries join together. First-order streams, those of Drainage Order 1, are unbranched. For convenience, they are defined as those streams which are shown as unbranched on Ordnance Survey maps, scale $\frac{1}{4}$ inch to 1 mile (1:250000), in moist climates [20]. Second-order streams are formed by the confluence of two first-order ones, third-order ones from two second-order ones, and so on (Fig. 4.5). There are, of course, far more streams of a low than of a high drainage order. If any two of the parameters of drainage order, the number of streams of each order, the mean length of stream of each order or the average drainage area of streams of each order are plotted against each other on semi-log graph paper (graph paper with one logarithmic and one arithmetic scale) then a straight line is obtained [20]. An anomaly should be pointed out: if ten second-order streams join a third-order one, the final stream remains of the third order, while if two-second order streams join immediately before flowing into the third-order one, these form a fourth-order stream. The upper reaches

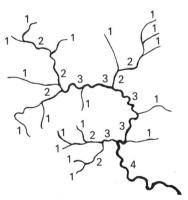

Fig. 4.5. Stream order.

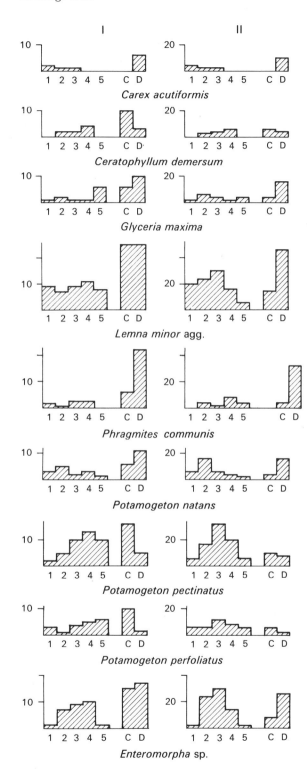

Fig. 4.6. Species best associated with dykes, drains and canals. Histograms I, on the left, show percentage occurrence in each type of stream; histograms II, on the right, show number of occurrences in each type of stream. Abbreviations: 1–5, drainage orders; C, canals and wide drains; D, dykes.

75

of the smallest streams marked on the $\frac{1}{4}$-inch map are often dry for most of the year when they are in the lowlands, but in the mountains perennial flow often occurs higher than the channel is marked on the map. This is probably a consequence of the falling water table in the lowlands (see Chapters 1 and 20): formerly perennial flow extended much farther upstream. However, these two types of first-order stream differ greatly in flow pattern and its associated variables, and thus in associated species.

The species best correlated with dykes and canals are shown in Fig. 4.6, and are of course those also best correlated with still shallow water on soft substrates (see Chapter 2 and below) (*Potamogeton natans*, however, is a more unexpected member of this group). Those plants best correlated with canals are easily washed away in flowing water, and, unlike the tall emergents best correlated with dykes, die in drought.

Species tending to decrease with increasing drainage order are shown in Fig. 4.7. These include the fringing herbs. Species which increase with increasing drainage order, shown in Fig. 4.8, include the plants of both large swift rivers (e.g. *Ranunculus*) and large slow ones (e.g. *Nuphar lutea*). The final group in Fig. 4.9 shows little association between distribution and drainage order.

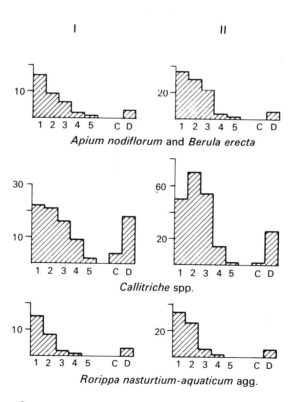

Apium nodiflorum and *Berula erecta*

Callitriche spp.

Rorippa nasturtium-aquaticum agg.

Fig. 4.7. Species decreasing with increasing drainage order. Details as for Fig. 4.6.

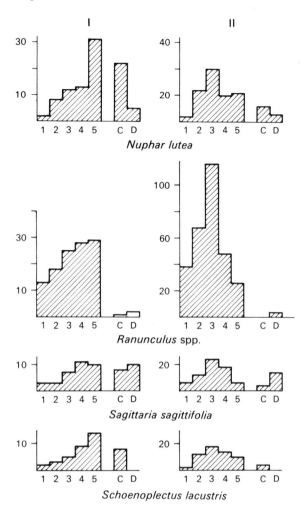

Fig. 4.8. Species increasing with increasing drainage order. Details as for Fig. 4.6.

DEPTH

Shallow watercourses often dry out in summer. Emergents and semi-emergents can usually tolerate such periods, but floating and sub-merged plants usually die quickly if dried. (Exceptions include the aerial forms of *Callitriche* and *Ranunculus* in summer-dry chalk streams.) Short droughts leave living parts of the plant in the soil or on damp mud, and these are able to grow again when the water returns, but droughts long enough to dry out the rhizomes and roots in the top soil are far more damaging, as new growth must come either from deep underground rhizomes or from propagules from outside the community (Plate 4). Some plants can survive in puddles remaining in the lower places, and these can form the nucleus for recolonising the stream. In hot climates a seasonal drought can be the equivalent of winter in

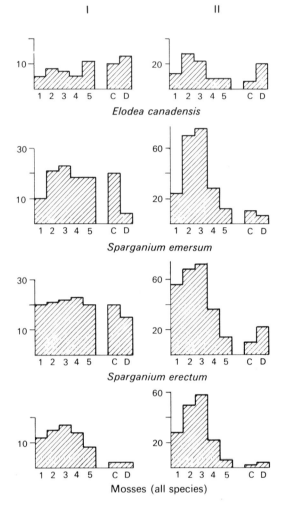

Fig. 4.9. Species showing little association with drainage order. Details as for Fig. 4.6.

temperate zones, as the time of year when the plants are not growing. It has been noted that drying canals in the Sudan for $3\frac{1}{2}$ months had little effect on the final plant growth [21]. In general, however, repeated summer droughts remove submerged and floating plants. A few species can colonise new places quickly, and grow quickly (e.g. *Lemna minor* agg.), and these can take advantage of temporary waters.

Shallow water, rather surprisingly, often harms submerged plants. They grow badly and often die, even though the shoots are in water and have space for growth (so this is a different phenomenon to death from drying). Deep water also causes problems for growing plants. For example, very little light reaches the channel bed (see Chapter 7), and plants may be unable to grow because of this. Also, in still and slow waters particularly, the concentration of dissolved gases may be very different at the top and bottom of deep waters because there is little mixing of the water and the effects of the air do not fully reach the

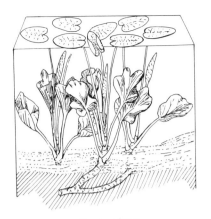

Nuphar lutea

Fig. 4.10

bottom. Most plant-bearing parts of British rivers are less than 1.25 m deep, and at least up to this depth hydrostatic pressure does not appear to affect plant growth. Deep water is necessary, though, for the full development of the larger aquatics such as *Ranunculus fluitans*, whose clumps may reach over 5 m long, 2 m wide and 1 m deep, and *Nuphar lutea*, whose leaves can spread through 1.5 m of water (Fig. 4.10). Thus, although most species will grow as smaller plants in shallower places, the provision of space for the full development of submerged plant parts is an important character of deeper water.

Depth can be a limiting factor in the distribution of tall emergents. Some of these plants may be able to live submerged in the water (e.g. *Epilobium hirsutum*) but most cannot because water is physiologically unsuitable for the whole of the shoot. The leaves of *Phragmites communis*, for example, are unable to photosynthesise under water and thus die, so this plant usually grows where at least a third of the shoot is above water level [22]. Another, *Phalaris arundinacea*, needs the whole shoot to be above water for at least part of the summer if it is to grow well. Even the very widespread *Sparganium erectum*, which does have a submerged form, is sparse and infrequent when permanently submerged (see Chapter 3 for the vulnerability of the submerged form). Most fringing herbs, in contrast, grow well when submerged, provided other habitat factors are suitable (chalk streams are the most suitable). However, these plants, being bushy, have a high hydraulic resistance to flow (see Chapter 3), and if they cannot anchor firmly – if, for example, they are growing in silt – they are washed away in storm flows. Emerged plants, therefore, may be excluded from deep water for either physiological or morphological causes.

The histograms of plant distribution in relation to water depth (Figs. 4.11–4.15), show the depth as that of the main part of the river in

Fig. 4.11. Species most closely associated with shallow water. Histograms I, on the left, show percentage occurrence at each depth; histograms II, on the right, show number of occurrences at each depth. Abbreviations for depth categories: 30, up to 30 cm deep; 75, 35–75 cm deep; 120, 80–120 cm deep; +, over 120 cm deep; T, turbid, probably deep.

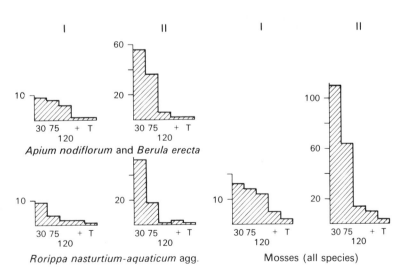

Apium nodiflorum and *Berula erecta*

Rorippa nasturtium-aquaticum agg.

Mosses (all species)

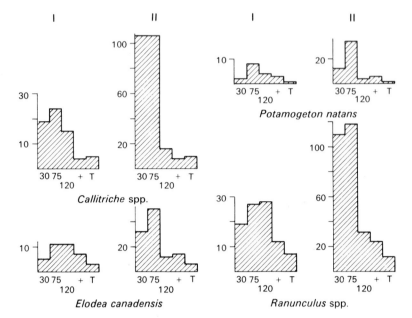

Fig. 4.12. Species most closely associated with fairly shallow water. Details as for Fig. 4.11.

Callitriche spp.

Potamogeton natans

Elodea canadensis

Ranunculus spp.

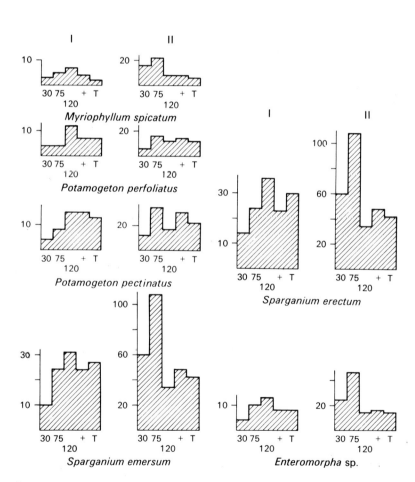

Myriophyllum spicatum

Potamogeton perfoliatus

Potamogeton pectinatus

Sparganium erectum

Fig. 4.13. Species most closely associated with fairly deep water. Details as for Fig. 4.11.

Sparganium emersum

Enteromorpha sp.

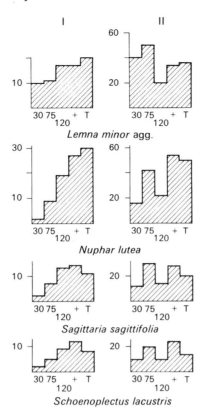

Depth

Fig. 4.14. Species most closely associated with deep water. Details as for Fig. 4.11.

Lemna minor agg.

Nuphar lutea

Sagittaria sagittifolia

Schoenoplectus lacustris

summer. Consequently the correlations are with river size rather than the actual water depth in which the plants are growing. Emergents recorded in deep rivers are growing in shallow water at the edges and even submergents are often confined to shallow bands at the sides.

The species best correlated with shallow streams are the fringing herbs and mosses (Fig. 4.11). Fringing herbs are commoner in the lowlands, where storm discharges are less, and mosses in the highlands, where stable boulders they can attach to are more frequent. The plants best correlated with somewhat deeper streams (Fig. 4.12) are submerged or floating ones. *Callitriche* and *Ranunculus* require high light and fairly stable substrates, and diminish in deep water for this reason. Some *Ranunculus* cannot tolerate very shallow water. The correlations for individual species are:

> Best correlated with water 35–75 cm deep: *Ranunculus aquatilis, Ranunculus calcareus, Ranunculus peltatus*
> Best correlated with water 80–100 cm deep, though abundant stands are best correlated with depths of 35–75 cm: *Ranunculus fluitans, Ranunculus penicillatus*

In fairly deep streams, the best-correlated species come from varied habitats (Fig. 4.13). Although *Sparganium erectum* is placed in this

81

group, in deeper watercourses it is found only at the edges. *Myrio-phyllum spicatum* is a plant of fairly shallow water, growing at the sides of deeper channels, and this is reflected in the correlation of good growth and water under 75 cm deep. The plants best correlated with deep rivers (Fig. 4.14) include also those characteristic of slow flow and silty substrates (see Chapter 2), and some that are indifferent to water depth (e.g. the free-floating *Lemna minor* agg., and *Polygonum amphibium*, whose floating shoots anchor in the bank). Finally, Fig. 4.15 shows species not belonging to the other groups. These include the tall emerged mono-cotyledons of ditches and stream edges. Plants of the bank are of course not causally related to the depth of the channel.

Fig. 4.15. Species poorly associated with water depth. Details as for Fig. 4.11. The tall emergents growing in shallow water either in shallow channels or at the edges of deeper ones also come into this category: *Carex acutiformis*, *Phragmites communis* and *Phalaris arundinacea*.

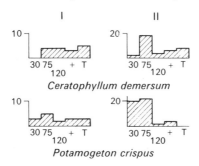

WIDTH–DEPTH ASSOCIATIONS

In becoming larger, watercourses usually become both wider and deeper. Channel width depends on the discharge carried by the channel and on the amount of sediment transported. In any set of circumstances, therefore, width is a fixed variable, although it can be varied somewhat by man, or by erosion or silting. Width is only of minor importance to plants (see above), but it is one of the most important parameters indicating the discharge. Depth is of much greater importance to plants but, because it fluctuates greatly, is less satisfactory to estimate for the season as a whole.

Some width–depth results are shown in Table 4.1. The plants best correlated with narrow shallow streams are the fringing herbs and mosses best correlated with shallow water. In the next group are the plants typical of moderate channels. These include both the plants best correlated with rather shallow water and ones best correlated with somewhat narrow channels (see above). The third group is of plants tending to occur in somewhat narrow though often deep channels. This is the first habitat parameter which links *Myriophyllum spicatum* and *Potamogeton crispus*, two species with many resemblances in their field distribution in the lowlands. The species in the widest, deepest channels naturally include some characteristic of deep waters.

When width is linked to depth, the species groups found are more meaningful as diagnostic groups than those for either width or depth separately. The groups in Table 4.1 are of species of streams of increasing size. The different characters of a watercourse must therefore be studied both singly and in combination to understand why plants grow in the habitats in which they do: singly because each factor acts separately, and in combination because there is more than one factor influencing each habitat.

TABLE 4.1. *Width–depth associations*

1   Concentrated in shallow and narrow channels

    Apium nodiflorum   *Rorippa nasturtium-aquaticum* agg.

    Berula erecta   Mosses

2   Concentrated in channels of moderate depth and width

    Callitriche spp.   *Phragmites communis*

    Elodea canadensis   (*Ranunculus* spp.)

    Phalaris arundinacea   *Sparganium erectum*

3   Concentrated in deep though not wide channels

    Ceratophyllum demersum   *Potamogeton crispus*

    Glyceria maxima   *Potamogeton natans*

    Myriophyllum spicatum

4   Concentrated in wide and very deep channels

    Nuphar lutea   *Sagittaria sagittifolia*

    Polygonum amphibium

5   Wide-ranging

    Lemna minor agg.   (*Ranunculus* spp.)

    Potamogeton pectinatus   *Schoenoplectus lacustris*

    Potamogeton perfoliatus   *Sparganium emersum*

5

Flow patterns and storm damage

Chalk streams have very stable flows, with much of their water coming from ground water springs. Other lowland streams have less stable flows, upland streams vary more and the mountain streams have the most variation in flow, with their large spate flows and lack of springs.

Different plants are damaged differently by storm flows. Some break in the water, and this sort of loss is quickly replaced. Others tend to be uprooted and washed away. Larger plants can shelter and protect smaller or less firmly anchored ones. Storm damage depends also on the type of storm flow and its duration, the longer the storm flow lasts, the greater the harm that is done. In the upper reaches very swift flow scours, breaks, and pulls out plants in the channel. Lower on the river force is usually somewhat less, but storm flows increase water depth, so that badly anchored plants are washed from the edges of the channel. The speed, force and duration of storm flows vary from place to place and from storm to storm, and thus plant damage is also variable.

River plants are influenced by both the normal and the more extreme flows of their habitat (Chapters 1, 2 and 3). Communities may be determined, for instance, by the presence or absence of severe storm flows. Drought flows can be important if they substantially alter turbulence or silting, but generally have a lesser and more localised effect (for drying effects, see Chapter 4).

Storm flows are important in their frequency as well as their intensity, the frequency of spates which can be tolerated by any particular species depending partly on the rate at which it can regrow after damage (Chapter 2). Because plants present in a natural flow regime are those suited to that regime, they can usually recover from storm and drought flows within a few weeks or, rarely, months, and it is only when (as happens every few decades or centuries) storm flows are exceptionally severe that it may take several years, or even decades, to make the damage good. Flow regimes alter very slowly, except with human interference such as flow regulation and water abstraction. On a smaller scale, though, plant patterns within one site may be much affected by storm or drought flows, as individual small populations are washed away and then replaced by other species, or the space they occupied left bare. Storm flows may differ in their effects as the result of differing amounts of debris they carry and the exact position of current paths, but will always damage a declining population, where plants are poorly anchored, more than they will a vigorous healthy population. The rising phase of a storm discharge does more damage than the falling phase, discharge being equal, because it comes first and therefore removes the weaker shoots, and carries the debris which accentuates the damage by pulling on the plants.

The water regime of a stream is of extreme importance to the vegetation, affecting both the species present and their abundance. As has already been seen in Chapters 2 and 3, water movement acts in many different ways. This chapter looks at the wider aspects, and considers the annual patterns of flow and the overall damage done to vegetation by storm flows, and so links up to the overall influence of topography on vegetation described in Chapters 11, 12, 13 and, to a lesser extent, Chapter 6, and gives some of the evidence for predicting the effects of man's interference with the water regime which is discussed in Chapter 20. Flow comprises a complex of habitat factors. Each must be understood singly before the total effect can be appreciated, but this inevitably leads to some overlap in the information given in each section concerned with flow.

FLOW REGIMES IN DIFFERENT STREAM TYPES

Chalk streams have the most stable flows. This is because much of their water comes from springs, which reduces the fluctuations found in streams fed directly from irregular rainfall. Chalk is relatively porous, acting as a storage reservoir and so the annual peak of flow occurs in March or April after this reservoir has been refilled by the winter rains (Fig. 5.1*a*). In streams other than chalk ones the proportion of run-off water entering the channel increases and that of ground water decreases, so that the flow becomes less stable and more spatey. With less underground storage, the annual peak of flow moves earlier, occurring as early as November in the mountains [23]. Soft sandstone has a somewhat porous bedrock also, but less spring water in the streams. Soft clay, on the other hand, is much less porous, so more run-off from the land reaches the streams and the flow is less stable (Fig. 5.1*b* of a mainly clay stream). In clay streams, the still greater instability of flow is matched by the instability of the easily eroded substrate. Much silt is washed in from the land, and deposited silt can be moved in storms. On the hard rocks instability and spateyness increase much more, because the ground is not porous and the rainfall is higher (Fig. 5.1*c*) [23]. The last group of watercourses is those with little flow: the dykes and drains of alluvial plains and the canals. The flow regime is more stable and more predictable, being determined mainly by flood gates, pumps, etc. and storm flows have little effect.

In chalk streams, storm flows are mild, allowing species to survive and regrow. As a result of the springs, perennial flow occurs nearer the source of the stream than is the case for other lowland stream types, and thus submerged and floating plants are also able to grow farther upstream. Soft sandstone streams are similar in having little storm damage, but have less water upstream. The instability of substrate and flow in

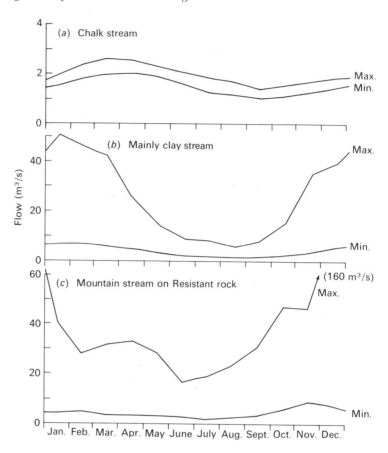

Fig. 5.1. Seasonal changes in flow patterns. Maximum and minimum mean flow values for: (*a*) chalk stream (R. Lambourn at Shaw); (*b*) a mainly clay stream (R. Great Ouse at Bedford); and (*c*) a mountain stream on Resistant rock (R. Dee at Bala). (Simplified after [24]).

lowland clay streams means they are susceptible to summer drought near the source, so few submerged or floating plants grow near the source except in local pools. In the middle reaches plants are often sparse because of unstable substrates, and in large rivers they tend to avoid the centre for the same reason, or because the water is deep. Other stream types are more likely to have plants right across the bed (if they occur at all) as the water tends to be clearer and the substrate more stable.

In the uplands and mountains, the swifter normal flows and the spates are both controlling influences on plant distribution. The spates have great force in the hills, where they can reduce a complete cover of thick vegetation to a few sparse plants, so the common plants in these areas are ones which do not erode easily and which can grow quickly between spates. In highland streams, flatter reaches have less force of water than steeper ones, so where the substrate is stable the flatter parts have more vegetation (probably of different species) and suffer less storm damage. If, though, these flatter parts have an unstable substrate of silt deposited in normal flows and disturbed during spate, then few plants can survive, and these tend to be ephemerals (potentially temporary species, e.g. *Callitriche* spp., *Lemna minor* agg.) or plants anchored to the banks (e.g. *Glyceria maxima, Polygonum amphibium*).

Minor changes in flow may cause minor variations in plant pattern, and conversely changes in plant pattern and growth cause minor variations in flow pattern. For example, plants growing better in moving water (e.g. *Ranunculus*) are benefited by increased current – unless the turbulence is high enough to cause substantial battering. Such small changes are happening continuously, and may benefit or harm individual plants while contributing to the overall stability of the plant community.

TYPES OF STORM DAMAGE

Any plant can be damaged by storm flow, though different species are likely to be harmed in different ways. In Table 5.1 the first group (1*a*) comprises those species whose early damage is usually to parts in the water by battering and velocity pull. If the shoots grow quickly, long-term damage is unlikely in normal storms. When plants which are normally able to anchor securely to firm substrates grow on soft soils, they are more liable to scour and damage. Recovery then depends on the presence and the growth rate of fragments, and is improved if rhizomes and roots are left in place (e.g. *Callitriche* spp.). A subsection of the group (1*b*) is the tall emergents whose shoots or leaves are likely to be damaged when bent over by the water or by debris carried in the current. Although individual shoots can be killed in this way, there is usually little total or permanent damage to the plant populations. Bank plants may also be included here, for they are sometimes low enough on the bank to be flooded after storms. Permanent damage is less than to channel plants, though, since they are flooded for less time. If, however, they are torn from the bank by velocity pull (see Chapter 3), the banks themselves may be damaged by erosion. The best protection against this is for the banks to be colonised by deep-rooted and firmly anchored plants which have a low hydraulic resistance to flow (for example, short grasses and alder trees). In the second group of plants in Table 5.1 the usual initial damage is that the whole plant is immediately uprooted and washed away. This may be the result of erosion or velocity pull (see Chapter 3 and below). Most of these species are bushy and poorly anchored, and typically occur in lowland streams and at the edges of small hill streams where danger of uprooting is least. Small losses are tolerable, but heavy losses would eliminate the species from a site. The third category in Table 5.1 is of plants which are often damaged when clods bearing these species slip into the water. This is most likely to happen if the banks are much trampled by anglers or cattle. Minor storm damage can also result from other processes such as abrasion (see Chapter 3).

TABLE 5.1 *Types of storm damage*

1(*a*) Species commonly torn (broken) above the ground

   Callitriche spp.   *Ranunculus* spp.

   Myriophyllum spp.   *(Schoenoplectus lacustris)*

   Nuphar lutea   *Sparganium emersum*

   Potamogeton crispus   *(Zannichellia palustris)*

   Potamogeton pectinatus

1(*b*) Species with long leaves commonly bent above the ground

   Carex spp.   *Sparganium erectum*

   Schoenoplectus lacustris

2 Species commonly uprooted from the ground

   Agrostis stolonifera   *Myriophyllum* spp. (in more severe storms)

   Apium nodoflorum   *(Nuphar lutea*, with local severe damage)

   Berula erecta   *Ranunculus* spp. (in more severe storms)

   Callitriche spp. (in more   *Rorippa amphibia*
   severe storms)

                      Rorippa nasturtium-aquaticum agg.

   Ceratophyllum demersum   *(Sparganium erectum*, with local severe damage)

   Elodea canadensis   *Veronica anagallis-aquatica* agg.

   Epilobium hirsutum   *Veronica beccabunga*

   Mentha aquatica   *Zannichellia palustris*

   Myosotis scorpioides

3 Species commonly falling into the stream with clods of earth eroded from the banks

   Agrostis stolonifera   *Glyceria maxima*

   Carex spp.

In small mountain streams the flow is usually too swift for flow-susceptible plants to grow in the channel, so storms cannot have much extra effect. On the edges, though, fringing herbs may grow well (see Chapter 13) in normal flows, though they are damaged in spates. Damage is thus concentrated on the edges of the streams. The other extreme is seen in silted lowland rivers with large deep-rooted plants. Here the vegetation is mainly at the sides, where water force is lower and, because the water is shallower, more light will reach the river bed. Damage may be greater at the centre or at the sides, depending on the stream, but it seldom affects the tall emergents at the edges.

Plant vigour affects the degree of damage they suffer. For example, small or unhealthy plants are less firmly anchored than large robust ones (see p. 54) and plants which are partly damaged are often more susceptible during subsequent storms. Also, if soft silt or sand has been consolidated by a root weft and the weft is then broken, the whole carpet of plants can be rolled up and eroded away. Similarly if river plants are cut, this exposes the loose sediment that has accumulated under the plants, and this may be washed away together with the plants rooted in it. Plants that have been partly uprooted often have more hydraulic resistance to flow, and so are more easily removed by velocity pull. For instance, *Berula erecta* has stolons which bear clusters of leaves before they bear anchoring roots. If these stolons are waving free in the current, the parent plant is more likely to be swept away than if it had no stolons.

Storm damage is increased by particles in the water, whether in the form of small particles like silt, or large ones like branches. The small particles cause abrasion (see Chapter 3). They occur throughout the water, but in general do less harm than the occasional large particles, which can pull much vegetation away as they sweep by.

Shoals may accumulate during low flows and then be colonised by quick-growing plants (Plate 7). Shoal and plants may then be washed away during the next severe storm. This is part of the constantly changing, yet stable, pattern of river vegetation. Shoals move downstream during high flows, the size of the discharge needed to move them depending on the size of the particles which constitute the shoal. During the falling phase of a storm flow some of the particles may be deposited, and the level of the stream bed thus raised.

There are several other factors which affect storm damage. For example, the main current in one storm may be in a different position to that in a previous storm – because of differences in water depth or in the size of plants or other obstructions – so the site of the damage may also vary. Rarely, places usually receiving little water force will receive most of the current, and severe local damage will result. Also, species may interact. An easily moved plant (e.g. *Rorippa amphibia*) may grow into large clumps when sheltered by a taller, firmly anchored species (e.g. *Sparganium erectum*) (Fig. 5.2). Then, during a storm flow, the easily moved plant will be washed away, but in the process may

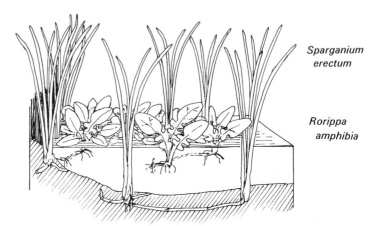

Sparganium erectum

Rorippa amphibia

Fig. 5.2

damage its protector, which in turn may be bent and broken and have its hydraulic resistance to flow increased by the small plants tangled in it, thus rendering it more likely to be uprooted by velocity pull.

The good growth of *Lemna minor* agg. depends on the absence of swift flows, whether these are storm flows or the normal discharge. It grows quickly, and so takes advantage of temporary suitable habitats, such as those sheltered by large leaves (of e.g. *Nuphar lutea*, *Sparganium erectum*) or indentations in the river bank (Fig. 5.3). Even small alteration in the flow, e.g. from light rain, can bring moving water to these places and wash off most of the *Lemna minor* agg., while severe storms cause more damage and can completely remove the species from the site. If there are any plants remaining these can re-form the population. Some streams which have intermittent periods of negligible flow (usually in late summer, or from the use of flood gates) may be completely covered with *Lemna minor* agg. at these times. Fringing herbs can grow very quickly in, for instance, small clay streams, where the water is shallow, competition absent, and nutrient status high because rich silt is present and accumulating (see Fig. 5.4; see also Chapter 12). Under these conditions plants can grow from a few fragments to clumps *c.* 4 m × 2 m in perhaps 4 months. Because these large clumps accumulate silt the shoots become rooted mainly in this silt and lose contact with the hard

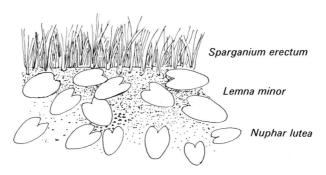

Sparganium erectum

Lemna minor

Nuphar lutea

Fig. 5.3

Apium nodiflorum

Fig. 5.4

bed below, which means they are then poorly anchored. They also have a very high hydraulic resistance to flow, because of their large bushy shoots, and receive the maximum force of the water, because they grow right into the centre of the narrow channel. Thus it is not surprising that such clumps are washed away in the first major storm. A few plants may be left, particularly if these are in havens (see Chapter 3), or new clumps may develop from fragments washed in from upstream. In small clay streams this cycle usually takes (2–)4–6(–10) months, depending partly on the plants' growth rate, and partly on the frequency of severe storms. Where the stream bed is harder, and there is not sufficient deposited silt for the plants to become detached from the bed or to grow rapidly because of the nutrients in the silt, the cycle is slower or non-existent. Chalk and mountain streams tend to have permanent populations of fringing herbs, often of small shoots which are only slightly damaged in the usual storm flows, and do not usually have either the luxuriant clumps or the wholesale losses which are so characteristic of the small clay streams.

DIFFERENT STORM FLOWS

The water force in a stream depends on the discharge, i.e. the mass of water, and the slope of the water surface. The steeper the slope, the greater the force, and the greater the turbulence, and for a given discharge a greater water force means shallower water, and vice versa. The slope of the water depends mostly on the slope of the channel, which is obviously greater in the hills than on the plains, but also depends on the flow at the time concerned, since the water surface is not exactly parallel to the channel bed. The water force affecting the plants is also influenced by the hydraulic depth of the water above them, i.e. the area of the cross-section of the water divided by the length of the wetted perimeter.

When rainfall is heavy and catchment conditions are suitable, the discharge may increase very quickly, leading to a steeply sloping water surface. If discharge increases more slowly during a rising flood the water surface will slope more gently. Given an equal discharge, the steeper flow will have the greater force of the two and do the greater damage. If the slope is the same in two storm flows, the one with the deeper water will have the more force and do the greater damage. The duration of the storm flow is important also, for plants are seldom damaged suddenly, and a storm discharge lasting several days will do more harm than one of equal force lasting less than a day. The poorly anchored plants, frail shoots, etc. will be removed in the first few hours and the degree of loss of semi-resistant parts then depends largely on the duration of a water force strong enough to damage the plants.

The actual damage due to storm flows of known intensity has been recorded from a few streams. One such example is a small chalk (–clay) stream with fringing herbs at the sides and in the centre where this was shallow and gravelly, and *Callitriche* growing in the deeper, slower areas with finer sediment. The effects on the vegetation of three storms are tabulated below.

Peak	Duration	Water slope	Damage
Storm 1			
10 cm above normal flow (vulnerable plants on bank touched only above 8 cm)	*c.* 3 days	Just flatter than in normal flow	(*a*) Up to and including peak flow (1 day). Removed or partly uprooted fringing herbs shallow-rooted on fine soil (*Veronica beccabunga* and *Rorippa nasturtium-aquaticum* agg. particularly). Carpet and root weft of fringing herbs in centre broken and partly rolled up (*Veronica anagallis-aquatica* agg. particularly). Some *Callitriche* pulled from silt, and broken off in water. (*b*) Subsiding flow to 7 cm above normal. A little further damage. (*c*) Subsiding flow below 7 cm above normal (4½ days). Shoots straighten, root wefts unroll, etc.
Storm 2			
25 cm above normal flow	*c.* 1 day	As steep as in normal flow	(*a*) Up to and including peak flow. A little removal of fringing herbs, and *Callitriche*. Fewer shoots bent, fewer shoots exposed. (*b*) Subsiding flow below 7 cm above normal. Recovery.
Storm 3			
25 cm above	*c.* 3 days	—	(*a*) Up to and including peak

92

flow. Considerable damage, particularly to *Callitriche* and *Rorippa nasturtium-aquaticum*, agg. some small stands being halved. *Veronica* spp. less damaged than in storm 1 as they were mainly in firmer substrates. (*b*) Subsiding flow. Recovery.

Damage throughout was mainly through velocity pull, pulling plants from the soil and breaking *Callitriche* within the water. A little damage came from erosion, in removing *Callitriche* from finer soil (in storms 1 and 3 mainly) and *Rorippa nasturtium-aquaticum* agg. (in storm 3). The change in the sediment pattern between storm 1, which was in early summer, and storm 3, which was in late summer, is reflected in the greater damage to *Veronica* spp. in storm 1, and *Rorippa nasturtium-aquaticum* agg. in storm 3. In both instances the finer substrates were the most damaged. The storm water was very silty in the earlier stages of the storm flow, but the amount of general removal (rather than disturbance) of the finer soil was too little to cause much damage. It can be seen from the data that damage depended mainly on the duration and depth of the storm flow, though the detailed effects depended also on the sediment pattern at the time of the storm.

Another stream studied was a side channel of a river in a flood plain, which had an almost flat water surface in both normal and storm flows. In storms the water level rose and the water movement increased very slightly, though not enough to cause damage by velocity pull, let alone by erosion or battering. The data from several storms are tabulated below:

Peak	Duration	Damage
10–30 cm above normal level	1 to several days	Some leaves of *Nuphar lutea, Sparganium erectum*, etc. bent in water, some broken. *Rorippa amphibia* growing in shelter at the side (some not anchored at all, some poorly anchored in soft silt), floated to new water surface and moved downstream.

In such circumstances as these, storm damage is minimal, and apart from a little damage by water movement, deep flooding, whether for a few hours or a few days, does no effective harm to channel or bank plants.

In a shallow medium-sized clay stream that was investigated, the substrate was of mixed particles, somewhat consolidated, with a layer of silt above. The principal species present were:

Myriophyllum spicatum   *Schoenoplectus lacustris*

Nuphar lutea   *Sparganium erectum*

The data from two storm flows can be summarised as:

Discharge	Damage
Storm 1	
2½ times normal flow	Removed some *Myriophyllum spicatum* and *Sparganium erectum*.

Storm 2
4 times normal flow

Removed half the *Myriophyllum spicatum*, washing away most small plants and reduced shoots *c.* 2 m long to *c.* 1 m. Tore or broke over half the *Nuphar lutea* leaves, but there was no noticeable loss of rhizomes. Emerged parts of *Schoenoplectus lacustris* and *Sparganium erectum* were much bent or broken, and a little rhizome uprooted.

In this case velocity pull was responsible for most of the damage, but erosion, abrasion and flood-bending also caused some harm. The damage done in storm 2, which occurred during the summer, was all made good in about 2 months. The loss to *Myriophyllum spicatum* was potentially the most serious, partly because this was the only species to lose much rhizome, and partly because the horizontal rhizomes, by which it spreads, grow in only very limited conditions. In the year in question, growth was possible both in spring, when the initial population developed, and in mid-summer, when the recovery took place. If the second growth had not been possible the population could have been seriously reduced by later storms, and if these had been severe the population might even have been lost. Thus it can be seen that the interaction of biology, damage and storm frequency is important (also see p. 67).

In the three examples cited above and the one described on p. 96, velocity pull was the most important cause of damage in two, battering in one, and the fourth had negligible damage anyway. These were all lowland streams without either spates or temporary shoals of silt, etc., and the data were recorded in summer, which is the time of year when the plant cover best protects the soil from erosion. Erosion is the most important cause of damage to sites where silt, sand or gravel accumulate during low flows. These include plant clumps which, as sediment accumulates, become rooted only in the sediment. Erosion and velocity pull are jointly responsible for most of the autumn wash-out damage, and much spate damage in the hills.

Exceptionally severe storms occur every few decades or centuries and cause severe erosion, with the loss of rhizomes and roots. It may take years, perhaps more than a decade, for the vegetation to recover fully from scour severe enough to cause an unstable substrate (e.g. R. Tees [25]). In a silting flow, silt can be deposited on the bed to form a new suitable substrate for plants within a year or two, but if a firm gravel or stone bed is broken then the recovery time is a lot longer.

DOWNSTREAM VARIATIONS IN STORM DAMAGE

The parameters of storm flow affecting plants are the water force, turbulence and depth, and the particles carried, all of which vary along a river.

The upper reaches of lowland streams are now often dry except during storms, and so the storm flows bring force and turbulence to plants which are otherwise not exposed to either (depth is irrelevant). The plants are necessarily emergents, and are mostly bushy. They therefore offer much hydraulic resistance to flow and, as most of them have not developed in flow, they may be insecurely rooted or rooted in loose soil. Quite frequently, tall dense emergents are able to grow near the source of a stream where the storm flow is very little (e.g. *Epilobium hirsutum, Solanum dulcamara*), whereas a little farther downstream (where summer storm flow is greater) plants are both sparser and shorter, and the species found are those offering less hydraulic resistance to flow and often those which have the ability to recover quickly after damage (e.g. *Apium nodiflorum, Veronica beccabunga*).

In the hills, upper reaches of streams usually have a normal flow that is swift and turbulent. The turbulence is usually enough for the additional turbulence of storm flows to have little extra effect, and here the importance of the storm flow lies in its increased force and the increased depth which brings this force to fringing plants growing at the edges of the channel. Channel plants are usually few, but in the infrequent streams where the substrate is stable in the usual spates and normal flow is not too fierce, channel plants may grow thickly (e.g. *Callitriche* spp., *Myriophyllum alterniflorum, Ranunculus aquatilis*). Fringing herbs cannot accumulate much silt, since not much silt is carried in the water, and as they are anchored to a firm substrate they are less easily swept away than in the lowlands. All fringing herbs, however, are bushy and will be swept away if the water force is sufficiently great, so will only grow well in those few mountain streams with an unusually low water force and some accumulation of sediment.

In the middle reaches of both lowland and hill streams the increased turbulence brought by storm flows is important. Plants are tangled and battered and locally pulled from the soil, and temporary and minor damage is common. The increased force is relatively less, though, than in the steeper upstream reaches. At the margins, water force is usually low and hardly affects tall emergents. However, the edges are usually silted and so the fringing herbs, which are shallow-rooted, do not anchor well. When the water level rises these plants are flooded, their soil softens, and because of their bushiness and insecure anchorage they are swept away by water without much force. This is a major reason why fringing herbs decrease downstream, except in chalk streams where the water level is more stable and there is less silting.

The substrate stability in middle reaches of streams varies. Chalk streams usually have a firm stable bed and low storm flows. There is thus little scour and plants typically grow all over the channel. Clay streams, on the other hand, have more silt, a less consolidated hard floor and greater storm flows. In these there can be much scour and few plants in the channel centre. In the hills, rivers with deposited, unconsolidated gravel are very unstable and few or no plants are able to grow, good vegetation only being found when the gravel is consolidated or is stabilised by near-by boulders.

In the lower reaches of rivers, in or near flood plains, there is seldom much force or turbulence in storm flows, and their main effect is to raise the water level, floating off poorly anchored plants from the edges, bending leaves, etc. They frequently deposit silt, though, and this may have a smothering effect (see Chapter 3). The total damage is little.

SEASONAL VARIATION IN STORM FLOWS

Most river plants die down in winter, so their above-ground parts cannot be harmed by battering or velocity pull in winter storms. The soil may be churned up and washed away, though, uprooting the underground parts of some species (e.g. *Sparganium erectum*). Some plants remain green but have much smaller populations or shoots (e.g. *Nuphar lutea, Rorippa nasturtium-aquaticum* agg.) while others stay quite large in winter (e.g. *Berula erecta, Ranunculus fluitans*). Most *Ranunculus* species are winter-green, and so offer hydraulic resistance to flow in winter storms. They are unusually resistant to erosion, and in the mountains they are exposed to fierce water in spates. Their second main habitat is chalk streams, where winter storms are the least.

Most aquatics start active growth between March and May. Shoots usually grow before roots, so a storm flow may do more damage in spring than in summer when the shoots are anchored. Also, there is more bare soil in spring, so erosion is more probable. Summer storms are the least likely to cause erosion as the protecting plant cover is greatest. Most die-back is between September and December. Dying shoots are battered and washed away more easily than healthy ones. Tree leaves fall in autumn, and large mounds can accumulate in the streams, to be decomposed and be washed farther downstream.

In summer, the exact timing of damage may be important. In Lincolnshire, a stream has been studied [19] which is *c.* 3 m wide, with a usual discharge of 10–40 cusecs (cubic feet of water per second). It is dominated by *Potamogeton pectinatus*, a species growing very quickly in early summer and dying back in autumn. A storm flow of 100 cusecs during rapid growth did a little damage which was quickly replaced. A later storm of 750 cusecs removed perhaps 80% of the vegetation, and although the plants were still growing slowly, the damage happened too late to be replaced. Consequently the late summer vegetation was only about a quarter of that normally expected. The next year, however, the plants grew well. The stored food in most aquatics seems to be enough to initiate more than one year's growth.

Several river plants have a strongly seasonal growth cycle. Damage during rapid growth is fully replaced, but later damage is not, and if the damage recurs several years running it may lead to the decrease or loss of the population (e.g. *Phragmites communis, Potamogeton pectinatus, Sparganium emersum*). Another group of species can grow at any season, other conditions being satisfactory. In these damage can always be followed by regrowth, but at some times of the year, e.g. mid-winter in Britain, growth is slow (e.g. *Callitriche* spp., *Elodea canadensis*). There is also variation in the seasons in which plants can spread quickly horizontally, colonising unoccupied ground. Examples of typical behaviour are:

Any season: *Callitriche* spp.

Early spring: *Berula erecta*

Early summer: *Groenlandia densa, Myriophyllum spicatum, Zannichellia palustris*

Late summer: *Apium nodiflorum* (frequent), *Hippuris vulgaris* (hot summers)

Early autumn: *Schoenoplectus lacustris*

In species with horizontal spread, the timing of the damage can affect its outcome. If it occurs before the spread the losses are quickly made good, provided some plants remained. If, on the other hand, the damage comes at the end of the annual growth period, the stand is left poor and sparse and liable to further damage and perhaps extinction in later storms.

6

Width–slope patterns

The width and slope of a stream, unlike the water speed, depth, etc., remain constant from day to day and nearly so from year to year. Most of the flow, substrate, sedimentation, and depth characters of a stream are summarised in, and can partly be deduced from, the width–slope pattern.

Topography affects the fierceness of flow and thus the distribution of plants. In the mountains the steeper and narrower streams have no large plants, but almost all upland and lowland streams can potentially contain vegetation. Luxuriant vegetation, however, is confined to certain parts of the width–slope pattern, its distribution depending on rock type and topography.

Different species are found on different width–slope patterns (referred to here as the pattern of the species), and they may occupy some parts of this pattern only in selected habitats – for instance in deeper water, or in places liable to spate, etc. The distributions can usually be interpreted in terms of the habit of the plant. The diagrams in this chapter can be used for predicting the changes in plant distribution that will occur after alterations to the river, such as flow regulation.

The total size of a stream channel is defined by its width and depth. The shape and size in turn depend on the discharge, the amount of bed sediment transported, the substrate type and the toughness of the channel banks. The movement of sediment on the channel bed partly depends on the channel slope. The slope of the river bed, and of the water surface, are usually fixed by the hilliness of the land, and minor man-made alterations, such as flood gates, do not affect the overall pattern.

Width and slope are usually constant, while discharge, sediment and depth may vary from day to day. It is therefore useful to consider width and slope as independent variables. The width used here is approximately that of the channel near a road bridge, and the slope is that on the 1 inch to 1 mile Ordnance Survey map (1:63 360) above the road bridge. This gives the general slope but not the minor variations: many streams, particularly hill ones, have alternating stretches of faster steeper flows and slower, deeper flatter reaches, and these are not measured separately here.

The roughness of a straight channel depends on the size of the particles of the bed and the banks, the shapes of these (ripples etc. on the bed, undulations in the bank), the river plants and the water depth. If discharge and, consequently, depth much increase, then since slope and width hardly vary roughness has to decrease. This means that vegetation is removed. Therefore, although flow and substrate inter-actions damage and remove plants in different ways and in different quantities in different conditions, this plant loss falls within the general physical equation determining roughness.

The width–slope pattern summarises most of the physical characteristics of a river which affect plants, so variations in discharge, substrate, loose sediment, depth and the height from hill top to stream bed can be deduced in part from the width–slope patterns. Since most plant nutrients come from the silt and mud, nutrient preferences can also be deduced, if the flow pattern is known. If slope is fixed, a fixed amount of turbulence can come either from a shallow slower stream or a deeper swifter one. Plants which require a silting flow or an eroding flow will grow only in particular combinations of depth and flow.

REGIONAL PATTERNS

Regional patterns of plant distribution are shown in Figs. 6.1–6.4. In each diagram one line surrounds those stream sites which have angio-sperms (ignoring those with only temporary grass clumps or temporarily flooded grass at the edges). When a second outline is present, this indicates sites without angiosperms. Streams are steeper towards the left of each diagram and wider towards the top, so that sites at the upper left have the fiercest flow and those at the lower right the least water movement.

The patterns in Fig. 6.1 all have a similar general shape. Fig. 6.1(*a*), (*b*) and (*c*) are of mountainous regions. In each there is a band of

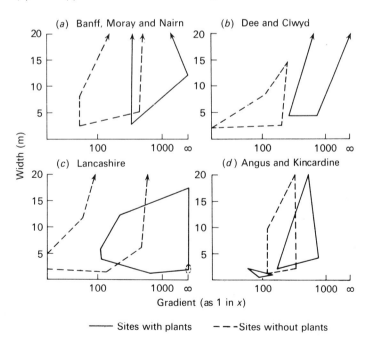

Fig. 6.1. Regional width–slope patterns: very mountainous regions. The sites shown in these patterns are from the regions named, but are not necessarily representative of the topography of the whole of these regions. The gradient (or slope) is taken from 1 inch Ordnance Survey maps and is plotted on a semi-log scale, so that 100 represents a gradient of 1:100, 1000 a gradient of 1:1000, and ∞ a gradient substantially less than 1:1000. The channel width is plotted on an arithmetical scale.

99

streams to the left without angiosperms, because the fall in height from
the mountain top to the stream channel is great, the water runs off with
great force, and the flow in this outer band of sites is too fierce for these
plants to be able to grow. The substrate is affected by the swift flows
too. Even over to the right, in the downstream reaches, the substrates
are mainly gravel or stone, except in flood plains. If the particles are
consolidated into firm beds many species can grow there, but unconsolidated gravel and stone are most unsuitable for plant growth because
they are moved in spates, disturbing and squashing the below-ground
parts of the plants, which are thus confined to sheltered and stabilised
places or are absent altogether. Silt may occur in local sheltered places,
and plants requiring either higher nutrient levels or a fine substrate for
rooting are confined to these areas.

Where the patterns of the sites with and without angiosperms overlap, there are several possible causes for this. For example, the streams
may have alternating stretches of swifter water without plants and
quieter water with them, and of the sites recorded some may be on the
swifter and some on the quieter reaches. Alternatively the sites without
plants may have fiercer flow because the hills beside them are higher;
or the sites without plants may be polluted, shaded, recently dredged,
etc.

If a mountainous district includes some lowland regions as well, the
lowland sites with plants will be placed on the diagram near the
mountainous ones without them (Fig. 6.1*d*). This is because the lowland
regions have only a small fall from hill top to stream channel compared

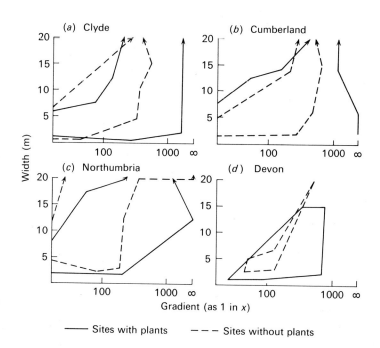

Fig. 6.2. Regional width–slope
patterns: mountainous
regions. Details as for Fig. 6.1.

—— Sites with plants   – – – Sites without plants

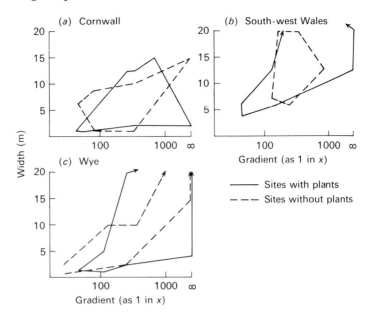

Fig. 6.3. Regional width–slope patterns: uplands. Details as for Fig. 6.1.

with the mountainous regions, and so in the same place on the width–slope pattern flow may be fierce or gentle depending on this fall from hill to channel. In the very mountainous region of northern Scotland, where gradients are very steep and falls from hill to valley are great, most streams do not bear angiosperms. Plants occur in those few channels where the gradients are flatter or – more often – the falls from hill to valley are less. *Juncus articulatus* occurs in fiercer flow than the other species. In flat watersheds the bog streams, with much plant growth, have narrow and nearly flat channels.

Fairly mountainous areas, where the flow is less fierce and the hills less steep than those illustrated in Fig. 6.1, are shown in Fig. 6.2. Smaller streams in this group may resemble those in the previous group, but the total pattern differs. Sites with plants occur almost throughout the pattern, plants can grow even in some of the steepest sloping streams. However, most of the sites without plants are in these smaller steeper streams, which are likely to have shallow turbulent water, or greater falls from hill top to stream channel, or to be polluted, etc. Almost all flat or downstream sites have plants, whether these are growing in the channel, on shoals, or as bank emergents.

Upland regions have lower hills (Chapter 1 and 13), less force of water and usually finer substrates (Fig. 6.3) than mountainous areas. More silt is produced from sandstone than from the Resistant rocks and more can accumulate in upland than in mountain areas. Since upland areas are often sandstone, their increased silt is the result of both these factors. An upland region usually includes some mountain and lowland streams, which unfortunately obscure the pattern. Species which cannot tolerate the extreme flow and substrate regimes of the mountains, but

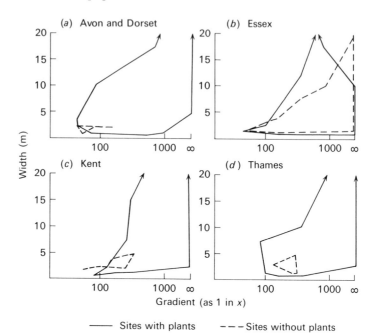

Fig. 6.4. Regional width–slope patterns: lowlands. Details as for Fig. 6.1.

——— Sites with plants   − − − Sites without plants

can withstand some spate, are able to grow in the upland streams and as the mountain species can potentially occur here too, these streams more often bear plants, are more species-rich, and more often contain dense vegetation than the mountain streams. If silt is deposited downstream and is usually undisturbed in spates, vegetation can be abundant, though silt which is frequently disturbed bears few plants. As flow becomes less fierce, other factors (such as shade, pollution, recent dredging, etc.) become relatively more important as the overall controlling factors of plant distribution. It is only where flow alone is fierce enough to exclude plants from a site that the other possible limiting factors are irrelevant.

Patterns of plant distribution in lowland streams are shown in Fig. 6.4, and it can be seen that sites with angiosperms occur throughout the diagram. In this region the storm flows are of much lower force, allowing plants to grow even in the steepest sloping streams. In general species requiring slow flows and fine substrates can do well, though chalk streams have less silt than those on other rock types. In clay streams plants may be excluded as a result of a combination of unstable substrate and unstable flow (see Chapters 5 and 12; shown in Fig. 6.4). For example, Essex is entirely clay, and in the pattern from here no sites are empty solely because of the width–slope combination. Pollution, shading, channel management practices, etc., may all exclude plants from sites. Shading is of greater importance here than it is in the hills: for example, small brooks often run beside, and are shaded by, hedges (Plate 3).

DISTRIBUTION OF DENSE VEGETATION

The development of a luxuriant river vegetation depends on the interaction of flow regime and geology. In lowlands, catchments often vary in geological type, so although small streams often have catchments of only one rock type, large rivers are often from two or more rock types. The distribution of sites full of plants (on a simple subjective assessment) on soft lowland rocks is shown in Fig. 6.5(a). These are in places with the least fierce flows: on flatter and wider sites. This is because in mild flows silt accumulated downstream, and if it is not churned up is able to bear luxuriant vegetation. The upper reaches of clay and sandstone streams have beds which are easily eroded and, particularly on clay soils, much silt enters the streams and can build up in plant clumps, in shoals, etc. However, this is then disturbed in storm flows and consequently these upstream sites are seldom full of vegetation. On chalk, in contrast, the bed is firm, silting is less and a perennial flow extends farther upstream, and thus the steeper streams can be full of vegetation.

The distribution of 'full' sites on hard rocks is shown in Fig. 6.5(b). The outline for Resistant rocks is to the left and above that for soft rocks, that is, on wider and steeper streams. The boulders of steeper hill streams do not move easily, even in spates, and help to stabilise gravel near them, which if it is stabilised and firm can allow the development of a good vegetation. Thus some hill streams on hard rock can bear abundant plants on steeper slopes than can streams on a soft rock. Lower on hill streams, however, the substrate is often deposited unstable gravel or stones, so plants are sparse. In the lowlands, the best downstream habitat for good growth of vegetation is the flatter narrower rivers. In the highlands this habitat is poor, since the substrate tends to be unstable gravel rather than stable silt, but in larger highland rivers, with their greater discharges, the substrates tend to be firmer, with

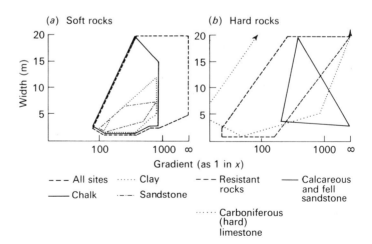

Fig. 6.5. Stream sites full of vegetation. Details as for Fig. 6.1.

little deposition and a consolidated bed that is seldom moved in spates, so here plants are able to grow thickly.

The smaller streams of flatter country on Resistant rocks (Fig. 6.5*b*) are often full of plants. Water force is low here, as in other lowlands. On the hard sandstones the pattern of full sites is intermediate between that of the Resistant rocks and that of soft sandstone (Fig. 6.5*a*). The full sites on hard limestone (Fig. 6.5*b*) form a larger pattern than on any other rock type. The more extreme points on the pattern are from places without fierce spates and a regulated flow. The very steep sites are from the Cheddar Gorge, which has no spates, and the very wide ones are from the Peak District, where the flows are upland rather than mountainous. Limestone forms a fertile rock, but in mountainous sites it is, as always, the fierce flow that is the factor controlling plant distribution.

SPECIES DISTRIBUTION

Different plant species are found in areas with different width–slope patterns. Figs. 6.6–6.11 show the width–slope patterns of various species derived from the vegetation of around 3500 sites throughout Britain. Because many species are very sparse a study of this size may be incomplete, and further study could well add a new habitat (e.g. a new rock type or flow regime) and thus a part to the width–slope pattern. When a few sites stand outside the general pattern, these discrepancies are usually attributable to interference with the flow regime, for instance, a species may occur in quiet water above a weir in a stream otherwise too steep for this species, and such sites have usually been excluded from the diagrams. Most species are more frequent in the lowlands than the mountains, and consequently the diagrams are more accurate for the lowlands.

Some plants occur in much the same width–slope pattern whether they occur in, say, lowland Essex or highland Aberdeenshire. Variations in the flow pattern as a result of topography are then irrelevant to the distribution. Most of the patterns do differ with topography, though, and these variations are described below. The patterns can often be subdivided according to stream depth, the narrower and steeper sites being shallower than the wider and flatter ones. Wider rivers have a faster flow than narrower ones on the same channel slope. Deeper streams have a faster flow type than shallower ones, other factors being equal. Thus if a plant requires a slow flow type, it grows in deeper water in wider than in shallower streams. Such species require a certain flow-and-substrate regime. If, on the other hand, a plant is almost confined to a single depth category, it is probably influenced more by depth than by flow regime.

Plants typical of upper reaches have width–slope patterns like those illustrated in Fig. 6.6. They are concentrated in narrow streams and shallow water.

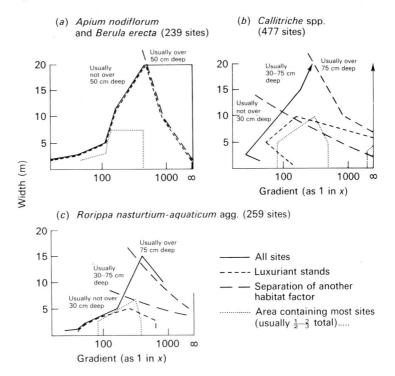

(a) *Apium nodiflorum*
and *Berula erecta* (239 sites)

(b) *Callitriche* spp.
(477 sites)

(c) *Rorippa nasturtium-aquaticum* agg. (259 sites)

——— All sites
- - - - - Luxuriant stands
— — Separation of another
habitat factor
············· Area containing most sites
(usually $\frac{1}{2}$–$\frac{2}{3}$ total).....

Width (m)

Gradient (as 1 in *x*)

Fig. 6.6. Species of narrower
and shallower streams.
Details as for Fig. 6.1.

Apium nodiflorum and *Berula erecta*, grouped together (Fig. 6.6*a*) are most frequent in the steepest streams, though most stands of good performance are placed only a little left of centre of the pattern. In small brooks clumps are often sparse because of the storm flows, which in the hills are spatey and in the lowlands scour the unstable, often dry, substrates. In small lowland brooks without scour, though, taller species can dominate. Abundant populations are mainly found in lowland streams with a firm medium-grained bed (typically chalk streams), for here the plants can anchor and sedimentation is little. In wider streams they are confined to the margins, as a result of the centre having too strong a flow, the water being too deep or too turbid, or the substrate being too coarse or unstable for these plants. The edges usually have fine soil, so that plants are easily washed away in storm flows (see Chapter 5). This means the populations in such streams are often temporary, developing from fragments washed down from upstream, and being themselves swept away in due course.

Rorippa nasturtium-aquaticum agg. has the same overall pattern, but its sites are more frequent in the centre of the diagram, and most of its well-grown populations are in steeper streams. It is excluded from most wide and deep rivers for the same reasons as *Apium nodiflorum*. *Rorippa nasturtium-aquaticum* agg. frequently grows very quickly, growing over any other short or submerged species present. It can cover small channels but, as most shoots are not anchored or poorly anchored, most of the stand is washed away in the next major storm. With this growth habit it does best in the shallowest channels with least fluctuation in depth. It does not grow well when submerged, and so avoids deeper water when it cannot grow on top of other plants. In contrast, *Berula erecta* grows well submerged, rarely shows explosive growth, and does not grow over and smother other fringing herbs. It does best somewhat downstream of *Rorippa nasturtium-aquaticum* agg.

The *Callitriche* spp. (Fig. 6.6*b*) are mainly *Callitriche obtusangula* and *Callitriche stagnalis*, with some *Callitriche hamulata* in more acid places and some *Callitriche platycarpa* in slower waters. There is rather more *Callitriche* in the slightly wider streams, and it is most frequent in steeper ones. Its well-grown, long-lived populations are most often found in somewhat shallow water with a fair flow and a medium-grained firm substrate, or else in somewhat shallow still waters in dykes and canals. *Callitriche* species do, however, occur sparsely in most other stream types, sometimes in only temporary populations (for example, on silt shoals liable to disturbance).

Myriophyllum alterniflorum, though not illustrated, probably belongs with this group of species, although it does occur in wide rivers also. Its main association seems to be with soft water rather than with physical factors.

The next group of plants, shown in Fig. 6.7, are those which occur mainly or only in faster flows. In the range studied, depth is of less importance as a controlling factor. Their pattern is from the lower left to the upper right in the width–slope diagrams, i.e. from steep streams to wide flat streams.

Ranunculus calcareus is most frequent, and most often luxuriant, in the steeper, narrower end of its range. It is almost confined to limestone. It usually prefers a considerable volume of water and so does not occur in the very steep streams at the far left of the diagram (if the artificial deep pools of the Cheddar Gorge are excluded). It has a similar range in the lowlands and the hills.

Ranunculus fluitans requires a stable substrate and, for good growth, a large volume of water. It therefore avoids sites where water volume is small (which includes the left of the diagram). It occurs in both lowland and hill sites, but in flatter reaches it is found in narrower streams in the lowlands, and wider ones in the hills. This separation of the width–slope pattern between habitats with and without spates has two causes. First, rivers are spatey in the hills, and spatey rivers tend to be wider for the same volume of flow. A second and more important factor in the distribution of *Ranunculus fluitans* is the difference in the distribution of suitable substrates for its growth between the lowlands and the hills. In lowland streams stable firm substrates tend to be found in medium-width stretches, while in the hills the more spatey flow means that these medium-width streams often have unstable gravel substrates and wide ones are more likely to have consolidated beds.

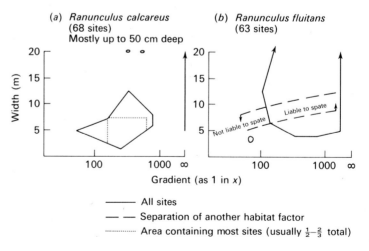

Fig. 6.7. Species mainly in swifter flow. Details as for Fig. 6.1.

——— All sites

— — Separation of another habitat factor

·········· Area containing most sites (usually $\frac{1}{2}$–$\frac{2}{3}$ total)

The plants of Fig. 6.8 occur similarly from small narrow to wide flat streams, that is, have a similar band from lower left to upper right of the diagram, as in Fig. 6.7. But they also have a band along the right-hand side, which means they grow in flat channels too, both wide and narrow. The sites the plants avoid are the slow-flowing and somewhat narrow streams, where there is unstable accumulating silt in which short-rooted plants cannot anchor and by which low-growing ones can be smothered. Some of this group also avoid the wider more gravelly streams (Fig. 6.8).

Myriophyllum spicatum is most frequent in the steeper and narrow part of its range, and the wider sites are in hill rivers. Its morphological characteristics are that it does not have a consolidating root weft, its rhizomes can spread over a wide area and its shoots are bushy. Thus it is probably absent from the low-central part of the diagram because it is vulnerable to velocity pull (see Chapter 3) and cannot remain in, and take full advantage of, small areas of suitable substrate; it is also not able, by means of a root weft, to improve the substrate to its own advantage.

Ranunculus aquatilis and *Ranunculus penicillatus* have a broadly similar pattern, and both show a separation between hill and lowland sites (see *Ranunculus fluitans*, above). Both form firm root wefts in stable coarser substrates. *Ranunculus aquatilis* grows in still water in dykes and canals, but there it is likely to be floating and delicate and this form could not survive in flow. Most of its records in narrow streams are in the lowlands, where the flow, particularly from chalk, tends to be stable. Small hill streams have a more variable discharge than lowland ones, and perhaps become too shallow in summer for these species to occur. In mountain streams *Ranunculus aquatilis* is typically found in upper middle or middle reaches, with *Ranunculus fluitans* below, where discharge and depth increase. Where both occur, *Ranunculus aquatilis* grows in the shallower faster parts and *Ranunculus fluitans* in the deeper slower ones. *Ranunculus penicillatus* seldom grows in small streams, and does not grow in the narrow still channels of the dykes, etc. The lowland form has longer leaves.

The *Potamogeton natans* pattern may be incomplete, as the species is infrequent. The lower position of the pattern on the diagram shows that it grows in slower waters than the preceding species. The lower reaches and still-water sites (on the right) are mostly lowland, and the wider and steeper ones (on the left) are mainly highland. It needs a fairly soft soil for rooting, though it can grow between stones. It is able to consolidate soil somewhat, though it does not have a dense root weft.

Polygonum amphibium is also infrequent, so may also have a width–slope pattern that is incomplete. The pattern is only for the form with floating shoots (not the emerged form), and as these are anchored in the bank, water depth is irrelevant. It typically occurs in slow waters, though it can tolerate spates by regrowing from the bank. Most sites on the left of the diagram are lowland clay streams, and on those on the right canals and hill streams.

Zannichellia palustris is more frequent in the steeper part of its range, and does not grow in deep water. It grows where some silt or sand will be deposited, the plants accumulating this and stabilising the sediment by means of their root weft. This root weft is broken by storm flows, trampling, etc., and the sediment, along with the plants, can then be washed away. If sediment is deposited very quickly, however, the plant cannot stabilise it, and again both sediment and plant will be eroded. Alternatively the plant grows in sites such as dykes where there is no scour. It has a small habitat range, mainly in the lowlands.

Enteromorpha sp. which is a large alga (not shown in the figure) has a

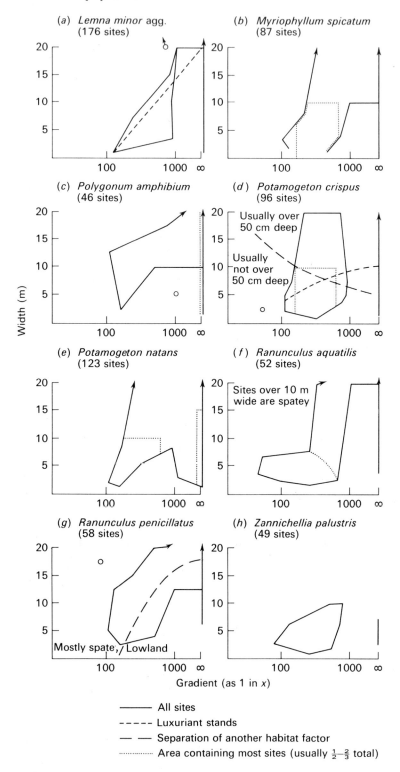

Fig. 6.8. Species of a central
pattern. Details as for Fig. 6.1.

rather similar pattern, but occurs in a wider habitat range. It is most frequent, and most often luxuriant, in dykes and drains and is seldom able to
grow abundantly in flow because the strands are easily broken and washed
downstream.

Potamogeton crispus is spread over most sites except those in the steepest
narrowest streams, and in the areas with least stable substrate. It is most
frequent in the centre of its range and is rare in flat narrow channels, though
it is in these flatter sites that most well-grown populations are found. On a
fairly firm medium-grained substrate it can grow well, forming a dense root
weft which keeps the soil stable below it: really unstable or coarse substrates
are unsuitable for its root weft. Its steepest sites are usually on clay, where
silting occurs furthest upstream. Its anchoring root weft, and its ability to
stay in discrete patches on suitable substrates, probably account for its wide
distribution and its frequency in flatter areas. It is bushy when growing well,
giving a high hydraulic resistance to flow, but it is brittle and easily broken
above ground by faster flows.

Lemna minor agg. is washed into streams from lakes, ponds and sheltered
places upstream. It becomes lodged in any sheltered place, and if conditions
are suitable it can grow very quickly. Stands developing during low flows
are swept off during high ones. In the wider and steeper rivers the sites are
mainly lowland, because water movement at such sites in the hills is too great
for it.

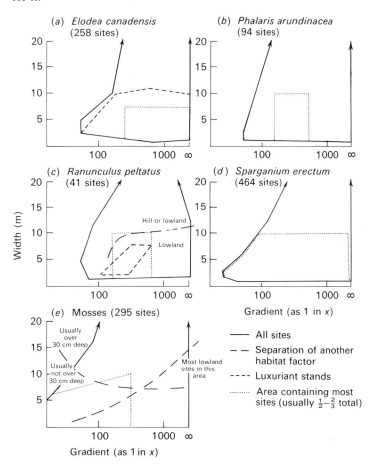

Fig. 6.9. Species of wide
pattern. Details as for Fig. 6.1.

109

The species of the next group, shown in Fig. 6.9, have wide overall patterns, though in other respects may vary considerably. Mosses, as a group, are concentrated in narrow steep streams (cf. Fig. 6.6). Most sites in such streams (at the far left of the diagram) are from the hills, because this size of stream in the lowlands is often dry in summer or is covered with plants tall enough to shade out mosses.

Elodea canadensis, when luxuriant, is commonest in the unstable silting habitat (lower right of the diagram) most avoided by the species illustrated in Fig. 6.8. It grows in the more sheltered places, and because it can quickly regrow from fragments does well in temporary as well as long-term habitats, and can spread widely in the hot summers in which it grows best. The pattern does not have a clear division on stream depth, and this probably reflects the wide range of places in which it can occur, at least temporarily. It seldom grows in the widest rivers. Its roots are short and straight and need soft soil, so are easily pulled up, and this is probably why the species avoids steep mountain streams.

Sparganium erectum is the commonest British river plant. It grows in shallow water and soft soil, and so can be found both in the centre of slow silty channels and at the sides of deeper ones. Luxuriant stands are mostly found in the small siltier channels (left of centre in the diagram).

Phalaris arundinacea is widespread. It does not tolerate permanent flooding and so grows in shallow channels which usually dry in summer, and where

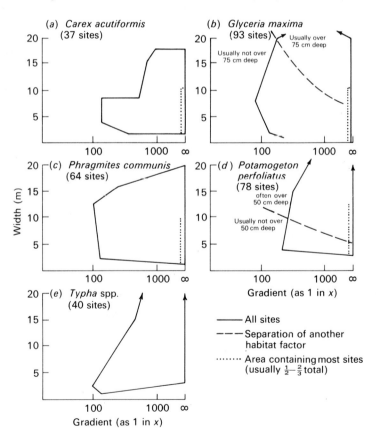

Fig. 6.10. Species occurring towards the right. Details as for Fig. 6.1. The tall emergent monocotyledons, *Carex acutiformis*, *Glyceria maxima*, *Phragmites communis* and *Typha* spp., grow only in the shallow water with little scour, so are confined to the sides of deeper or more flowing channels.

flows are not swift enough to erode it from unstable substrates. As often happens, it tolerates long-term submergence best in chalk streams. It is also found on gravel and stone banks and shoals in spatey hill streams – where it is dry in low flows and flooded in high ones, and where it can anchor to the gravel. Elsewhere it occurs on edges and banks, sometimes growing down into the water. Fragments can become established at least temporarily in other places.

Ranunculus peltatus has the widest range of the *Ranunculus* species. In common with the others it has a tough anchoring rhizome and root weft. Like *Ranunculus aquatilis* it can grow in still water as a delicate, often non-rooted form and like *Ranunculus fluitans*, *Ranunculus aquatilis* and *Ranunculus penicillatus*, in flatter places it grows in wider channels in the hills than in the lowlands. It frequently occurs in small chalk streams upstream of *Ranunculus calcareus*.

Blanket weed (long trailing filamentous algae, most often *Cladophora*) could also be placed in this group. It is most frequent, and most frequently luxuriant, in streams less than 10 m wide. In these surveys the wider rivers recorded were mainly in the hills and so tended to be too swift for blanket weed.

The group of plants shown in Fig. 6.10 have a similar but smaller pattern, which does not extend into the steeper or wider sites. Most of the tall emerged monocotyledons are included here.

The tall emerged monocotyledons require shallow water, fine soil and little scour. *Carex acutiformis* avoids spates, growing in shallow dykes and at the edges of other lowland channels, particularly on chalk and alluvium. *Glyceria maxima* is also mainly a lowland species, and luxuriant stands are mostly found in fairly flat and narrow watercourses, though it frequently grows along the edges of larger channels. *Phragmites communis* is distributed similarly, though it is less common in streams and much more frequent in dykes. It tolerates even less scour than the others. *Typha* spp. are less common.

Potamogeton perfoliatus grows well in silty substrates with little flow, and in firmer substrates in moderate flows. It has a deep rhizome system, no root weft, bushy shoots which can grow large, and usually grows slowly. These characters exclude it from the steeper and wider channels.

There are only a few records for *Oenanthe fluviatilis*, but it probably belongs in this group. It is typically found in lower parts of chalk streams which have eutrophic influence, so physical factors may be of lesser importance than nutrient status to this plant. It is rather bushy, and forms a root weft in fairly firm substrates.

The main species characteristic of slow flows are shown in Fig. 6.11. They have basically similar patterns, well into the flatter reaches to the right, and with sites concentrated to the right, even if they are dense in the lower centre also. Stands of good performance avoid the upper left of the range; i.e. the more extreme flows.

Sparganium emersum extends the furthest into steep streams. It sometimes grows well even in moderate flow, though luxuriant growth is mainly in the flatter sites at the right of the diagram (except in the part avoided by the species of Fig. 6.8). Its straight deep roots need soft soil and grow best in fine particles. Its leaves are easily damaged in flow.

Nuphar lutea, *Potamogeton pectinatus* and *Sagittaria sagittifolia* have similar patterns. All have deep straight roots and tolerate both silting and disturbance of the upper silt (except that *Potamogeton pectinatus* can be swept away in winter and spring; see Chapters 3 and 5). They cannot tolerate much battering. The pattern is clearly divided on depth in Fig. 6.11 and this confirms that they have a preference for a definite substrate and flow type. In the steeper parts of the range *Potamogeton pectinatus* is more likely to occur in faster flows and *Nuphar lutea* and *Sagittaria sagittifolia* in slower ones, with *Sparganium emersum* being intermediate.

Nuphar lutea is the commonest in the deeper waters. Most of its abundant stands are in flatter streams, and it is infrequent in streams over 15 m wide. Its rhizome system is large and deep, but it can be uprooted in severe low-land storms, and, not surprisingly, is absent from mountain rivers. In upland streams it occurs only in the flatter parts (to the far right of the pattern) where deep soft soil may accumulate. As the plants are large they grow better with a large volume of water. *Nuphar lutea* and *Sagittaria sagittifolia* both extend into narrow steep clay streams in Sussex, where quiet silting stretches reach far upstream. *Nuphar lutea* also has a secondary distribution on Resistant rocks, in small streams in nearly flat country. Here the sediment is much poorer in nutrients, coming from the neighbouring moorland.

Potamogeton pectinatus can grow well in moderate flow, anchoring in soft deep soil by its complex rhizome system (it does not have a root weft). It is

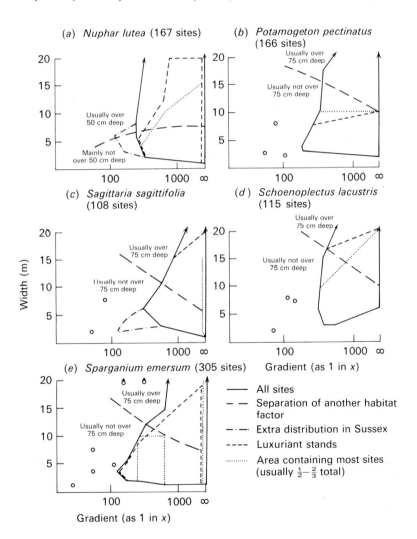

Fig. 6.11. Species of slower flow. Details as for Fig. 6.1.

severely battered in faster flows and cannot anchor properly to coarse or firm substrates, so well-grown stands are usually found in narrower streams and it seldom occurs in channels over 15 m wide. It grows somewhat into the hills, but the upland and mountain sites are, of course, the flatter sites to the right of the pattern.

Sagittaria sagittifolia is commonest in drains, canals and larger dykes (bottom right of the pattern). It is damaged by battering, but is less easily uprooted than *Nuphar lutea*. It extends further into the hills than *Nuphar lutea* but less far than *Potamogeton pectinatus*.

Schoenoplectus lacustris has a similar pattern to the other species except that it avoids the habitats with little flow and much silting (bottom right of the pattern). Its roots are straight and often deep and it has dense, nearly superficial rhizomes. This means it is able to anchor both in deep silt, by using its roots, and in stable gravel, with its rhizome mat – an unusual characteristic. Its above-ground parts are versatile also, because there are both strap-like flexible submerged leaves and large stiff emerged shoots; although storm flows break the stiff shoots the submerged leaves are only harmed if there is much turbulence. The proportion of submerged leaves increases in faster flows, where the emerged shoots are more liable to damage. It is found mainly in the lowlands, but can grow well in fast flows without much spate. There is little separation with depth on the width–slope pattern, probably because of the morphological diversity. Bearing in mind this diversity, however, the pattern is small. Its avoidance of mountain streams can be attributed to spate losses through erosion and velocity pull (the latter mostly affecting the emerged shoots), but it is also rare in canals, drains and dykes. In large rivers with stretches of both fast and slow water, as above and below weirs, *Schoenoplectus lacustris* is typically common in the faster flow and *Sagittaria sagittifolia* in the slower.

The last category of width–slope pattern, Fig. 6.12, contains only one member, *Ceratophyllum demersum*. This is large, with much hydraulic resistance to flow and no roots for anchorage, and it grows too slowly to be an ephemeral like *Lemna minor* agg. It is therefore confined to the flattest places, to still waters and to lowland streams with a slope of less than 1 : 1000. It provides a striking final example of the value of width–slope patterns in elucidating the ecology of river plants.

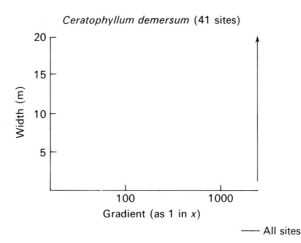

Fig. 6.12. Species of the far right. Details as for Fig. 6.1.

USE OF WIDTH–SLOPE PATTERNS

The width and slope of a channel can be obtained very easily from a visit to the site and a large-scale map. In general, a species will occur within its pattern as illustrated in the preceding figures. It has been seen that the patterns are explicable in terms of the biology and ecology of the plant and thus form a valid measure of plant distribution. They can, therefore, be used to predict changes in plant distribution in rivers which are changing or are likely to change. For example, if the future depth, normal flow pattern and spateyness are known, the likelihood of any species remaining in or invading a particular site can be assessed (see Chapter 20). Predictions from the patterns are not complete, though, as chemical factors may change as well as physical ones and these are only partly accounted for on the overall patterns. The most important chemical factor is the nutrient regime which is determined partly by the amount of silt present and partly by the nutrient status of that silt (see Chapter 8). If pollution is present, or the water is turbid, or the site has recently been dredged, etc., these factors must also be taken into account, and may result in species not being found at sites within their width–slope pattern that they would otherwise be able to tolerate.

7

Light

Plants need light for photosynthesis, and so their growth can be decreased by lack of light. Because the light reaching submerged plants is lessened by passing through the water, their growth (production) is usually much less than that of emerged plants (see Chapter 9). Light can also be reduced by the shading effect of trees, etc. above the stream, and of the river plants themselves.

SHADING FROM ABOVE

Trees, shrubs and tall bank plants frequently overhang the water, and shade it (Plates 2, 3, 24 and 25; Figs. 10.14 and 10.15). The field distribution of river plants in relation to shading is shown in Table 7.1. Species which have green shoots in winter are those which tend to be harmed the most by shade. Of the common species of river plants *Ranunculus* spp. are the most light-demanding and *Sparganium emersum* is the most shade-tolerant. Most lowland streams were shaded before the woods were cleared for farming. (See Chapter 1.) Shading is also a simple way of preventing the excessive plant growth (see Chapters 19 and 21) which can cause flooding. A lowland stream is safe for flood prevention if it is not more than a quarter full of vegetation (see Chapter 19). To keep a *Ranunculus* stream safe from flooding the overhanging trees should let through at most *c.* 60–70% of full sunlight if the shade is uniform (see note in Table 7.1), or *c.* 40–55% of full sunlight in a dapple shade. At the other extreme, a *Sparganium emersum* stream needs *c.* 40–55% of full sunlight if the shade is uniform and less than 30% of full sunlight in a dapple shade.

LIGHT UNDER WATER

When light reaches the water surface, between 5% and 25% of it is reflected back from the surface and does not enter the water. The proportion of sunlight reflected increased as the sun gets lower in the sky. About 6–8% of the diffuse sky light (not the full sunlight) is reflected and this varies much less with the position of the sun [26]. Once below the surface, light is absorbed as it passes through the water. These are three causes of this loss: (1) the water itself; (2) turbidity, due to the particles in the water; and (3) colouring matter in the water. Of these, (1) is stable but (2) and (3) vary greatly. They can be changed in various ways; for instance, river water is more turbid than that of lakes, since rivers drain water from the land, and so plants cannot grow as far

TABLE 7.1 *Shade tolerance*

Divided into groups, from the most tolerant species in Group 1, to the least tolerant in Group 6.

1. Shade-tolerant species

Sparganium emersum

2.

Callitriche spp.   *Sparganium erectum*

Phragmites communis (probably)

3.

Alisma plantago-aquatica   *Elodea canadensis*

Ceratophyllum demersum (probably)   *Myosotis scorpioides*

4.

Apium nodiflorum   *Nuphar lutea*

Berula erecta   *Potamogeton crispus*

Glyceria maxima (probably)   *Sagittaria sagittifolia*

5.

Rorippa nasturtium-aquaticum agg.   *Zannichellia palustris*
                                      (probably)

Veronica anagallis-aquatica agg. (probably)

6. Light-requiring species

Ranunculus spp.   *Veronica beccabunga*

Group 1 occurs often in medium continuous shade, and is common in medium discontinuous shade. It occasionally occurs in heavy discontinuous or dapple shade. Group 6 is sparse in medium dapple shade and in light continuous shade.

Shade types are defined as the percentage of full sunlight reaching 1–2 m above water level, between mid-June and mid-September, i.e.

Heavy shade	10–40 % full sunlight
Medium shade	45–65 % full sunlight
Light shade	70–80 % full sunlight

below the water level in rivers as in lakes. Plants are usually considered to need at least 1% of full daylight in order to be able to survive [26].

In clear streams some plants can grow at about 2 m below the surface, but they are frequently restricted to shallower water. In lakes, they can grow below the level where they can be seen from the surface, but because streams are more turbid and more coloured, the drop in light is much sharper and, effectively, if plants are present they can be seen from above the water. There are, of course, some exceptions. Pale and large particles hinder viewing more than they hinder the passage of light, and a good deal of vegetation can grow unseen in paper mill effluents and, to a lesser extent, where boats stir up silt.

During storm flows the water is usually turbid from silt being washed in from the land and moved downstream. However, storm flows are infrequent enough during the plants' growing season (i.e. the summer) for the decrease in photosynthesis that they produce probably to have little overall effect on the vegetation (see Chapters 3, 5 and 6 for the other effects of storm flows). Variations in turbidity during normal flows are more important and the productivity of the light-sensitive *Ranunculus* can vary over a two-fold range with these changes.

Turbidity is influenced by rock type [27]. Turbidity is least in chalk streams since chalk dissolves in water and few eroded particles are left to be washed off the land. Resistant rocks also erode very slowly and usually bear clear streams, though material can enter them from drift or soil above the rock. The silting from sandstone is greater, and clay streams are the most turbid, with many small particles in the water. The lower reaches of clay streams, and drains, contain the highest proportion of vegetation with leaves on or above the water. These leaves are not harmed by turbidity.

Small algae (phytoplankton) also make water turbid, as do bacteria and microscopic animals. Algae are present in all streams, but in swift-flowing waters the algae are washed along quickly, while slow flows allow rapid reproduction of the algae and so the water may become turbid. In suitable slow rivers, phytoplankton can be very dense and comprise most of the plant life in the river (see Chapter 9), and the lower reaches of clay rivers, and other slow lowland streams, can be very turbid from phytoplankton. Turbidity can also be increased by human interference, both directly, e.g. by coal mine effluents, and indirectly, as when agricultural practices increase the silt being washed from the land (see Chapter 22).

Water may be coloured green-grey in clay streams, particularly in winter. The frequent coloration, however, is brown, the water being stained with peat or humus from the land around. Brown water occurs in some dykes on organic alluvium, and in streams from peaty heath, moorland or blanket bog catchments, such as the New Forest and the northern moors. The only region of Britain with many very dark brown streams is, though, the far north of Scotland, on both Resistant rocks and sandstone. Mountain streams are clear where there is little peat in the catchment, e.g. the Lake District (for North America see Chapters 15 and 16).

PLANTS WITHIN WATER

Turbid and coloured water cannot, of course, harm the growth of plants whose leaves are mainly above water level, such as the emergents and most free-floating plants (e.g. *Lemna* spp.). The rooted floating plants receive full light on their floating leaves, but the shoots must first grow up through the water. Plants such as *Nuphar lutea*, which have large rhizomes and much stored food in the soil, can grow well in turbid water, since the young leaves can grow to the surface using food from these reserves.

The tolerance to turbidity of various plants is shown in Table 7.2.

TABLE 7.2   *Tolerance to turbid water*

1. Most tolerant of turbid water

    Ceratophyllum demersum   *Polygonum amphibium*

    Lemna minor agg.   *Sagittaria sagittifolia*

    Nuphar lutea   *Schoenoplectus lacustris*

2. Intermediate

    Callitriche spp.   *Potamogeton pectinatus*

    Myriophyllum spicatum   *Sparganium emersum*

    Potamogeton natans   *Sparganium erectum*

3. Least tolerant of turbid water

    Elodea canadensis   *Ranunculus* spp.

    Potamogeton perfoliatis   Mosses

Since shading affects the whole plant whereas turbidity affects only those parts in the water, there are differences between Tables 7.1 and 7.2. The plants most tolerant to turbidity either have parts above the water surface, or are characteristic of deep water (*Ceratophyllum demersum*) or both (e.g. *Nuphar lutea*). The species least tolerant of turbidity all grow submerged, the mosses and *Elodea canadensis* usually growing close to the substrate so that they receive least of the light entering the water. *Ranunculus* spp. are also the least tolerant of shading. The depth distribution and light requirements of several *Potamogeton* spp. have been studied [28, 29] and it has been found that the less light the species needs, the deeper it grows. These species tolerant of low light have the lowest respiration per unit area of leaf, and so can stay alive and grow with the least photosynthesis.

Plants within water

In still water the plants may be able to adjust their growth habit in order to decrease self-shading and shading by taller water plants. In currents, however, the habit is determined by the flow, not the light (see Chapter 3), though within these limits as much of the plant as possible will grow in the light [26], partly because leaves that receive too little light for photosynthesis often fall. The taller a plant can grow, whether within or above the water, the more chance it has of obtaining light if there is competition between the plants. However, while competition may be important in shallow still waters, it has little effect in swift deep ones where flow is again the dominant controlling factor (see Chapter 10).

8

Nutrients

River plants (except for free-floating ones) take up nutrients from both soil and water, the soil usually being the more important source. Most soil nutrients are in the silt particles, so the soil texture affects its nutrient status. As well as the stable silt, temporary silt deposited in, on, or under plant shoots can be an important source of nutrients, particularly to plants in somewhat nutrient-deficient habitats.

The silt from different rock types differs in nutrient status. The flow regime controls silt deposition, and different rock types differ in the amount of silt entering the channels. These three factors control the nutrient pattern of stream types, and each stream type has a characteristic vegetation which can be linked to this nutrient pattern. Streams on Resistant rocks are often oligotrophic, those on chalk and sandstone tend to be mesotrophic, and clay streams are often eutrophic.

River plants need, for their growth and development, both the gasses oxygen and carbon dioxide (or another form of dissolved carbon: bicarbonate), and dissolved mineral nutrients such as calcium, potassium, phosphate and nitrate. The minerals are taken up from the water and the soil, and oxygen and carbon from the water, the air and (for carbon) the soil. When nutrients are absorbed, the area immediately surrounding the site of uptake (e.g. the water around each leaf) is depleted of nutrients, so uptake becomes progressively more difficult. However, water movement brings in fresh supplies of nutrients, so that moving water is, for plants, richer in nutrients than still water. It seems probable that flow is more important for the carbon supply to submerged plants than for the mineral supply, because carbon is one of the nutrients most likely to be limiting for the growth of submerged shoots [30]. All animals and plants use oxygen for respiration, while photosynthesis uses carbon dioxide and produces oxygen. The presence of green plants usually increases the total oxygen supply of the river because of the oxygen released during the day when the plants are photosynthesising; this improves the habitat for fish. In some circumstances, though, plants can actually decrease oxygenation, as when there is much decaying plant material (the micro-organisms which effect the decomposition use oxygen in respiration), or when much of a plant clump is too shaded by its upper shoots to photosynthesise, or when the water becomes very warm.

120

NUTRIENT UPTAKE

Water plants can obtain their mineral nutrients from the soil or from the water. Free-floating plants which are near the water surface (e.g. *Lemna minor* agg.) necessarily obtain all their nutrients from the water, while tall emerged plants on the banks (e.g. some *Phragmites communis*) must obtain all their minerals from the soil. Other groups can use both soil and water nutrients, though under experimental conditions [31] a plant's entire nutrient intake may come from either the water or the soil, and, at least for short periods, this is adequate for growth. Nutrients can be taken up by roots, stems and leaves, and uptake by roots and by shoots may be independent, each being unaffected by the nutrients surrounding the other (shown for phosphorus [32]).

Soil contains more nutrients than the same volume of water, and within the soil the water present contains more nutrients than the water above ground (e.g. 6–20 times per unit wet weight of silt). The nutrients are held in the silt, so as roots take up nutrients from the soil water they are replenished from the silt. (The concentrations of exchangeable nutrients in wet silt are around three to six times higher than those in the soil water alone, exchangeable nutrients being defined as those which are extractable from the soil with ammonium acetate and are considered to be available for uptake by plants.)

Since the soil is richer in nutrients than the water, this is the main source [31, 32, 33, 34]. However, in chalk stream water there is more nitrogen and phosphorus than is needed for the growth of dominant *Ranunculus*, and the amount of phosphorus removed from the water is the same as that held in the *Ranunculus*. *Ranunculus* can also grow well, at least for short periods, with nutrients supplied only from below. Low levels of water nutrients may lead to poor plant development in experiments [31], and it is suggested that development is poor below *c.* 1 ppm of nitrate-nitrogen and *c.* 30 μg of phosphate-phosphorus [36]. If it is the soil and not the water nutrients which are very low, more roots may develop in the water above ground, which increases the amount of nutrients taken up from the water [32] and the temporary silt. In streams, water nutrients often seem adequate for survival but not for good growth. For instance, plant fragments can live for many months in the water, staying healthy but small and not growing well or luxuriantly, e.g.

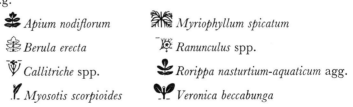

Apium nodiflorum	*Myriophyllum spicatum*
Berula erecta	*Ranunculus* spp.
Callitriche spp.	*Rorippa nasturtium-aquaticum* agg.
Myosotis scorpioides	*Veronica beccabunga*

These are common in the floating fringe of middle reaches of chalk streams (see Chapter 10, Fig. 10.2), and also occur elsewhere. If, and only if, the fragments become anchored, they can increase in size and start to spread. (The floating fragments usually have roots, so this poor growth is not due to lack of growth hormone developed in roots.) Soil nutrients are essential for normal growth.

It seems that only part of the plant need be in a nutrient-rich place. For example, if one part of a bed is silty and another is nutrient-poor (e.g. clean-washed sand), a rhizomatous plant which can grow well only in the silty part can, at least in experiments, move nutrients along its rhizomes so that shoots can grow well in the nutrient-deficient area. As rhizomes are usually short-lived, the shoots will eventually be cut off from the high nutrient source, and further growth in the deficient area will be unsatisfactory.

Submerged plant parts can, as mentioned above, all obtain nutrients from the soil or the water, but the morphology and habit of the plant may determine which of its parts are more important for uptake. Leaves and stems may have a thick water-repellent fatty cuticle, in which case mineral uptake is difficult, or this cuticle may be very thin or negligible (Table 8.1). Woody rhizomes and rootstocks do not, of

TABLE 8.1 *Amounts of cuticle in different species*

1. Very little or negligible cuticle

Callitriche spp.   *Potamogeton crispus*

Ceratophyllum demersum (leaf)   *Potamogeton lucens* (leaf)

Elodea canadensis (leaf)   *Rorippa nasturtium-aquaticum* agg.
                                                  (some leaves)

Groenlandia densa (leaf)   *Sagittaria sagittifolia* (submerged
                                                  leaf)

Hottonia palustris (leaf)   *Schoenoplectus lacustris* (submerged
                                                  leaf)

Myriophyllum spicatum (stem)   *Zannichellia palustris*

Nuphar lutea (submerged leaf)

2. Little cuticle

Apium nodiflorum   *Myosotis scorpioides*

Ceratophyllum demersum (stem)   *Nuphar lutea* (submerged petiole)

Elodea canadensis (stem)   *Ranunculus aquatilis* (leaf)

Groenlandia densa (stem)   *Ranunculus peltatus* (broad leaf)

Hottonia palustris (stem)   *Rorippa nasturtium-aquaticum* agg.
                                                  (some leaves, stems)

TABLE 8.1 (*cont.*)

⊤ *Lemna minor* agg.	⌐ *Sparganium emersum*
⋆ *Lemna polyrhiza*	*Sparganium erectum* (submerged leaf)
Mentha aquatica (leaf)	*Veronica beccabunga*

3. More cuticle

Agrostis stolonifera	*Potamogeton lucens* (stem)
Carex acutiformis	*Potamogeton natans*
Hydrocharis morsus-ranae	*Ranunculus aquatilis* (stem)
Mentha aquatica (stem)	*Ranunculus peltatus* (stem)
Myriophyllum spicatum (leaf)	*Rorippa nasturtium-aquaticum* agg. (emerged stem)
Nuphar lutea (floating leaf)	*Schoenoplectus lacustris* (emerged)
Phalaris arundinacea	*Sparganium erectum* (emerged leaf)

course, absorb much nutrients. Many aquatics can, under experimental conditions, grow well initially with few or no roots, but roots are the parts best adapted for nutrient uptake. In general, deep-rooted species (see Chapter 3) have all their roots in the ground, while shallow-rooted ones often have them above ground as well. Roots in the water reach much the same nutrient supplies as the lower shoots there, but roots in the soil grow well beyond the rhizomes, reaching different nutrient supplies. When soil nutrients are inadequate, roots may develop more above ground [32], increasing the proportion of nutrients obtained above ground. The development of individual roots is also influenced by soil type. If a plant grows best in eutrophic places (e.g. *Sagittaria sagittifolia*, *Sparganium emersum*), its roots are longer and wider in silt than in a soft mixed-grained substrate.

Uptake can vary in different seasons. Roots of winter-dormant plants usually start growing in spring or early summer, some weeks after the shoots have emerged. Tall emergents such as:

Phalaris arundinacea	*Schoenoplectus lacustris*
Phragmites communis	*Sparganium erectum*

absorb little through their thick rhizomes or aerial parts, so effective nutrient uptake occurs only when living roots are present. This seasonal uptake can be important in relation to pollution. Following the winter flooding by sea water of a bed of *Phragmites communis*, spring growth of the shoots was normal, but when the new roots grew and were influenced by the salt, the shoots developed badly and many died. If, in

123

contrast, a pollutant can both enter and be lost from a habitat while it cannot be absorbed by the plants, no harm results.

In conclusion, mineral nutrients necessary for plant growth, and pollutants damaging to this, can be taken up by plants from both the soil and the water. In general, soil is more important because it contains more nutrients and usually more roots. Water uptake is probably needed for the optimal growth of submerged plants and is usually enough to keep plant fragments alive (these contain nutrients before being detached) though not enough for their normal growth. In some circumstances water uptake is more important than soil uptake, as when poisonous pollutants are in the water but hardly affect the soil, or when the soil is particularly nutrient-deficient and water nutrient-rich. Dependence also varies with habit, tall emergents being necessarily almost entirely dependent on soil nutrients, while submerged, rooted floating, and short semi-emergents obtain nutrients mainly from the soil (though the proportions vary with the habitat and, presumably, the species concerned) and free-floating plants near the surface depend on water uptake.

SOURCES OF NUTRIENT SUPPLY

Water

When stream water is derived from springs it contains the same nutrient concentrations as the ground water in the aquifer from which it originated. When stream water is derived from rainfall running off the land, however, it contains dissolved substances picked up from the soil. Thus nutrients in run-off are usually related to the rock type of the catchment, unless, as in acid bogs, the soil differs from the underlying rock or there is much human interference with the catchment. The last main source of water nutrients is from effluents from sewage works, etc. (see Chapter 22). There is some exchange of nutrients between soil and water, so long-term changes in chemical habitat factors tend to influence both, though in the short-term nutrient concentrations in water tend to fluctuate more than those in the soil. As mentioned above, the effective nutrient supply to the plant is increased when the water is moving.

Stable substrate of the channel bed

The substrate likewise contains and supplies nutrients. It is composed of particles of different sizes and of varying proportions (see Chapters 2 and 3). Silt particles, which contain most nutrients, occur around and between larger particles, as well as in silt beds, etc. From this nutrient-rich silt the water within the soil becomes nutrient-rich also (see above). The total store of nutrients is greatest in a silt bed and least in clean-washed sand, or the clean-washed stone beds of mountain streams, etc. Substrates with little silt can be nutrient-deficient for plants (though sands and gravels which appear non-silty may in fact contain considerable amounts of silt on and between the particles). In addition, substrates of similar texture, derived from different sources, may differ in both the concentrations and the ratios of the nutrients they contain (see below). The nutrients usually reflect those of the rock types of the catchment (with the same reservations referred to above, for water).

There is no clear separation between the substrates described here and the temporary silt discussed in the following section, but the distinction is useful ecologically because although both supply nutrients, one does so on a long-term the other on a short-term basis. Both are derived from erosion of the channel bottom, falling material from the channel banks, sediment washed in from the catchment and, in downstream reaches, sediment transported from upper reaches. As loose silt is moved the most easily, and is very nutrient-rich, it can have an independent effect.

Temporary and unstable silt

Where the stable substrate is nutrient-rich, supplying the nutrients needed for the luxuriant development of the plants present, temporary silt is irrelevant, but elsewhere it is of great importance. Plants contain stores of nutrients which can be moved to new parts of the plant and supply it for some while, and in a low-nutrient habitat good growth and development can be achieved with temporary silt, which allows intermittent good uptake and accumulations of nutrient reserves.

Stream water carries fine particles, especially during storm flows, and these are deposited where flow is checked (see Chapters 2 and 3). When silt is deposited on a plant part the nutrient supply to, and uptake of, that leaf, stem or root will be much increased. Silt is caught most easily on individual leaves if these are large (e.g. *Nuphar lutea*) or hairy (e.g. *Myriophyllum spicatum*), or, as in *Elodea canadensis*, they are made hairy by filamentous algae growing on them. Shoots near the ground, particularly if within a clump of at the side of a channel, receive the least flow (Chapter 3) and so frequently collect silt. This silt may be washed off soon after a storm or may remain for some time.

Plant clumps accumulated silt (Fig. 3.2). For example:

Apium nodiflorum	*Ranunculus* spp.
Callitriche spp.	*Rorippa* spp.
Elodea canadensis	*Veronica* spp.
Myriophyllum spp.	*Zannichellia palustris*
Potamogeton pectinatus	

This silt may be in firm hummocks, or in the form of silty water trapped between shoots. Changing flow patterns and storms move this silt, thus renewing nutrient supplies. The above-ground roots (see above) are mainly in this silty area, and usually belong to silt-trapping species.

When plants are newly established, their few and small shoots cannot trap silt well. If the nutrients of the stable substrate are inadequate, therefore, growth will be greatly stimulated once the plant is able to trap silt. Frequently species cannot grow, or grow well, unless bands of temporary silt are present or can be accumulated by the plants. These species include, on coarse or low-nutrient substrates:

Callitriche spp. *Myriophyllum* spp.

Elodea canadensis *Potamogeton crispus*

Groenlandia densa *Zannichellia palustris*

An extreme example was seen in clean-washed sandy gravel of a chalk brook, where frequent cutting had made the vegetation dwarf, shoots being very short and small. The plants included:

Apium nodiflorum *Ranunculus peltatus*

Callitriche spp. *Veronica beccabunga*

These remained dwarf for over a year after the cutting ceased. Then the flow pattern changed and silt was deposited along the sides of the channel. Within a few months the plants were growing well, spreading quickly, and had shoots of normal size.

Silting decreases the light reaching the leaves, thus decreasing photosynthesis. This may be harmful if most leaves are covered, but is immaterial when most photosynthesis is carried out by leaves higher in the water (e.g. *Nuphar lutea*). (The non-rooted *Ceratophyllum demersum* usually has silt around the lower shoots and may indeed have some shoots buried in silt.)

Organic matter

Decomposing organic matter is provided by dead river plants and animals, trees and other plants which overhang the channel, dropping leaves, etc. into the water and material washed in from the land. When in the form of small particles, it forms a part of the temporary silt (as well as of the stable substrate). Decaying organic matter produces both mineral nutrients and organic substances. On peat lands, of course, peat particles are washed into the channels, and this is particularly important for plants where acid peat particles are deposited in a stream. Apart from the peat lands, the silt of chalk streams tends to have the highest proportion of organic matter, probably because chalk streams contain much plant life, and because decomposing chalk leads to little mineral silt. Some organic substances may enhance, or be necessary for, the good growth of aquatics. This is an additional reason for the chemical importance of silt for the plants.

PLANT DISTRIBUTION IN RELATION TO WATER NUTRIENTS

Plant distribution and luxuriance are determined by a complex of chemical, physical and biotic factors, and the individual nutrients in the water, or indeed the total nutrients there, are only a small part of this complex. Plants are likely, in fact, to be better correlated with flow type or depth than with any or all water nutrients. As the soil is usually the more important nutrient source, there are unlikely to be waters so low in nutrients that plants cannot grow well. Also, waters too nutrient-rich for any plant growth are unlikely to occur without human interference. Within these extremes, different species have different nutrient ranges with which they are best correlated, and within these, narrower ranges in which they can grow luxuriantly. Water nutrients fluctuate during the year, so many samples must be analysed in order to obtain a good estimate of the average nutrient content of the water. In some circumstances, seasonal variations may be crucial. (In ponds, the rootless *Ceratophyllum demersum* can occur if there is high nitrogen present for part of the year only [37]). Water analyses of rivers and large brooks, supplied by the (former) River Authorities and River Purification Boards have been correlated with plant distribution, and Table 8.2 summarises the typical nutrient preferences of some widespread species. (The ranges correspond to the highest peak of the left-hand histograms in Chapters 2 and 4.) The division of plant species into groups according to nitrate-nitrogen and phosphate-phosphorus levels in the water corresponds quite well with those groups postulated in Chapter 1 (pp. 12f). When the plants are classified according to levels of chloride, nitrite-nitrogen and total dissolved solids, the correlation is less good, and there is no such relation in the groups with common requirements

TABLE 8.2   *Plant distribution in relation to water chemistry*

The species are listed under the concentration with which they are best correlated. Concentrations are in parts per million (ppm) unless otherwise stated. Water analyses supplied by River Authorities and River Purification Boards (see text).

1.   Alkalinity (as calcium carbonate)
Below 50   Mosses

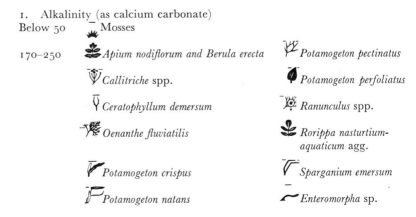

170–250   *Apium nodiflorum* and *Berula erecta*   *Potamogeton pectinatus*

Callitriche spp.   *Potamogeton perfoliatus*

Ceratophyllum demersum   *Ranunculus* spp.

Oenanthe fluviatilis   *Rorippa nasturtium-aquaticum* agg.

Potamogeton crispus   *Sparganium emersum*

Potamogeton natans   *Enteromorpha* sp.

TABLE 8.2 (*cont.*)

Over 250	*Callitriche* spp.		*Schoenoplectus lacustris*
	Lemna minor agg.		*Sparganium emersum*
	Nuphar lutea		*Sparganium erectum*
	Rorippa nasturtium-aquaticum agg.		*Zannichellia palustris*
	Sagittaria sagittifolia		
Poorly correlated	*Elodea canadensis*		*Myriophyllum spicatum*

2. Biological Oxygen Demand

2.5–4	*Apium nodiflorum* and *Berula erecta*		*Ranunculus* spp.
	Callitriche spp.		*Rorippa nasturtium-aquaticum* agg.
	Lemna minor agg.		*Sagittaria sagittifolia*
	Myriophyllum spicatum		*Schoenoplectus lacustris*
	Nuphar lutea		*Sparganium emersum*
	Potamogeton crispus		*Sparganium erectum*
	Potamogeton natans		*Zannichellia palustris*
	Potamogeton pectinatus		Mosses
	Potamogeton perfoliatus		*Enteromorpha* sp.
over 4	*Sagittaria sagittifolia*		

3. Chloride

Below 15	Mosses		
15–35	*Apium nodiflorum* and *Berula erecta*		*Ranunculus* spp.
	Elodea canadensis		
35–65	*Callitriche* spp.		*Potamogeton natans*
66–100	*Nuphar lutea*		*Schoenoplectus lacustris*
	Oenanthe fluviatilis		*Sparganium emersum*
	Potamogeton pectinatus		*Sparganium erectum*
	Sagittaria sagittifolia		*Enteromorpha* sp.
over 100	*Ceratophyllum demersum*		*Potamogeton pectinatus*

TABLE 8.2 *(cont.)*

Lemna minor agg.

Poorly
correlated   *Myriophyllum spicatum*   *Zannichellia palustris*

Rorippa nasturtium-aquaticum agg.

4. Nitrate-nitrogen

Below 1   *Myriophyllum spicatum*   *Ranunculus* spp.

1–3   *Lemna minor* agg.   *Potamogeton perfoliatus*

Myriophyllum spicatum

3–6   *Apium nodiflorum* and *Berula erecta*   *Rorippa nasturtium-aquaticum* agg.

Nuphar lutea   *Sagittaria sagittifolia*

Oenanthe fluviatilis   *Sparganium erectum*

Potamogeton perfoliatus

Over 6   *Nuphar lutea*   *Sparganium emersum*

Potamogeton pectinatus   *Sparganium erectum*

Sagittaria sagittifolia   *Zannichellia palustris*

Schoenoplectus lacustris   *Enteromorpha* sp.

5. Ammonia-nitrogen

Below 0.1   *Apium nodiflorum* and *Berula erecta*

0.1–0.3   *Apium nodiflorum* and *Berula erecta*   *Potamogeton pectinatus*

Callitriche spp.   *Ranunculus* spp.

Elodea canadensis   *Rorippa nasturtium-aquaticum* agg.

Lemna minor agg.   *Sagittaria sagittifolia*

Myriophyllum spicatum   *Schoenoplectus lacustris*

Nuphar lutea   *Sparganium emersum*

Polygonum amphibium   *Sparganium erectum*

Potamogeton crispus   Mosses

Potamogeton natans   *Enteromorpha* sp.

Over 0.3   *Lemna minor* agg.

TABLE 8.2 (*cont.*)

6. Nitrite-nitrogen

Below 0.05   *Elodea canadensis*   *Ranunculus* spp.

             Myriophyllum spicatum   Mosses

0.05–0.1   *Callitriche* spp.   *Sagittaria sagittifolia*

Over 0.1   *Lemna minor* agg.   *Schoenoplectus lacustris*

             Nuphar lutea   *Sparganium emersum*

             Potamogeton pectinatus   *Sparganium erectum*

7. Phosphate-phosphorus

Below 0.3   *Apium nodiflorum* and *Berula erecta*   *Potamogeton natens*

             Callitriche spp.   *Ranunculus* spp.

             Elodea canadensis   Mosses

0.3–1.2   *Lemna minor* agg.   *Potamogeton pectinatus*

1.2–3.0   *Nuphar lutea*   *Sparganium emersum*

             Sagittaria sagittifolia   *Sparganium erectum*

             Schoenoplectus lacustris

Over 3.0   *Myriophyllum spicatum*   *Enteromorpha* sp.

8. pH (in units of pH)

Below 7.5   *Elodea canadensis*

7.5–8.0   *Apium nodiflorum* and *Berula erecta*   *Ranunculus* spp.

             Lemna minor agg.   *Sagittaria sagittifolia*

             Nuphar lutea   *Sparganium emersum*

             Potamogeton crispus   *Sparganium erectum*

Over 8.0   *Nuphar lutea*   *Schoenoplectus lacustris*

             Potamogeton crispus   *Sparganium emersum*

             Potamogeton pectinatus   *Sparganium erectum*

             Potamogeton perfoliatus   *Enteromorpha* sp.

Poorly   *Callitriche* spp.   *Potamogeton natans*
correlated

TABLE 8.2 (*cont.*)

⦿ *Ceratophyllum demersum*　　　　　　⦿ *Rorippa nasturtium-aquaticum* agg.

⦿ *Myriophyllum spicatum*　　　　　　⦿ *Zannichellia palustris*

⦿ *Oenanthe fluviatilis*

9. Total dissolved solids

Below 350 ⦿ *Apium nodiflorum* and *Berula erecta*　　⦿ *Potamogeton natans*

⦿ *Elodea canadensis*　　　　　　⦿ *Schoenoplectus lacustris*

⦿ *Myriophyllum spicatum*　　　　　　⦿ *Sparganium erectum*

⦿ *Polygonum amphibium*

350–500 ⦿ *Callitriche* spp.　　　　　　⦿ *Rorippa nasturtium-aquaticum* agg.

⦿ *Potamogeton perfoliatus*　　　　　　⦿ *Zannichellia palustris*

⦿ *Ranunculus* spp.　　　　　　⦿ *Enteromorpha* sp.

Over 500 ⦿ *Ceratophyllum demersum*　　　　⦿ *Potamogeton pectinatus*

⦿ *Lemna minor* agg.　　　　　　⦿ *Sagittaria sagittifolia*

⦿ *Nuphar lutea*　　　　　　⦿ *Schoenoplectus lacustris*

⦿ *Potamogeton crispus*

Poorly correlated ⦿ *Oenanthe fluviatilis*　　　　⦿ *Sparganium emersum*

10. Total hardness (calcium carbonate plus magnesium carbonate)

Below 100 ⦿ *Myriophyllum spicatum*　　　　⦿ *Ranunculus* spp.

⦿ *Polygonum amphibium*　　　　　　⦿ Mosses

100–350 ⦿ *Potamogeton natans*　　　　　　⦿ *Enteromorpha* sp.

Over 350 ⦿ *Lemna minor* agg.　　　　　　⦿ *Schoenoplectus lacustris*

⦿ *Nuphar lutea*　　　　　　⦿ *Sparganium emersum*

⦿ *Potamogeton pectinatus*　　　　　　⦿ *Sparganium erectum*

⦿ *Sagittaria sagittifolia*　　　　　　⦿ *Enteromorpha* sp.

Poorly correlated ⦿ *Callitriche* spp.　　　　⦿ *Potamogeton crispus*

⦿ *Elodea canadensis*　　　　　　⦿ *Potamogeton perfoliatus*

for alkalinity, pH and total water hardness. (There were, however, no analyses from dystrophic or oligotrophic streams with acid peat in the channel; see soil analyses below.) Ammonium-nitrogen and Biological Oxygen Demand are the two parameters most widely accepted as indicators of water quality for domestic use, but they are generally irrelevant to plant distribution. (There are too few records of the major cations – calcium, magnesium, potassium and sodium – to determine correlations. For soil data, see below.) Water nutrients cannot account for all trophic variations between species. For instance, middle reaches of swift gravelly stretches of part-limestone lowland streams may have:

Ranunculus spp.   Fringing herbs

while in the slow silty stretches alternating with these, typical species include:

Myriophyllum spicatum   *Schoenoplectus lacustris*

Within a single river, R. Tweed, plant distribution has been correlated with five nutrient parameters (Table 8.3; [15]). There is some general correspondence between the nutrients and the trophic preferences of the plants which are described elsewhere in this book (alkalinity, Table 8.2, is often well correlated with conductivity, Table 8.3). Of course, only a few rock types are included in the system.

TABLE 8.3   *Plant distribution in relation to water chemistry in R. Tweed*

1. Negatively correlated with conductivity and calcium

   Myriophyllum alterniflorum

2. Negatively correlated with conductivity

   Ranunculus aquatilis agg.

3. Not correlated (with conductivity, calcium, phosphate-phosphorus or nitrate-nitrogen)

   Callitriche spp.   *Potamogeton pectinatus*

   Elodea canadensis   *Potamogeton × suecicus*

   Potamogeton × olivaceus   *Ranunculus circinatus* + *hybrid*

   Potamogeton × salicifolius   *Ranunculus penicillatus* var. *calcareus*

4. Correlated with one of the following: conductivity, calcium (ca), or nitrate-nitrogen (NO_3-N)

   Potamogeton lucens   *Potamogeton perfoliatus* (NO_3-N)
   (conductivity)

   Potamogeton natans (Ca)   *Potamogeton pusillus* (NO_3-N)

TABLE 8.3 *(cont.)*

5.   Correlated with two or three of these parameters

Potamogeton berchtoldii *Zannichellia palustris* (conductivity, Ca,
(conductivity, Ca)   NO$_3$-N)

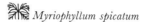*Ranunculus fluitans* × ?
(conductivity, NO$_3$-N)

6.   Correlated with conductivity, calcium, phosphate-phosphorus and
nitrate-nitrogen

Myriophyllum spicatum   *Potamogeton crispus*

Data from [15].

NUTRIENT PATTERNS IN SILT

Different rock types differ in the nutrients present in the silt of their
channels. The star diagrams in Fig. 8.1 summarise the results of a
country-wide survey of nutrient levels at approximately 450 sites. (The
conclusions in Chapters 2, 4 and 6 were based on at least 1000 sites, and
so are more accurate.) On the star diagrams, the line joins the average
concentrations of each nutrient. The error marks on the diagrams (the
standard error of the mean) show how variable the records are. If
the error mark is close to the line, the nutrient levels are similar in the
samples analysed. (When the error is not marked at all, as in Fig. 8.4, it
is too large to be shown on a diagram of this size, and the average is a
less reliable measure of the nutrients found.) Table 8.4 is complement-
ary to Fig. 8.1, listing the hardness ratio of each rock type. The hardness
ratio is the combined calcium and magnesium present in the site
divided by the combined sodium and potassium [38] and has proved
very useful in interpreting lake vegetation in terms of nutrient status for

TABLE 8.4   *Hardness ratio of silt from different watercourse types*

Rock or soil type	Hardness ratio
Peat dykes	1.4 (0.2)
Silt dykes	2.0 (0.6)
Soft sandstone streams	2.5 (0.2)
Resistant rock streams	2.6 (0.3)
Clay streams	3.6 (0.3)
Hard sandstone streams	4.9 (0.6)
Chalk streams	4.9 (0.8)

The hardness ratio is $\dfrac{\text{calcium} + \text{magnesium}}{\text{sodium} + \text{potassium}}$.

In brackets after each value is the standard error of the mean, a
measure of the variability of the results.

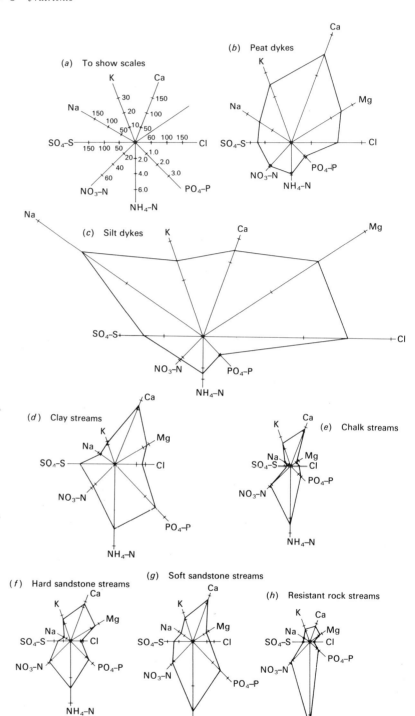

Fig. 8.1. Star diagrams of nutrient patterns for different stream types. The scales shown in (*a*) give nutrient levels (in p.p.m.) in water extracted from silt. The lines join the average values for each nutrient. The two small marks indicate the standard error of the mean, a measure of the variability of the results. Where no such marks are shown, the error is too large to be placed on a diagram of this size. Ca, calcium; Mg, magnesium; Cl, chloride; PO_4-P, phosphate–phosphorus; NH_4-N, ammonia–nitrogen; NO_3-N, nitrate–nitrogen; SO_4-S, sulphate–sulphur; Na, sodium; K, potassium.

plants. Both ratios and quantities of nutrients are important in determining how plants react to the chemical regime. The total nutrients present depend both on the nutrient levels in the silt and on the amount of silt present. Ratios are independent of the amount of silt. If these two parameters are used together, a coherent pattern of nutrient regimes and plant communities emerges.

In Fig. 8.1 and Table 8.4, the first division is between alluvial dykes (on peat and silt) and streams. Dykes have a very low hardness ratio, low ammonia-nitrogen, and very high calcium, magnesium, sodium, potassium, chloride, and, to a lesser extent, sulphate-sulphur. This combination of low hardness ratio and high nutrient levels is unusual. Peat dykes at the most mesotrophic end of the range, may have as the principal species:

Hottonia palustris   *Utricularia vulgaris*

Potamogeton alpinus

and these presumably occur in this nutrient-rich environment because of the low hardness ratio. Silt dykes have higher nutrient levels, and a slightly higher hardness ratio. They do not have the plants especially characteristic of nutrient-poor sites, and may have species commonly found in eutrophic (clay) sites, such as (see Chapter 14):

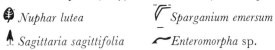

Nuphar lutea   *Sparganium emersum*

Sagittaria sagittifolia   *Enteromorpha* sp.

Dykes contain an unusual variety of plant communities (see Chapter 14) and this could be partly due to their unusual nutrient regime.

Clay streams are the most nutrient-rich, their silt containing the most calcium, magnesium, sodium, phosphate-phosphorus, chloride and sulphate-sulphur; they have a medium hardness ratio. The characteristic species of their most eutrophic downstream reaches (see Chapters 1, 11 and 12) are described as eutrophic.

Chalk has a high hardness ratio (more because the sodium content is low than because the calcium is high) and magnesium, phosphate-phosphorus, chloride and sulphate-sulphur are very low. This combination of low nutrients and high hardness ratio is here termed mesotrophic, and the species characteristic of upper and middle reaches of chalk streams are described as mesotrophic. There is some overlap of these species with mesotrophic dykes, e.g.

Lemna trisulca   *Ranunculus aquatilis*

There is more overlap between chalk brooks and silty upper brooks on Resistant rocks (which are nutrient-rich for their rock type), with soft sandstone streams, and with the upland brooks on hard sandstone (see Chapters 11–13). Hard sandstone is the rock type nearest in nutrient status to chalk, being similar in most nutrients and in hardness ratio. However, it has more magnesium, phosphate-phosphorus, chloride, sulphate-sulphur and less potassium, and so is of slightly higher nutrient status. This is especially true in the lower reaches, as hard sandstone produces much more inorganic silt than chalk (see below). Soft sandstone has a very low hardness ratio, but high levels of calcium, potassium, sodium and phosphate-phosphorus, and fairly high levels of chloride and sulphate-sulphur. The final nutrient status can presumably be considered similar to that of chalk, but it is different in that it has a low hardness ratio and high nutrients, instead of the reverse. The vegetation is mainly mesotrophic, but with more eutrophic plants than occur on chalk. These are presumably influenced by the high nutrient levels (as well as by the greater amount of silt; see below).

Streams on Resistant rock are very low in most nutrients (particularly so in calcium, phosphate-phosphorus, chloride and sulphate) and the hardness ratio is as low as on soft sandstone. These streams are the most nutrient-poor, being low in both nutrients and hardness ratio. The country on Resistant rocks varies from arable land to blanket bog, channels on blanket bog containing the least nutrients of all.

The average nutrient levels of a few blanket bog channels are (in p.p.m.):

Calcium 1.5	Magnesium 2	Sodium 25
Potassium 3	Nitrate-nitrogen 4	Ammonia-nitrogen 4
Phosphate-phosphorus 0.4	Chloride 35	Sulphate-sulphur 8
	Hardness ratio 0.6	

(Nitrogen levels are as high as in other stream types.)

The dystrophic species of this habitat include:

Drosera anglica   *Littorella uniflora*

Eriophorum angustifolium *Menyanthes uniflora*

When there is both acid peat and mineral sediment on the channel bed, the nutrient levels are slightly higher and the typical oligotrophic species include:

Callitriche hamulata   *Juncus bulbosos*

Eleocharis acicularis   *Myriophyllum alterniflorum*

Eleogiton fluitans

The star diagram in Fig. 8.4(*a*) shows the nutrient pattern in which such species occur. Oligotrophic habitats are usually streams on fairly flat moorland or bog, but can be swift highland ones with very little sediment, or upper non-silted reaches of streams on acid soft sandstone, e.g. in the New Forest. Streams with less or no peat have the higher nutrient status shown in Fig. 8.1(*h*).

Interestingly, nitrogen levels vary little between stream types, but vary considerably from place to place within them. This is most probably because decomposing organic matter produces considerable nitrogen in all channels, but the amount of dead plants or animals in

TABLE 8.5 *Manganese contents of silt from different watercourse types*

Values are in p.p.m. and the figure in brackets is the standard error. Methods as for Table 8.6.

Peat dykes	0.5 (0.2)
Silt dykes	0.6 (0.1)
Chalk streams	0.5 (0.3)
Clay streams	0.5 (0.1)
Soft sandstone streams	1.0 (0.2)
Resistant rock streams	3.5 (2.0)
Hard sandstone streams	3.5 (1.1)

any one site will vary. Table 8.5 shows that manganese tends to be higher on hard than on soft rocks.

When silt is analysed from channels from mixed rock types, such as clay and soft limestone, or hard sandstone and Resistant rock, its nutrient status is intermediate between that of the two types considered separately.

Using the combination of nutrient levels and hardness ratio, the comparative nutrient status of the rock types is:

Low   **High**

Resistant with peat	Resistant	Chalk Sandstone	Clay

NUTRIENT STATUS OF DIFFERENT STREAM TYPES

Trophic status depends both on the quality and the quantity of the nutrients present, and differences in the nutrient status of silt between rock types are enhanced by the different amounts of silt produced:

Low   **High**

Hard limestone Resistant rock	Chalk	Soft sandstone	Hard sandstone	Clay

It is the flow regime of each reach, however, that determines how much of this silt is actually deposited in the channel, and is thus available to influence plants.

Passing downstream along a river, flow generally becomes slower, and silt tends to increase. It is temporary silt that is often important for plant nutrition in upper and middle reaches, while silt in the middle and lower reaches is stable. Also, the nutrient status of the silt, sometimes in relation to most of the major nutrients, often increases downstream, particularly in rivers on Resistant rocks. Downstream increases in anions or conductivity in river water, though, can usually be attributed to changes in rock type or the entry of sewage effluents rather than to

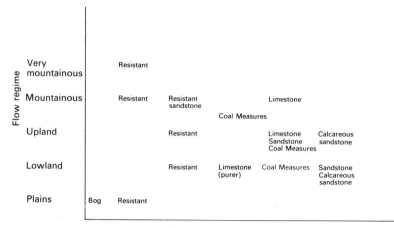

Fig. 8.2. Stream types on hard rocks.

silt. As a result of these changes in nutrient status the vegetation of British streams shows a consistent shift towards species of more eutrophic preferences as one passes from upper to lower reaches (also see Chapter 11 and, for North American streams, Chapter 17). There are three stream types in which downstream eutrophication is particularly marked. The first of these is streams on Resistant rock rising on blanket bog, which when they receive mineral silt from the more fertile land downstream may change from being dystrophic to mesotrophic channels. Secondly, on hard sandstone, mountain brooks are clean-washed, with negligible silt and, sometimes, oligotrophic plants. Farther down, however, flow may be checked sufficiently to allow thick nutrient-rich silt to accumulate on the bed. Thirdly, streams rising on a nutrient-poor rock type and flowing on to a nutrient-rich one will show a sharp increase in nutrient status.

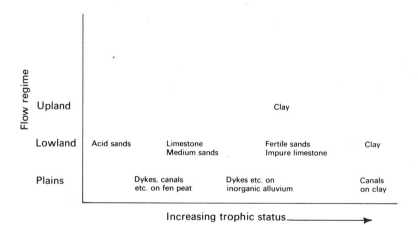

Fig. 8.3. Stream types on soft rocks.

The classification of British stream types, as shown in Figs. 8.2 and 8.3, is by nutrient status and flow type. The nutrient factors, summarised here, are:

Nutrient status of silt (and water)
Amount of silt entering the channel
Amount of this silt which can remain in the substrate

(Coal Measures, calcareous sandstone and hard limestone streams are included in Fig. 8.3, even though the nutrient data for these types are incomplete.)

Flow, of course, affects plants directly (see Chapters 2–6), as well as affecting the silt deposition, and other factors also differentiate between stream types. Some, such as the stability of the hard bed, are described elsewhere in this book (see Chapter 3). A minor but important factor may be the slope (and stability) of the bank. For instance, soft sandstone brooks, in contrast with those on chalk or hard sandstone, often have banks which are too steep for abundant fringing herbs at the sides (Fig. 10.2).

The importance of Figs. 8.2 and 8.3 is that each category has its own characteristic vegetation type, recognisable and classifiable as such (see Chapters 11–14). The main controlling factors of British stream vegetation are flow type and geology, and these can be interpreted in terms of topography, water inflow, silting and nutrient status (and other minor factors such as bed stability, turbidity, etc.).

DISTRIBUTION OF SPECIES IN RELATION TO SILT NUTRIENTS

Dystrophic and oligotrophic species are confined to a narrow range of habitats, with a narrow and characteristic nutrient status. Dystrophic nutrient levels are given above. Fig. 8.4(*a*) is a star diagram showing the low nutrient levels of the sites in which oligotrophic species occur (for nitrogen, see above). Fig. 8.4(*b*), for sites with mesotrophic species, is a larger pattern, and that for eutrophic species (Fig. 8.4*c*) is larger again. These species groups can be identified and separated on nutrient status. However, mesotrophic and eutrophic species have wide and sometimes overlapping ranges, and there are also species groups intermediate between these (see Chapters 1, 12, etc.) and it is difficult to get clear separations on nutrient regimes (particularly on such a small number of samples). Star diagrams for some species are shown in Fig. 8.5. Contracted patterns (narrow ranges of these nutrients which vary considerably within the mesotrophic–eutrophic range) are illustrated in (*a*) and (*b*) (*Ranunculus* spp. and mosses), and wide patterns with high nutrients in (*c*) and (*d*) (*Alisma plantago-aquatica* and *Enteromorpha* sp.). (Mosses and *Enteromorpha* both have at least some access to silt nutrients.) Species with particularly low values for a single nutrient are shown in (*e*) and (*f*) (*Myriophyllum spicatum* for magnesium, and *Potamogeton crispus* for phosphate-phosphorus), and with particularly high

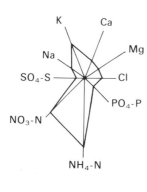

(a) Oligotrophic species

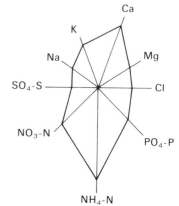

(b) Mesotrophic species

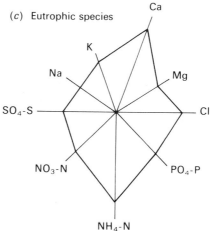

(c) Eutrophic species

Fig. 8.4. Star diagrams of nutrient patterns for different species groups. Details as for Fig. 8.1. (a) Represents all sites containing oligotrophic species, (b) all sites containing mesotrophic species, and (c) all sites containing eutrophic species.

values of one nutrient in (g) and (h) (*Potamogeton pectinatus* for chloride, and *Sparganium erectum* for ammonia-nitrogen).

Plants also vary in their distribution according to hardness ratio. Using the average of all sites in which any given species was found, the following generalisations between hardness ratio aad type of species can be made:

Hardness ratio	Types of species
0–1	Dystrophic, oligotrophic and dyke species
1–2	Mainly oligotrophic and dyke species
2–4	A varied group, including eutrophic species
4+	Mainly limestone species, including fringing herbs

It again appears that silt nutrient content and hardness ratio jointly determine nutrient status for plants, so that, for instance, clay species require a medium hardness ratio but very high silt nutrient levels,

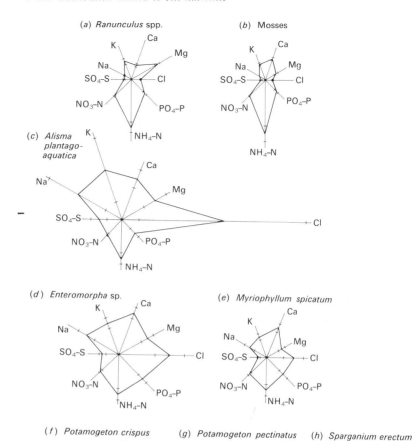

Fig. 8.5. Star diagrams of nutrient patterns for individual species (or aggregates). Details as for Fig. 8.1. except that in order to make the standard errors small enough to show on the patterns, the site with the highest nutrient levels in each instance has been omitted from the calculations.

while chalk species are characteristic of a high hardness ratio and low nutrient contents.

Plant distribution in relation to silt chemistry is shown in Table 8.6, and should be compared with the water chemistry date in Table 8.2. Both tables show mesotrophic and eutrophic plants only (apart from part of the aggregate – mosses). Sodium and chloride levels are important in separating these two groups (also see Fig. 8.4 and, for chloride, Table 8.2) and some separation is shown with magnesium, potassium and sulphate-sulphur, but not with the other nutrients. Differences in nitrate-nitrogen and phosphate-phosphorus levels in the water are more important ecologically than similar differences in silt.

141

TABLE 8.6 *Plant distribution in relation to silt chemistry*

The species are listed under the concentrations with which they are best related. Concentrations are in p.p.m. Data are of water extracted from silt and analysed on Atomic Absorption Spectrometer, Flame Photometer, and Hach Fieldlab DR/EL2.

1. Calcium

Below 50   Mosses

50–100   (*Carex acutiformis*)   *Potamogeton crispus*

        Ceratophyllum demersum   *Potamogeton pectinatus*

        Myriophyllum spicatum   *Ranunculus* spp.

        Phalaris arundinacea   *Sagittaria sagittifolia*

100–150   *Apium nodiflorum*   *Potamogeton crispus*

        (*Carex acutiformis*)   *Potamogeton pectinatus*

        Elodea canadensis   *Rorippa nasturtium-aquaticum* agg.

        Glyceria maxima   *Schoenoplectus lacustris*

        Lemna minor agg.   *Sparganium emersum*

        Mentha aquatica   *Veronica beccabunga*

        Nuphar lutea

150–300   (*Carex acutiformis*)   *Potamogeton perfoliatus*

        Mentha aquatica   *Schoenoplectus lacustris*

        Myosotis scorpioides   *Sparganium emersum*

        Potamogeton pectinatus

Over 300   *Phragmites communis*   *Enteromorpha* sp.

        Polygonum amphibium

Poorly correlated   *Alisma plantago-aquatica*   *Sparganium erectum*

        Callitriche spp.

2. Chloride

Below 20   *Callitriche* spp.   *Rorippa nasturtium-aquaticum* agg.

TABLE 8.6 (*cont.*)

	Carex acutiformis	Veronica beccabunga
	Elodea canadensis	Mosses
	Mentha aquatica	

20–40	Apium nodiflorum	Phalaris arundinacea
	Carex acutiformis	Potamogeton perfoliatus
	Elodea canadensis	Ranunculus spp.
	Myosotis scorpioides	Rorippa nasturtium-aquaticum agg.
	Myriophyllum spicatum	Sparganium erectum
40–60	Glyceria maxima	Rorippa nasturtium-aquaticum agg.
	Nuphar lutea	Sagittaria sagittifolia
	Phragmites communis	Schoenoplectus lacustris
	Potamogeton crispus	Sparganium emersum
60–80	Glyceria maxima	Enteromorpha sp.
	Nuphar lutea	
Over 80	Alisma plantago-aquatica	Phragmites communis
	Ceratophyllum demersum	Potamogeton pectinatus
	Lemna minor agg.	

3. Magnesium

Below 5	Apium nodiflorum	Phalaris arundinacea
	Callitriche spp.	Ranunculus spp.
	Carex acutiformis	Rorippa nasturtium-aquaticum agg.
	Myriophyllum spicatum	Veronica beccabunga
5–10	Elodea canadensis	Potamogeton pectinatus
	Lemna minor agg.	Sagittaria sagittifolia
	Nuphar lutea	Sparganium emersum
	Potamogeton crispus	Sparganium erectum

143

TABLE 8.6 *(cont.)*

10–20	*Glyceria maxima*		*Schoenoplectus lacustris*
	Nuphar lutea		*Sparganium erectum*
20–40	*Alisma plantago-aquatica*		*Schoenoplectus lacustris*
	Ceratophyllum demersum		*Sparganium erectum*
	Lemna minor agg.	–	Mosses
	Polygonum amphibium	–	*Enteromorpha* sp.
40+	*Ceratophyllum demersum*		*Phragmites communis*

4.   Nitrate-nitrogen

Below 3	*Alisma plantago-aquatica*		*Mentha aquatica*
	Callitriche spp.		*Phalaris arundinacea*
3–4	*Callitriche* spp.		
	Ceratophyllum demersum		*Potamogeton crispus*
	Mentha aquatica		*Potamogeton pectinatus*
	Myriophyllum spicatum		*Rorippa nasturtium-aquaticum* agg.
	Nuphar lutea		*Veronica beccabunga*
	Phragmites communis	–	Mosses
	Polygonum amphibium	–	*Enteromorpha* sp.
4+	*Elodea canadensis*		*Rorippa nasturtium-aquaticum* agg.
	Lemna minor agg.	–	Mosses

Poorly correlated	*Apium nodiflorum*		*Ranunculus* spp.
	Carex acutiformis		*Sagittaria sagittifolia*
	Glyceria maxima		*Sparganium emersum*
	Myosotis scorpioides		*Sparganium erectum*
	Potamogeton perfoliatus		

5.   Ammonia-nitrogen

Below 2.5	*Callitriche* spp.		*Polygonum amphibium*
2.5–5	*Ceratophyllum demersum*		*Lemna minor* agg.

TABLE 8.6 (*cont.*)

5–10

Mentha aquatica

Nuphar lutea

Phalaris arundinacea

Polygonum amphibium

Potamogeton crispus

Rorippa nasturtium-aquaticum agg.

Enteromorpha sp.

10+

Apium nodiflorum

Lemna minor agg.

Mentha aquatica

Myosotis scorpioides

Phalaris arundinacea

Sparganium erectum

Enteromorpha sp.

Poorly correlated

Alisma plantago-aquatica

Carex acutiformis

Elodea canadensis

Glyceria maxima

Myriophyllum spicatum

Phragmites communis

Potamogeton pectinatus

Ranunculus spp.

Sagittaria sagittifolia

Schoenoplectus lacustris

Sparganium emersum

Mosses

6. Phosphate-phosphorus

Below 1

Alisma plantago-aquatica

Ceratophyllum demersum

Elodea canadensis

Glyceria maxima

Nuphar lutea

Phalaris arundinacea

Polygonum amphibium

Ranunculus spp.

Rorippa nasturtium-aquaticum agg.

Sagittaria sagittifolia

Schoenoplectus lacustris

1–2

Mentha aquatica

Myriophyllum spicatum

Nuphar lutea

Phragmites communis

Polygonum amphibium

Rorippa nasturtium-aquaticum agg.

Sagittaria sagittifolia

2–3

Callitriche spp.

Myriophyllum spicatum

Schoenoplectus lacustris

Sparganium erectum

145

TABLE 8.6 (*cont.*)

	🌿 *Potamogeton pectinatus*	🌱 *Veronica beccabunga*
	✹ *Ranunculus* spp.	🌿 Mosses
3+	🌱 *Myosotis scorpioides*	🌿 *Rorippa nasturtium-aquaticum* agg.
	🌾 *Phalaris arundinacea*	🌿 *Sparganium erectum*

Poorly correlated

	🌿 *Apium nodiflorum*	🍃 *Potamogeton perfoliatus*
	🌱 *Carex acutiformis*	🌿 *Sparganium emersum*
	🌿 *Lemna minor* agg.	〰 *Enteromorpha* sp.
	🌿 *Potamogeton crispus*	

7.   Potassium

Below 10

	🌱 *Alisma plantago-aquatica*	✹ *Ranunculus* spp.
		🌱 *Veronica beccabunga*
	〰 *Polygonum amphibium*	🌿 Mosses
	🍃 *Potamogeton perfoliatus*	

10–20

	✳ *Elodea canadensis*	✹ *Ranunculus* spp.
	🌿 *Lemna minor* agg.	🌱 *Sagittaria sagittifolia*
	🌿 *Potamogeton pectinatus*	🌿 *Schoenoplectus lacustris*
	🍃 *Potamogeton perfoliatus*	🌿 *Sparganium emersum*

20–40

	🌱 *Alisma plantago-aquatica*	🌿 *Glyceria maxima*
	🌱 *Carex acutiformis*	🌿 *Lemna minor* agg.
	🌿 *Ceratophyllum demersum*	🌿 *Rorippa nasturtium-aquaticum* agg.

40+

	🌿 *Callitriche* spp.	〰 *Enteromorpha* sp.
	🌿 *Lemna minor* agg.	

Poorly correlated

	🌿 *Apium nodiflorum*	🌿 *Nuphar lutea*
	🌿 *Mentha aquatica*	🌾 *Phalaris arundinacea*
	🌱 *Myosotis scorpioides*	🌿 *Sparganium erectum*

TABLE 8.6 (*cont.*)

8. Sodium

Below 20	*Callitriche* spp.	*Ranunculus* spp.
	Mentha aquatica	*Veronica beccabunga*
	Phalaris arundinacea	Mosses
20–40	*Apium nodiflorum*	*Ranunculus* spp.
	Myosotis scorpioides	*Rorippa nasturtium-aquaticum* agg.
	Nuphar lutea	*Sparganium emersum*
	Potamogeton crispus	
40–60	*Alisma plantago-aquatica*	*Sagittaria sagittifolia*
	Elodea canadensis	*Schoenoplectus lacustris*
	Glyceria maxima	*Sparganium emersum*
	Nuphar lutea	
60–100	*Sparganium erectum*	
100 +	*Alisma plantago-aquatica*	*Potamogeton perfoliatus*
	Ceratophyllum demersum	*Sagittaria sagittifolia*
	Lemna minor agg.	*Enteromorpha* sp.
	Phragmites communis	
Poorly correlated	*Carex acutiformis*	*Potamogeton pectinatus*
	Myriophyllum spicatum	

9. Sulphate-sulphur

Below 10	*Myosotis scorpioides*	Mosses
10–30	*Apium nodiflorum*	*Rorippa nasturtium-aquaticum* agg.
	Callitriche spp.	*Schoenoplectus lacustris*
	Carex arundinacea	*Veronica beccabunga*
	Phalaris arundinacea	Mosses
	Ranunculus spp.	
30–80	*Ceratophyllum demersum*	*Potamogeton pectinatus*

147

TABLE 8.6 (*cont.*)

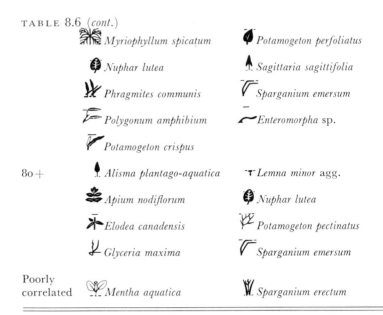

	Myriophyllum spicatum	Potamogeton perfoliatus
	Nuphar lutea	Sagittaria sagittifolia
	Phragmites communis	Sparganium emersum
	Polygonum amphibium	Enteromorpha sp.
	Potamogeton crispus	
80+	Alisma plantago-aquatica	Lemna minor agg.
	Apium nodiflorum	Nuphar lutea
	Elodea canadensis	Potamogeton pectinatus
	Glyceria maxima	Sparganium emersum
Poorly correlated	Mentha aquatica	Sparganium erectum

Plant nutrient contents

Oligotrophic species, and the grouping of all species from oligotrophic sites, contain very low levels of calcium, potassium, nitrogen and phosphorus. The wide-ranging mesotrophic and eutrophic species, in contrast, each have widely varying nutrient contents. However, if all plants sampled on each rock type are grouped, the values are much more stable.

In Resistant rock streams plants have very low calcium, potassium, nitrogen and phosphorus levels (though those from the more fertile sites contain higher levels than those from the oligotrophic ones); in chalk streams plants are low in magnesium; in both clay and sandstone streams plants have high levels of potassium, nitrogen and phosphorus, but plants on clay are rich in sodium while those on sandstone are rich in magnesium and low in sodium (also see [39]).

Plant communities and vegetation types are described in Chapters 12–14. They are closely related to rock (and soil) type, which, in turn, determine the nutrient status of the silt in the channel substrate. The association between species and bedrock is through the community type, and both the quality and the quantity of the silt can be linked with this. As far as individual mesotrophic to eutrophic species are concerned, the quantity of silt is too variable to be assessed and the species usually occur in a wide range of habitats, not well related to individual nutrients of the silt or water. The type of variation is shown well by considering communities containing three or more fringing herbs. Such communities typically occur in chalk brooks (high hardness ratio, low nutrient contents), upland hard sandstone brooks (similar to chalk but higher in some nutrients often lower in silt bands), silty sheltered mountain brooks on Resistant rocks (low hardness ratio, rather low nutrients but ample silt), some soft sandstone brooks with gently sloping banks (low hardness ratio, fairly high nutrients) and sheltered upland hard limestone brooks (chemically probably like chalk, with some silt present). All these are mesotrophic habitats, but are classified as such because of different combinations of nutrient characters, the link being with the rock type and silt content rather than with, say, calcium or phosphate levels separately. If these fringing herbs are considered as individual species rather than as a group, each has a much wider habitat range and thus a yet wider nutrient range. In other words, communities have smaller nutrient ranges than do their component species. It must be borne in mind though that nutrients are only one of the complex of factors determining plant distribution, and must not be considered in isolation from other factors such as flow.

There are also other complications which are not considered in Tables 8.2 and 8.6. For example, it appears that if plants have adequate nutrients for part of the time, e.g. from temporary silt, this is sufficient for continuous good growth. It is also possible for nutrients to be present in too high amounts. In particular, if several nutrients are jointly present in high concentrations in water they are more toxic than if only one is at a high level. Also, there is as yet no relevant evidence on relationships between carbon sources and the distribution of river plants.

9
Productivity

Productivity, the rate at which the vegetation increases, is greater for emerged than for submerged plants, mainly because the submerged plants receive less light. The total production of river plants in streams is very variable and depends on how much of the stream is occupied by plants at all and the relative proportions of emerged plants, submerged plants and algae.

The *production* of vegetation is the weight of new organic matter made by plants over a given period of time, *productivity* is the rate at which this organic material is formed and the *biomass* is the total weight of plants present at any one time. Productivity is very variable in different rivers, some having no vegetation, some a little and others being choked with plants. Among aquatics, most of the research so far has been on the largest amount of production – and the maximum biomass – which can occur, rather than on the variations between different plant species and the effect on production of differences in habitat such as substrate texture, trophic status, etc.

The main division of river plants according to production is between the emerged and the submerged plants. Emergents receive full sunlight, and because their lower parts are in water their production, unlike that of most land plants, is rarely hindered by shortage of water. Thus the production of tall emergents is usually higher than that of any other type of vegetation and they also have a high productivity. Submergents, in contrast, receive much less light, since a proportion of it is reflected or absorbed during its passage through the water above them, and they

TABLE 9.1   *Production of water plants in good conditions*

	Maximum biomass (kg organic wt per m²)	Annual net production (kg organic wt per m²)	Productivity (g dry wt per m² per day)	Carbon up-take (mg per g dry wt per hour)
Emergents	4–10	4–6	12–48	3–9
Submergents	0.4–0.7	0.5–0.8	2–10(—25)	2–10(—20)

Adapted from [40].
Data refer to temperate regions.
Emergents are tall species such as *Phragmites communis* and *Typha* spp. and do not include short emergents such as *Myosotis scorpioides*. Submergents include *Chara* spp. and *Myriophyllum spicatum*.

150

may often suffer from a shortage of carbon (see Chapter 8). Their production is therefore very low, averaging only about a tenth of that of the tall emergents (Table 9.1).

Production depends both on features of the plant's habitat, and on the rate at which green shoots can fix carbon and energy into organic compounds under optimum conditions. As can be seen from Table 9.1, the amounts of carbon fixed (per unit dry leaf weight per hour) in emergents and submergents are similar, so the difference in productivity between the two groups is a result of differences in habitats, not their biochemical efficiency [40]. There is too little evidence at present to determine whether variations in habitat within the two main groups cause variations in production.

The production in a chalk stream, Bere stream, has been studied [41] and the results are summarised in Table 9.2. The stream is fairly small, with a moderately swift flow and plants growing right across the channel. *Ranunculus calcareus*, a submergent, is the main dominant, growing most in spring and early summer. *Rorippa nasturtium-aquaticum* agg., a fringing herb, increases greatly in late summer, particularly at the sides, and partly overgrows the *Ranunculus*. More of the stream is covered by the submergent and in the stream as a whole its production is the higher, but the productivity of the fringing herb is much the greater and where the habitat is suitable for it the biomass of the vegetation is mainly *Rorippa nasturtium-aquaticum* agg. There are also algae present, growing and suspended in the water. Their production is about half that of the larger plants. As in many chalk streams, commercial watercress beds occur near the source, and fragments from these are washed into the stream. To a lesser extent organic material enters the stream as leaves, etc., from overhanging trees and bank plants. When the plants and animals die, their remains take some time

TABLE 9.2   *Production in Bere stream*

(a) *Production (metric tonnes organic matter per year) for the whole stream*

Submergents (mainly *Ranunculus*)	8
Emergents (mainly *Rorippa nasturtium-aquaticum* agg.)	5
Algae	6
Tree leaves, etc.	2
Bank plants, etc.	1
From watercress beds	38
In spring water	17

(b) *Annual net production (kcal per m² per year) at one site*

Submergents (mainly *Ranunculus*)	550
Emergents (mainly *Rorippa nasturtium-aquaticum* agg.)	1800
Algae (in water and on bed)	700

Adapted after [41].

TABLE 9.3 *Production^a in R. Thames*

Acorus calamus (tall emergent)	Site 1 23.7		Site 2 16.4
Nuphar lutea (rooted floating)	Site 1 0.1		Site 2 27.6
Tree leaves, etc.		79	
Algae in water (phytoplankton)		4388	

Adapted from [43].

^{a}Annual net production and input of tree leaves etc. (kcal per m^2 per year).

to decompose completely, and some of this dead organic matter is washed downstream.

A study has been made of production in R. Thames [42, 43], the results of which are shown in Table 9.3. R. Thames is a large slow eutrophic lowland river, with a very different vegetation from the chalk stream. The water is deep and turbid, and so the larger plants are able to grow only on the shallow edges. The slow eutrophic waters are very suitable for the growth of floating algae, though, and in the reach studied they accounted for most of the production. The organic matter contributed by the larger plants comes partly from the tall emergent *Acorus calamus* and the floating-leaved rooted *Nuphar lutea*, but mainly from tree leaves dropping into the water. Where trees overhang the water these have the double effect of increasing the tree leaf content of the site and decreasing the river plants as a result of shading. *Acorus calamus* grows in shallow water, while *Nuphar lutea* is usually in water 0.5–2 m deep, so that the proportions of each, and indeed the total biomass of the two, depends on the slope of the channel below water level.

10

Plant patterns

A few river channels have uniform vegetation from one bank to the other, but in most the plants found at the edges are different from those in the centre. Some channels have essentially similar species growing in much the same positions and to much the same size each year, while in others there are large changes from year to year, or even between early summer and late summer. This type of variations can be the result of differences in factors which are physical (e.g. flow, substrate texture), chemical (e.g. the nutrients in the silt) or biotic (e.g. shading by larger plants, grazing by animals). It is easier to list the causes of such patterning than it is to classify plant behaviour, and some phenomena, for example the pattern of fringing herbs, are due to more than one cause. A final limiting factor on vegetation patterns is that any one site will only support a specified number of species and only some of these will be capable of luxuriant growth (see Chapters 11–14).

This chapter brings together descriptions of the sort of plant patterns which can be seen within sites – the emerged plants at the sides of the channel and the submerged ones in the middle; the decrease in vegetation under trees; and the differences at the same site when it is observed in May and in August, for instance. Most of these patterns are governed by the factors such as channel depth, flow, substrate and shading which have already been discussed in previous chapters. However, the patterns due to storm flows, fully described in Chapters 5 and 3, are not repeated here, nor are e.g. mosaics due to substrate stability. Here we consider the patterns from the point of view of plant behaviour, rather than as an effect of a physical factor such as flow, and also introduce the study of river vegetation as a whole, which is continued in Chapter 11 with a discussion of the changes which occur along a river system, and in Chapters 12–14 with an examination of the different plant communities and vegetation of various stream types. The small-scale plant patterns described here occur within the larger-scale ones described later, and are irrelevant to the diagnosis of a site as being e.g. a chalk stream, an upper reach or a polluted habitat.

ACROSS THE CHANNEL

Cross-sections of channels vary in shape, from, at one extreme, deep canals with uniformly deep water between piled banks, to small chalk or hill streams with a shallow channel and gentle side slopes at the other (Fig. 10.1). River plants can potentially live on all angles of

(a) Canal

Fig. 10.1. Contrasted channel shapes. (*a*) Canal (deep water, piled sides). (*b*) Small chalk stream.

(b) Small chalk stream

substrate and at all water depths commonly found in Britain, though concrete or piled slopes cannot, of course, bear rooted plants. More importantly, plants often cannot grow in deep water, either because the substrate is unstable, or because in turbid water too little light reaches the bed (see Chapters 3, 4 and 7). Fig. 10.2 shows some typical variations in vegetation with the slope of the bank above and below water level, the slope of the bed, and the general size and flow regime of the channel. The plant communities themselves are described in Chapters 12–14.

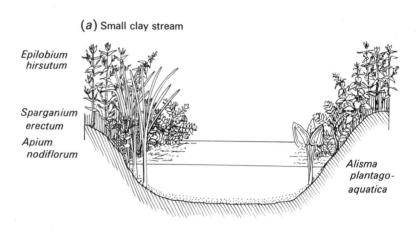

(a) Small clay stream

Epilobium hirsutum

Sparganium erectum

Apium nodiflorum

Alisma plantago-aquatica

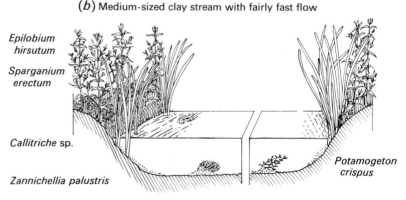

(b) Medium-sized clay stream with fairly fast flow

Epilobium hirsutum

Sparganium erectum

Callitriche sp.

Zannichellia palustris

Potamogeton crispus

Fig. 10.2. Sections of different stream types. (Some points of basic ecology illustrated here are not described elsewhere in this book.) (*a*) Small clay stream: silted bottom; steep banks; slow flow. (*b*) Medium-sized clay stream with fairly fast flow: gravelly bottom; steep banks; very sparse plants.

(*c*) Medium-sized clay stream with slow flow

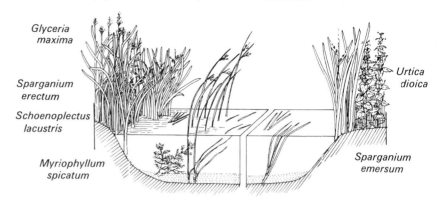

Glyceria maxima

Sparganium erectum

Schoenoplectus lacustris

Myriophyllum spicatum

Urtica dioica

Sparganium emersum

(*d*) Large clay stream

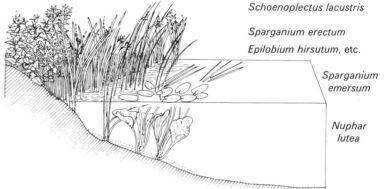

Schoenoplectus lacustris

Sparganium erectum

Epilobium hirsutum, etc.

Sparganium emersum

Nuphar lutea

Fig. 10.2 (cont.). Sections of different stream types. (Some points of basic ecology illustrated here are not described elsewhere in this book.) (*c*) Medium-sized clay stream with slow flow: silted bottom; large or low banks; much vegetation; tall monocotyledons in large patches or bands. (*d*) Large clay stream: centre too deep and turbid or unstable for plants, slow flow; low banks; tall monocotyledons in large patches or bands. (*e*) Small chalk stream: gravelly bottom; fairly fast flow; gently sloping banks. (*f*) Medium to large chalk stream: gravelly bottom; moderate flow; low banks; tall monocotyledons in narrow band only; submerged plants throughout the channel.

(*e*) Small chalk stream

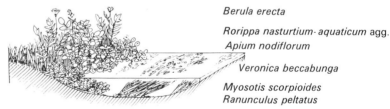

Berula erecta

Rorippa nasturtium-aquaticum agg.
Apium nodiflorum

Veronica beccabunga

Myosotis scorpioides
Ranunculus peltatus

(*f*) Medium to large chalk stream

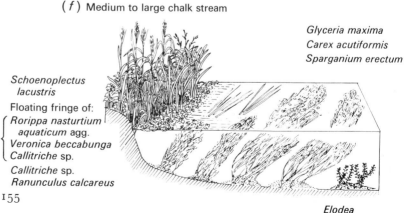

Glyceria maxima
Carex acutiformis
Sparganium erectum

Schoenoplectus lacustris

Floating fringe of:
{ *Rorippa nasturtium aquaticum* agg.
Veronica beccabunga
Callitriche sp.

Callitriche sp.
Ranunculus calcareus

155

Elodea canadensis

(*g*) Medium-sized soft sandstone stream

Epilobium
hirsutum, etc.

Callitriche sp.

Veronica
beccabunga

Ranunculus cf. aquatilis.

Potamogeton crispus

(*h*) Small hard sandstone stream

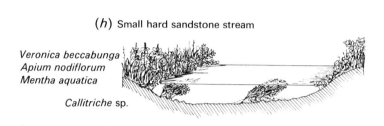

Veronica beccabunga
Apium nodiflorum
Mentha aquatica

Callitriche sp.

(*i*) Medium-sized hard sandstone stream

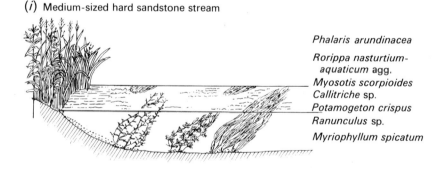

Phalaris arundinacea

Rorippa nasturtium-
aquaticum agg.

Myosotis scorpioides
Callitriche sp.
Potamogeton crispus
Ranunculus sp.
Myriophyllum spicatum

Fig. 10.2 (cont.). Sections of different stream types. (Some points of basic ecology illustrated here are not described elsewhere in this book.) (*g*) Medium-sized soft sandstone stream: sandy and gravelly bottom; moderate flow; steep or gently-sloping banks. (*h*) Small hard sandstone stream: sandy and gravelly bottom; swift flow; gently sloping banks. (*i*) Medium-sized hard sandstone stream: silted in sheltered parts; gravelly bottom in exposed ones; moderate flow; banks not steep. (*j*) Small hard limestone stream: coarse substrate with some fine soil in shelter; swift flow.

(*j*) Small hard limestone stream

Filipendula ulmaria
Mimulus guttatus

Veronica beccabunga

Rorippa nasturtium
aquaticum agg.

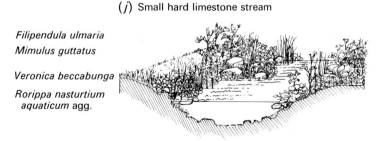

(*k*) Small Resistant rock stream with very fierce flow

(*l*) Small Resistant rock stream with less fierce flow

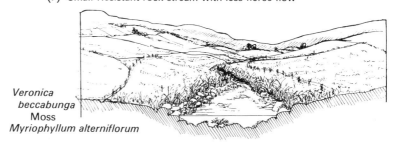

Veronica
beccabunga
Moss
Myriophyllum alterniflorum

Fig. 10.2 (cont.). Sections of different stream types. (Some points of basic ecology illustrated here are not described elsewhere in this book.) (*k*) Small Resistant rock stream with very fierce flow: coarse substrate; gently sloping banks; large fall from hill to channel and steeply sloping channel; no river plants. (*l*) Small Resistant rock stream with less fierce flow: as (*k*) but with less steep falls from hill to channel, and a less steeply sloping channel. (*m*) Large Resistant rock stream with very fierce flow: coarse bouldery substrate; very swift flow. (*n*) Medium to large Resistant rock stream with much less force of water than (*m*), in middle reaches: coarse, non-bouldery substrate; fairly swift flow; less steep hillsides, etc. (*Polygonum amphibium* is included to show habit only. It occurs in shelter, and as regards both flow and trophic status rarely overlaps with *Myriophyllum alterniflorum*.)

(*m*) Large resistant rock stream with very fierce flow

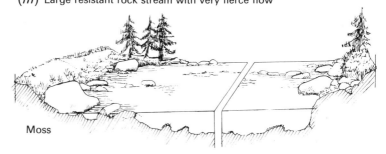

Moss

(*n*) Medium to large Resistant rock stream with much less force of water than (*m*)

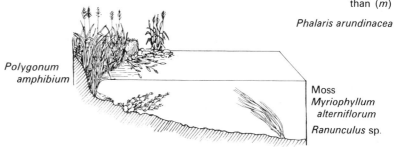

Phalaris arundinacea

Polygonum
amphibium

Moss
Myriophyllum
alterniflorum
Ranunculus sp.

157

(*o*) Dyke with water level lowered to
well below ground level

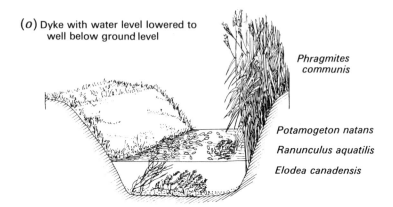

*Phragmites
communis*

Potamogeton natans

Ranunculus aquatilis

Elodea canadensis

Fig. 10.2 (cont.). Sections of
different stream types. (Some
points of basic ecology
illustrated here are not
described elsewhere in this
book.) (*o*) Dyke with water
level lowered to well below
ground level: negligible
flow: dredged sufficiently
recently to prevent growth of
tall monocotyledons on the
bed; one bank grazed, or
treated with maleic hydrazide
and covered with short
grasses, the other unmanaged
and dominated by *Phragmites
communis*; channel plants
included to show different
habits. (*p*) Dyke with water
level near ground level:
negligible flow; dredged
sufficiently recently to prevent
growth of tall monocotyledons
on the bed; fringing herbs
able to grow on the shallow
bank; other channel plants
included to show different
habits. (*q*) Canal: negligible
flow; slightly silted floor;
species-rich (see Chapters 14
and 21). The shallow ledge
forms a favourable habitat,
being shallow and affected
little by boats. Disused canals
may be full of plants, ones
that are used somewhat have
fringes of plants, and much
use results in the
disappearance of vegetation.

(*p*) Dyke with water level near ground level

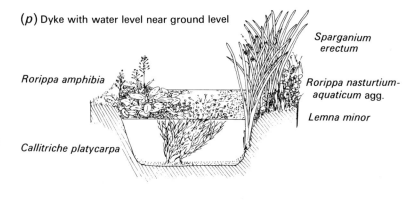

*Sparganium
erectum*

Rorippa amphibia

*Rorippa nasturtium-
aquaticum* agg.

Lemna minor

Callitriche platycarpa

(*q*) Canal

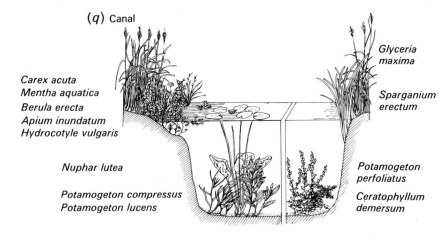

*Glyceria
maxima*

*Sparganium
erectum*

Carex acuta
Mentha aquatica
Berula erecta
Apium inundatum
Hydrocotyle vulgaris

Nuphar lutea

Potamogeton compressus
Potamogeton lucens

*Potamogeton
perfoliatus*

*Ceratophyllum
demersum*

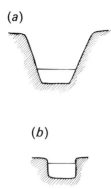

(a)

(b)

Fig. 10.3. A cross-section of a dyke (*a*) with the water level well below ground level, and (*b*) with the water level almost at ground level.

Let us first consider the channel banks and their plants. High banks may be man-made to prevent flooding, for example in the flood plains of large rivers, or they may be the result of a lowering of the ground water level, as in some fen dykes. Where wetlands have been drained for arable farming, the water may be as much as 2 m below ground level (Fig. 10.3*a*), but in little-drained wetlands ground level and water level almost coincide and banks will be very low (Fig. 10.3*b*). Lowland chalk streams have a very stable flow and also usually have very low banks. Sandstone banks are slightly higher, and clay ones are higher still (Fig. 10.2). The low banks associated with mountain streams means that they may flood land beside the river downstream unless they are artificially embanked.

Banks vary in steepness as well as in height, and this can influence vegetation. For example, narrow channels with steep banks 1–2 m high may be completely shaded by tall herbs on the bank, whereas if banks of the same height slope more gradually there may be enough light for plants to grow in the channel. In wide channels, however, channel vegetation is hardly influenced by the slope of the bank above the water.

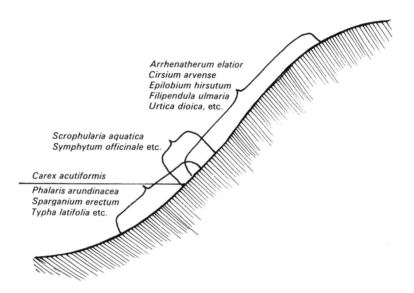

Arrhenatherum elatior
Cirsium arvense
Epilobium hirsutum
Filipendula ulmaria
Urtica dioica, etc.

Scrophularia aquatica
Symphytum officinale etc.

Carex acutiformis

Phalaris arundinacea
Sparganium erectum
Typha latifolia etc.

Fig. 10.4. The habitat preferences in relation to water level of the main groups of bank plants.

Bank plants in the lowlands are usually tall, but may be short when they are grazed (e.g. by sheep on the Romney Marsh), sprayed with herbicides (as in much of the Lincolnshire River Division) or regularly cut (as on some tow paths). In the mountains the plants are short. Scattered trees are widespread along lowland streams, and are often also found in valleys in the highlands. The main groups of tall bank plants are those of dry land, damp ground, and of habitats intermittently flooded (Fig. 10.4). Of those species in the last group, some

159

seldom grow far from the water (e.g. *Carex acutiformis*, *Sparganium erectum*), others often grow well up the bank (e.g. *Glyceria maxima*, *Phalaris arundinacea*), while yet others vary markedly in habitat (e.g. *Phragmites communis* grows well up the banks of dykes, but elsewhere is seldom far from the water).

At the edge of the channel, river plants, if present, may be tall emerged monocotyledons or fringing herbs. Rarely, other large broad-leaved plants may occur (e.g. *Alisma plantago-aquatica*, most frequent in somewhat small clay streams; and *Rumex hydrolapathum*, most frequent in canals), or small grasses may be prominent (as opposed to present but inconspicuous). When both tall and short plants can grow at the edge, the tall will shade and kill the short. However, the habitat preferences of the tall monocotyledons and the fringing herbs, while overlapping, do differ. The tall monocotyledons (except sometimes *Phalaris arundinacea*) require or prefer a soft fine substrate in which their deep roots and large rhizomes can anchor, and, being slow-growing perennials, need habitats which are stable, preferably for some years. They are also, being tall and deep-rooted, less likely to be harmed by fluctuating water levels and are thus more able to colonise steep slopes. The fringing herbs, in contrast, can invade temporary habitats, and can root in both fine and coarse substrates (though they can anchor more firmly in coarser ones). They offer much hydraulic resistance to flow and their root systems are short, so on fine soils they are likely to be floated off in storm flows (see Chapter 5). They require fairly gentle slopes for the proper development of clumps, though fragments can become lodged on steeper banks.

Within the channel itself it is possible for emerged, floating and submerged plants to grow, and some characteristic patterns are shown in Fig. 10.2. The basic habitat preferences of the tall monocotyledons and fringing herbs are the same as those described above for the channel edges. Tall monocotyledons are characteristic of shallow water and deep soft soils, with little scour but possibly some drying. This means they are usually found at the sides of watercourses, sometimes on stable shoals and occasionally right across the channel. They are commonest in clay streams and shallow dykes, where also they are most likely to dominate. Fringing herbs grow well in shallow water, at the sides of the channel or in the centre, particularly on firm gravel in moderate flow (Fig. 10.2), and especially in chalk and sandstone streams. Clay streams tend to have much fine sediment, so that clumps are easily washed away, while streams on Resistant rocks tend to have too fierce a flow and too little sediment for their good development. Fringing herbs grow well, and can develop into clumps, from fragments lodged on obstructions in the channel, though such clumps may be easily washed away.

Rooted submerged and floating plants may cover all the bed, as is quite common in small brooks. In a larger river, though, they are often confined to the shallower water at the sides (Fig. 10.2). These water-supported plants can be patterned across the channel in various ways (Fig. 10.2; and see below and Chapters 2, 3, 4). Free-floating plants are usually less important. They are necessarily commonest in still and slow waters, their pattern tending to depend on shelter, i.e. the wind and current in relation to the positions of plants, banks and other obstructions (Fig. 10.2). They may occur locally in faster flows, as when *Lemna trisulca* grows tangled with submerged shoots in small chalk streams. The third group of floating plants are those with floating shoots anchored to steep banks; and these are necessarily found close to these steep banks (Fig. 10.2).

DIRECTION OF FLOW

In still water plant parts will grow out in all directions, but in a current they will be patterned according to the direction of flow, the plants usually being longest in the direction of the current, shortest vertically, rooted mostly at the upstream end, and with shoots trailing downstream (Plate 5). Current, then, determines both the pattern of the population and what spaces between plants are available for colonisation. It similarly affects rhizome growth. In still and slow waters the rhizomes often grow in all directions (e.g. *Nuphar lutea*), while in faster flows growth is often most across the channel with only a little growth upstream and variable development downstream (e.g. *Berula erecta, Groenlandia densa, Myriophyllum spicatum, Potamogeton crispus, Ranunculus penicillatus*).

ALONG THE CHANNEL

In most natural channels bends produce patterns of erosion and sedimentation (Fig. 10.5), which in turn lead to the patterns of vegetation along the channel. The erosion on the bends may actually undercut the banks, and then clods fall into the water. If these bear species which can grow submerged (e.g. *Agrostis stolonifera, Epilobium hirsutum*) then the plants will start to grow where the clods fall, but will probably soon be washed downstream, where they may again lodge and grow. More importantly, on the channel bed itself places liable to the most erosion usually have different vegetation (e.g. Plate 5), but the differences do depend on the habitat. In the mountains, for instance, the areas that are eroded most may be bare, while sheltered ones have fringing herbs.

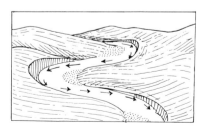

Fig. 10.5

Again, in quieter streams the fringing herbs may be on the most-eroded parts with tall monocotyledons in the sheltered places, and in very slow flows tall monocotyledons may be present throughout. For variations with fluctuations in flow and substrate regimes, see Chapters 2, 3 and 6.

PROTECTION

Often, plants are unable to grow in an open habitat because the flow is too fast or too turbulent, or the substrate too unstable, etc., though they can grow well in that site if local protection is provided from these unfavourable factors. Such protection may be given by a large plant. In faster flows current force is decreased just upstream of a clump while turbulence remains high, and species characteristic of somewhat slower flows may grow here (Fig. 10.6) (e.g. *Groenlandia densa, Potamogeton crispus, Zannichellia palustris* above *Ranunculus*). Some plants of slower waters which tolerate low light can grow as isolated shoots within large clumps, where they are sheltered (e.g. *Elodea canadensis* and *Potamogeton crispus* in *Ranunculus*).

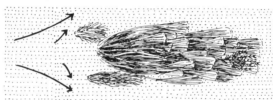

Fig. 10.6

Ranunculus penicillatus

Elodea canadensis

Zannichellia palustris

Emergents often accumulate silt between them, forming hummocks and providing shelter for other plants. These populations may be permanent, or temporarily growing in the shelter while recolonising

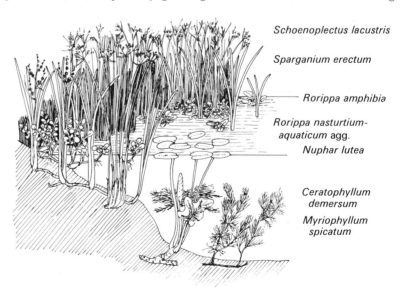

Fig. 10.7. Tall emergents sheltering short, poorly anchored emergents and submerged and floating plants (lower reach of clay stream, R. Great Ouse).

Schoenoplectus lacustris

Sparganium erectum

Rorippa amphibia

Rorippa nasturtium-aquaticum agg.

Nuphar lutea

Ceratophyllum demersum

Myriophyllum spicatum

from fragments (Fig. 10.7, Plate 7). Examples of these are *Schoenoplectus lacustris* and *Sparganium erectum* providing protection; *Nuphar lutea* and *Myriophyllum spicatum* as permanent populations in the shelter; *Ceratophyllum demersum* as a temporary submergent; *Rorippa amphibia* and *Rorippa nasturtium-aquaticum* agg. as temporary emergents.

Fig. 10.8

Any plant with parts at the water surface may trap viable propagules. Fringing herbs are often held and grow for a while, sometimes explosively, but their roots are seldom long enough to anchor in the soil and though they may survive mild storms they are likely to be dispersed in the next severe storm flow. Fragments of trailing plants can lodge well on firm small obstructions, such as stones or stakes, where they can remain with the ends trailing (Fig. 10.8; Plate 8) (e.g. *Myriophyllum* spp., *Potamogeton* spp., *Ranunculus* spp.). For fragments to become rooted to the ground, they must be near enough to the ground for a month or two. Even temporary plants may, however, reach over 1 m × 0.5 m before being washed away. Indentations in the stream bank, boulders and other obstructions can all provide local protection, allowing plants to grow which could not tolerate the main current.

SHADING FROM ABOVE

Channels are most often shaded by trees and bushes, though bridges and buildings may have local effects and, as has already been mentioned, tall bank herbs can shade narrow channels (see Chapters 7 and 21; Plates 2, 3, 24 and 25). Patterns of shading lead to patterns of plants, since although no river plants can live under heavy shade, some species are more shade-tolerant than others. For example:

Berula erecta – Ranunculus
  In light shade *Berula erecta* increases
Nuphar lutea – Sparganium emersum
  In light shade *Sparganium emersum* increases

In any one site shade patterns may vary with time, as when a tree falls and gives full light to a previously-shaded place. Plants will then invade and increase in size until the community becomes stable. In one example, fringing herbs and *Callitriche* invaded a place previously without plants, and within 2 years a stable pattern had formed.

BRIDGE AND WEIR PATTERNS

Bridge piers and other structures in the water may slow the water upstream, cause swirling and turbulence around them, and perhaps faster flows or deep pools downstream. Weirs also have slower water upstream and faster water downstream of the structure, and as they are often sited at bridges the effect of the two together is additive. Near large bridges, therefore, there is likely to be a range of flow, and consequently substrate, types. This means that such bridges are useful for studying the varieties of plant communities which can occur within the overall chemical regime of each stream (for examples, see Plates 9 and 10).

SEDIMENTATION AND PLANT CYCLES

Sedimentation varies both along the length of a stream and within individual sites. Where sedimentation varies, plants may vary also. In practice such plant patterns occur only where the sediment is silt or sand, since loose gravel and stones are unsuitable for plant growth. The deposited sediment may form shoals, on which plants can then grow, or in faster flow it may accumulate only inside plant clumps, not between them. Sand is deposited in faster flows than is silt because the particles are heavier. Silt is more nutrient-rich, and its presence in a clump may lead to shoots growing larger and denser, so some form of pattern can emerge from variations in the type of sediment. Also, deposition may vary across a channel, e.g. sand in *Berula erecta* in the middle, silt in *Rorippa nasturtium-aquaticum* agg. at the sides.

Shoals build up in shelter, and may be temporary, long-term or cyclical, depending mainly on the flow regime although plant cover can help to stabilise them. In deeper water they are first colonised by quick-growing submergents, while shallow ones are invaded by fringing herbs. Tall monocotyledons and more slow-growing submergents invade if the shoal persists. Regular disturbance or erosion leads to cyclical patterns. The cycle of fringing herbs, where this depends on silting, plant growth and storm flows, is described in Chapters 3 and 5.

Unusual sedimentation can alter plant distribution. A bare stretch of hard gravel bed observed for 5 years became silted during exceptionally low flows in one summer. *Zannichellia palustris* (with sparser other species) invaded and spread, and because of the extra trapping of silt by the *Zannichellia palustris* plants the silt became 5–20 cm deep. This depth was unstable and was swept away in the autumn storms, and as the general flow increased at that time, the area remained bare. Similar patterns occur with other species.

The true plant cycles depending on the morphology of the dominant plant, however, occur when the stream has a hard bed, the plants are short-rooted and sedimentation is sufficient for plants to become rooted only in the sediment. Species which move their rooting level completely into the sediment are washed away easily, while those staying more firmly in the gravel are less damaged in storms, so that differential damage and cycles result. Cycles of *Ranunculus* (which is firmly rooted) and *Rorippa nasturtium-aquatilis* agg. (which is poorly rooted) are des-

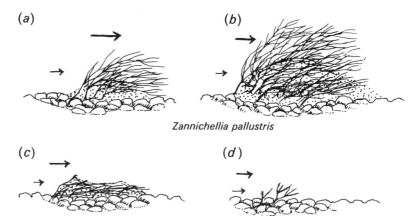

Zannichellia pallustris

Fig. 10.9. Development and erosion of hummock of *Zannichellia palustris.*

cribed in [35]. *Berula erecta* and *Zannichellia palustris* (which are sediment-rooted) may also show patterns with *Ranunculus*. In cycles, it is the age of the plant *patch* which is important, not that of the individual rooted shoot, because it is on the age of the clump (as well as flow and substrate regime) that the depth of sediment present depends. Sediment usually accumulates most in summer and wash-out is most in winter; damaging storms can occur at any season. Some cycles tend to be annual, controlled by the general growth and weather cycles (as in the firmly rooted *Ranunculus*; see [35]), while others are much more dependent on the minor fluctuations during the growing season, and several factors here affect the duration of the cycle.

Hummocks of sediment accumulate under *Zannichellia palustris* (Fig. 10.9). As these rise they become more exposed to damage from flow, trampling, etc. Erosion of the top leads to erosion of the whole hummock leaving plants on the firm bed to form the nuclei of new hummocks. This plant is summer-green, so cycles are ended in the autumn anyway. Carpets are formed by *Berula erecta* (Fig. 10.10). This is winter-green, so cycles may take a year or more. It spreads most in spring, so then many shoots are joined by living rhizomes and are difficult to move. Anchoring strength is less in autumn and winter, and storm flow of equal discharge will wash away more *Berula erecta* in autumn than in spring (see Chapters 3 and 5). The faster the sedimentation, the faster the cycle, provided the sediment can be stabilised by the root weft. Cycles can last anything between about 2 months and several years.

Cyclical changes are of much botanical interest, but are uncommon unless the ordinary changes between spring and autumn are included. They occur with a combination of shallow water, a hard bed, sediment deposited in plant clumps though not much outside them, no severe spates, and storm flows occurring only infrequently. Most sites are on chalk, with sandstone and chalk – clay ones ranking next. The cycles are ended by storm flows. This is in contrast with cycles on land, which

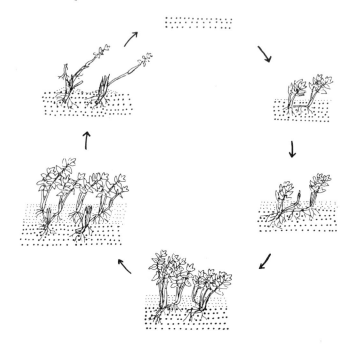

Fig. 10.10. Development and erosion of a carpet of *Berula erecta*.

are more usually determined by the growth cycle of the dominant plant. This again emphasises the control which flow exerts on river plants.

POOR ESTABLISHMENT

When plant patches are firmly established in a site, properly anchored and growing well, they are likely to remain for some while – perhaps for many years, if not disturbed by severe storm flows, dredging, trampling, etc. However, many plants are transients, so that plant patterns occur because of their coming and going. Poor establishment may be because the plants are not held in the substrate, or because the substrate is itself a temporary habitat.

Most river plants can grow from the fragments washed down in storm flows, provided these are of a suitable size. A few (e.g. *Potamogeton pectinatus*) need a rhizome portion, but most can grow well from just shoots, though some trailing species (e.g. *Ranunculus*) grow best from fragments from the lower part of the plant. Often portions only a few centimetres long are viable (e.g. *Callitriche*). Plants which can live in shallow water at channel margins, can often grow well there from fragments. These may become anchored, or they may form a floating fringe (Fig. 10.2*f*). A band of floating plants up to about 1 m wide is particularly characteristic of medium-sized chalk streams. Because they are not rooted in the nutrient-rich soil the plants remain small, but a fringe can survive until the next very severe storm flow and may last for 2 years or more, though trimmed and kept in place by the current.

(Because the plants are separate, some can be moved without affecting the rest.) This vegetation pattern contrasts with the large luxuriant single-species patches of rooted fringing herbs that are so characteristic of small clay streams (see Chapters 3 and 5).

Passing downstream, the lack of shallow water and increased silting and floods mean it is increasingly difficult for fragments to anchor, and so relatively more of the clumps of fringing herbs (*Callitriche*, etc.) are temporary. *Callitriche* spp. and *Elodea canadensis* both propagate well from fragments, and both are characteristic of temporary silt shoals. Large populations can form, be churned up by storm flows, and then re-colonise.

Other patterns caused by poor establishment are described elsewhere in this chapter, and in Chapter 3.

AGEING

Most aquatics propagate vegetatively, and although clumps and patches may move, the death of such patches (as opposed to individual shoots) from old age seems improbable. One aquatic, *Phragmites communis*, has been estimated to live over 1000 years in large marshes [44]. Some plant patterns are, however, related to the age of the plant. For example, most fringing herbs have below-ground parts which live less than a year. Thus if they are to persist at a site, new shoots must continually develop beside the old ones as they die. This regrowth may be outside or inside the area of the patch in the previous year, and so the patch may move. This alteration in distribution is more pronounced in the channel centre than it is at the edge. This is because the edges comprise a smaller habitat and so new growth is more likely to be within the area of old plants. It also follows that plants with long-lived rhizomes (such as *Nuphar lutea*) will move around any given site less. Ageing can affect the survival of emerged plants anchored in the bank but growing out into the channel. As mentioned in Chapter 3, they are able to remain in place while still joined to the bank by a strong rhizome, but are likely to be washed away once this rhizome dies.

SEASONAL AND ANNUAL CHANGES

Vegetation often varies from early summer to later summer, and from one year to another. These changes are due both to differences in the growing pattern of the plants, and to differences in habitat, flow pattern, temperature, etc. The changes described here are those of the growing seasons, not the differences between winter and summer forms.

In the south of Britain some species usually reach their maximum size in early summer, others in mid- or late summer. Local conditions may alter this pattern, however.

Early species: *Berula erecta, Callitriche* spp., *Hottonia palustris, Ranunculus* spp.

Late species: *Apium nodiflorum, Ceratophyllum demersum, Potamogeton* spp., *Rorippa nasturtium-aquaticum* agg., *Sagittaria sagittifolia*

Streams which contain plants from both these groups usually look very different in May and September (Fig. 10.11; [35, 41]).

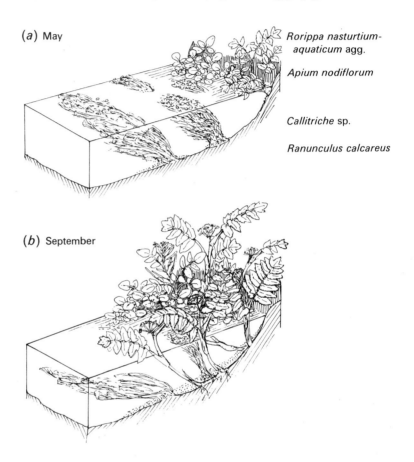

(*a*) May

Rorippa nasturtium-aquaticum agg.

Apium nodiflorum

Callitriche sp.

Ranunculus calcareus

(*b*) September

Fig. 10.11

    Some populations vary in position and luxuriance from year to year. Those which vary most are species which have intermittent rapid or explosive growth, or have some character, such as poor anchorage, which potentially causes large losses. Such species are frequently those that propagate well from fragments. Plant patches retaining their position and size may show the reverse characters, or may be confined to a substrate or flow regime occurring in only a small part of the stream (e.g. a silt shoal, or fast flow below a bridge). Patches restricted to certain substrates are particularly common in chalk–clay streams, as these tend to have variable substrates. When such mosaics occur, plant clumps remain stable because shoots moving outside them get swept away (see Chapter 3).

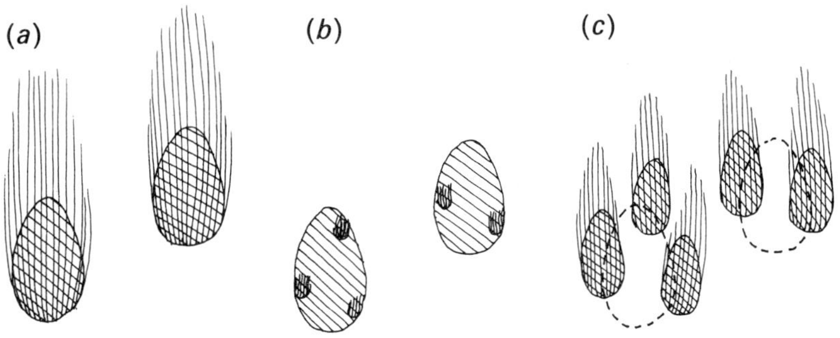

Fig. 10.12. Movement of *Ranunculus* clumps after cutting. (*a*) Uncut plants. (*b*) After cutting, with only a few viable tufts of shoots. (*c*) New plants.

Plants which form trailing clumps in moving water (e.g. *Callitriche, Oenanthe fluviatilis, Ranunculus*) usually keep their shape. This shape can be determined by internal factors or by small-scale variations in substrate etc. If substrate and other factors permit, though, the clump may move slowly. During the growing season the clump may be destroyed by cutting, trampling, etc., leaving scattered patches of shoots which can each form the nucleus of a new clump (Fig. 10.12). The position of these clumps may be different from that of the original one, and indeed clump and bare soil areas may be reversed, substrate and flow permitting. From year to year, therefore, the position and pattern of the plant clumps vary.

The growth of the plants themselves may cause seasonal variations in habitat. In shallow streams plants alter the current pattern as they grow during the summer, which leads to increased erosion in some places and greater shelter and sedimentation in others. These changes may then affect the existing plants and the position of any new ones.

Annual variations in water depth and flow may lead to annual changes in the plants. Channels drying in summer will lose their submerged plant parts and land species may invade; this change is reversed on flooding. If flow becomes slower in late summer, slow-flow species will show a relative increase (e.g. *Lemna minor* agg.), and faster-flow species a decrease (e.g. *Ranunculus* spp). The interaction between flow and substrate may cause annual variations in *Ranunculus*. In clay, or near-clay streams *Ranunculus* is often absent because the silt is unsuitable for anchorage and makes the habitat too nutrient-rich. In years of low flows, however, *Ranunculus penicillatus* may appear in those sites where more turbulent water and reduced silt allow the plant to anchor to the hard bed below and provide it with a lower nutrient supply. The *Ranunculus* appears in parts of the channel which are normally empty. Minor changes between one year and another can also be caused by temperature. For example, *Elodea canadensis* grows much better in exceptionally hot summers, and in 1975 often grew very well, particularly in N. Britain. It was more frequent and more luxuriant, extending its habitat range into faster flows where, because of the losses, it could not survive if growth was slow. Heat waves may kill plant parts in still water, where the temperature rises most (e.g. *Myriophyllum verticill-atum*).

COMPETITION

The overall controlling factors in the distribution of river plants are flow and substrate, and competition is, in consequence, much less important in river communities than in typical land communities. It does occur, however, and it alters the pattern of plant growth. The effect can be due to the mere presence of one species making the habitat less suitable for another species. For example, if a *Sparganium erectum* patch develops upstream of a *Ranunculus* clump, the *Ranunculus* grows badly because the *Sparganium erectum* shelters it from flow and may cause silt to be deposited on it (Fig. 10.13). It is also possible for tall plants to kill short ones by shading them, or for emergents to smother submergents by the silt they accumulate. So if both can grow well in any one place, it is the taller which survive. Tall emergents shade short ones, emergents shade floating and submerged species, floating plants shade submerged ones, and submergents high in the water shade those below (Fig. 10.14). Compensating for this, both flow and human interference remove mostly the taller plants, so the end result is that the overall plant pattern remains stable although the detailed arrangement in any one place may vary.

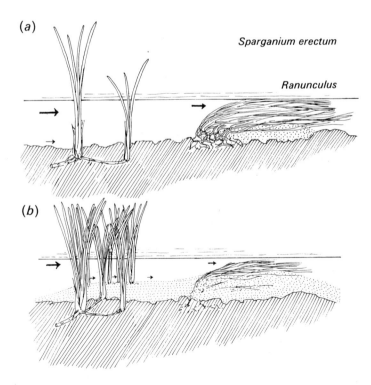

(a)

Sparganium erectum

Ranunculus

(b)

Fig. 10.13. *Sparganium erectum* damaging *Ranunculus*. (*a*) Establishment of *Sparganium erectum* upstream of a well-grown *Ranunculus* clump. (*b*) Thick clump of *Sparganium erectum* causing silt deposition in, and lack of flow to, the *Ranunculus* clump, which is growing badly.

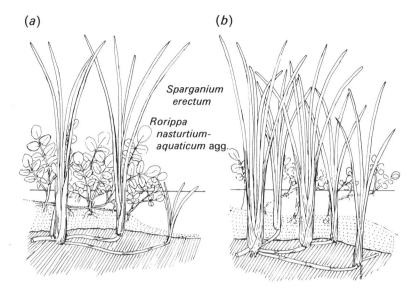

Fig. 10.14. *Sparganium erectum* shading and killing *Rorippa nasturtium-aquaticum* agg. (*a*) Early stage. (*b*) Late stage.

An interesting example of competition, in a form more characteristic of water plants than land plants, occurs between *Ranunculus* and *Berula erecta*. In chalk streams which flow too fast for the good growth of *Berula erecta*, this species can colonise the sheltered patch upstream of *Ranunculus* clumps. Once established *Berula erecta* can spread downstream, over

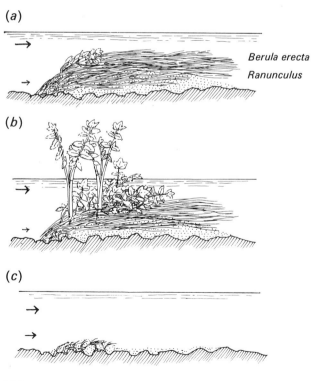

Fig. 10.15. *Berula erecta* shading and harming *Ranunculus*. (*a*) *Berula erecta* colonising a *Ranunculus* plant. (*b*) The well-grown *Berula erecta* shading *Ranunculus*. Most *Berula erecta* roots are in the *Ranunculus* plant or the loose sediment below this. *Berula erecta* weakens *Ranunculus* and increases its hydraulic resistance to flow. (*c*) *Ranunculus* is washed away, taking *Berula erecta* with it, as the latter is not rooted independently. The few remaining *Ranunculus* shoots can re-grow.

171

the *Ranunculus* and perhaps on the ground as well, tangling and rooting in the firmly anchored *Ranunculus* clumps. In due course, however, the shading effect of the *Berula erecta* weakens the *Ranunculus* and both species are washed away (Fig. 10.15).

MANAGEMENT

The large-scale effects of management are described in Chapters 20 and 21. Some effects of management on plant patterns are that killing large bands of tall monocotyledons at the sides of silting rivers (e.g. by dalapon) means that the silt accumulated by the plants is then eroded and submerged or floating plants invade instead; cutting part rather than all the channel leads to differential growth and invasion and some species (e.g. *Berula erecta*, *Ranunculus peltatus*) becoming dwarf under frequent cutting (see Chapter 8); and trampling by animals or man can lead to plant patterns influenced by disturbance.

11

Downstream changes

Passing downstream from the source a river normally gets larger, the flow slower and the substrate siltier; thus the vegetation changes also. In the hills, fringing herbs and Callitriche *tend to occur in shallow, fairly stable and fairly swift water in the upper reaches, and they decrease as water level fluctuations increase lower down. Tall monocotyledons and submerged plants increase in the lower reaches where there is more silt and less scour. In the lowlands, where the flow is less strong, there is more vegetation and submerged plants occur further upstream, up to the limit of perennial flow. Plants of slow flow are common and often luxuriant in the lower reaches, much more so than in the mountain streams.*

From the source to the mouth of a stream the habitat changes, and, of course, the plants change too. The main downstream changes are that the parameters of channel size, including width, depth and drainage order, increase; the proportion of fine sediment and the trophic status increase; water movement and the amount of light reaching the

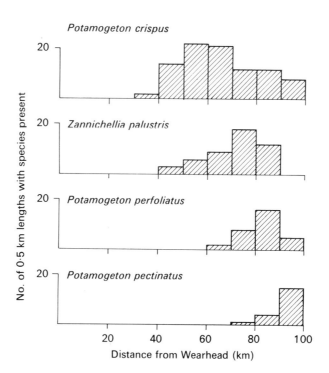

Fig. 11.1. Downstream changes in species frequency in R. Wear. (Adapted from [13].)

173

channel bed tend to decrease; and the nature of storm flows alters (see Chapter 5). Superimposed upon these general trends there may be marked variations in any one stretch of the stream as the result of changes in topography, geology, human interference, etc.

The changes in plants from swift upstream to slow downstream stretches in several British rivers were described in about 1930 [5, 6] and more recently some northern rivers have been studied [13, 14, 15, 45, 46]. Fig. 11.1 shows the downstream increases in species of slower flow or higher nutrient status, and decreases in the species of the reverse habitat.

This chapter centres round vegetation maps of whole rivers (Figs. 11.2–11.18). Each of these shows the river pattern and the species present at selected sites along the length (usually at road bridges). The rivers illustrate the vegetation of the main British stream types, showing a typical example of a lowland clay stream, an upland hard sandstone stream, etc. The actual plant communities of these stream types are described in Chapter 12 for streams on soft rocks (mainly lowland ones), and in Chapter 13 for streams on hard rocks (mainly upland and mountain ones). These three chapters thus form a unit, this chapter describing the downstream progressions, and the others the plant communities of each stage and of the stream types as a whole.

STREAMS OF DIFFERENT TYPES

No one river can show all the possible progressions of plants from source to mouth. Some typical progressions are shown in Figs. 11.2–11.18, which comprise:

Very mountainous, Resistant rock, R. Oykell (Fig. 11.2)
Mountainous, Resistant rock, R. Vyrnwy (Fig. 11.3)
Mountainous and plain, Resistant rock, R. Dart (Fig. 11.4)
Upland hard sandstone, R. Tone (Fig. 11.5)
Upland hard sandstone and clay, R. Leadon (Fig. 11.6)
Lowland chalk, R. Itchen (Fig. 11.7)
Lowland chalk, R. Darent (Fig. 11.8a–c)
Mountainous and lowland, limestone and clay, R. Derwent (Fig. 11.9)
Lowland sandstone, R. Meese (Fig. 11.10)
Lowland sandstone, R. Blythe (Fig. 11.11)
Lowland clay, R. Chelmer (Fig. 11.12)
Upland clay, R. Chew (Fig. 11.13)
Lowland clay and alluvium, R. Axe (Fig. 11.14)
Lowland chalk and fertile sandstone, R. Avon (Fig. 11.15)
Mountainous, Resistant rock and hard sandstone, R. Tweed (Fig. 11.16)
Mountainous, Resistant rock and hard calcareous rock, R. Clyde (Fig. 11.17)
Upland, Resistant rock and hard sandstone, R. Lugg (Fig. 11.18)

As a summary, Table 11.1 lists the characteristic species of upper and lower reaches of some different stream types.

The downstream changes described below are discussed firstly as they apply in general to streams of a particular type (mountain streams, etc.). The features considered can be seen from the river diagrams. Next, a very brief description is given of the rivers illustrated for each stream type (most of the necessary information being on the diagram). A typical stream is always given here, and when there is a particularly interesting unusual stream showing relevant features, this also is included and described. The full descriptions of the composition of river vegetation are given in Chapters 12 and 13 and are mostly illustrated in Fig. 10.2, and so are not repeated here.

TABLE 11.1 *Characteristic species of the upper and lower reaches of different stream types*

(Species in brackets tend to occur more towards middle reaches.)

1.   Resistant rocks
  (i) Upper reaches

   † *Juncus articulatus*   (Short-leaved *Ranunculus*, e.g. *Ranunculus aquatilis*)

   Myriophyllum alterniflorum   *Veronica beccabunga*

   Phalaris arundinacaea   Mosses

  (ii) Lower reaches

   (Elodea canadensis)   *Ranunculus fluitans*

   Myriophyllum spicatum   *Sparganium emersum*

   Potamogeton perfoliatus   *Sparganium erectum*

2.   Hard sandstone
  (i) Upper reaches

   Apium nodiflorum   *Rorippa nasturtium-aquaticum* agg.

   Callitriche spp.   *Veronica beccabunga*

  (ii) Lower reaches

   Callitriche spp.   *Ranunculus* spp.

   Elodea canadensis   *Sparganium emersum*

   Potamogeton crispus   *Sparganium erectum*

TABLE 11.1 (*cont.*)

3. Hard limestone
    (i) Upper reaches

    Mentha aquatica

    Mimulus guttatus

    Rorippa nasturtium-aquaticum agg.

    Veronica beccabunga

    Mosses

    (ii) Lower reaches

    Callitriche spp.

    Groenlandia densa

    Myriophyllum spicatum

    Ranunculus spp.

    Zannichellia palustris

4. Chalk
    (i) Upper reaches

    Apium nodiflorum

    Mentha aquatica

    Myosotis scorpioides

    Phalaris arundinacea

    Short-leaved *Ranunculus*, e.g. *Ranunculus peltatus*

    Rorippa nasturtium-aquaticum agg.

    Veronica anagallis-aquatica

    Veronica beccabunga

    (ii) Lower reaches

    Callitriche spp.

    Elodea canadensis

    Glyceria maxima

    Lemna minor agg.

    Myriophyllum spicatum

    Ranunculus calcareus

    Schoenoplectus lacustris

    Sparganium emersum

    Sparganium erectum

    Zannichellia palustris

5. Soft sandstone
    (i) Upper reaches

    Apium nodiflorum

    Callitriche spp.

    Epilobium hirsutum

    Phalaris arundinacea

    Sparganium erectum

    Veronica beccabunga

    (ii) Lower reaches

    Callitriche spp.

    Glyceria maxima

    Ranunculus spp.

    Sparganium emersum

TABLE 11.1 (*cont.*)

⊤ *Lemna minor* agg.

❶ *Potamogeton perfoliatus*

W *Sparganium erectum*

6. Clay
    (i) Upper reaches

❧ *Apium nodiflorum*

Ѱ *Callitriche* spp.

❧ *Epilobium hirsutum*

冊 *Phalaris arundinacea*

W *Sparganium erectum*

❧ *Veronica beccabunga*

    (ii) Lower reaches

✸ *Elodea canadensis*

⊤ *Lemna minor* agg.

❧ *Nuphar lutea*

☞ *Polygonum amphitium*

❶ (*Potamogeton perfoliatus*)

✿ *Rorippa amphibia*

Ⓐ *Sagittaria sagittifolia*

Ⓣ *Schoenoplectus lacustris*

𝒱 *Sparganium emersum*

W *Sparganium erectum*

⌒ *Enteromorpha* sp.

EXPLANATION OF FIGS. 11.2–11.8

These diagrams of the vegetation along whole rivers illustrate the general vegetation patterns of the main types of British rivers. (In Chapter 22 some characteristic effects of pollution are discussed.) Each site is recorded from a bridge, whose approximate position is marked on the diagram.

(1) Comparing near-by sites shows the type of variation found within the same community: a number of species are potentially present at each site, but only some of these occur at any one site.

(2) Comparing sites along the length of the channel shows the changes in species composition and species diversity from the source to the mouth.

(3) The definitions of lowland, upland and mountain streams, in terms of height of hill, fall from hill to channel and slope of channel, are given on p. 225. The rainfall varies, being less in the lowlands of south and east England, and greater in the north and west of the country. It is only in the far north of Scotland (R. Oykell, Fig. 11.2) that rainfall has much effect on the vegetation. The high rainfall there increases the mountainous nature (for plants) of the stream type.

(4) The effects of geological variations, in Figs. 11.15–11.18, are described only briefly in the text, but can be seen in more detail in the diagrams.

(5) At each site the widespread species are shown by the symbol used elsewhere in this book, and listed again below. Plants of doubtful identity (including possible hybrids) are given the symbol of the species they are closest to. Any other essential information, such as very recent dredging, is also given.

(6) At each site, the species present are listed in the following order: tall monocotyledons: fringing herbs; other emergents; water-supported higher plants; mosses and algae. Symbols in boxes indicate the species is present in quantity.

(7) Symbols in brackets are for sites on streams too small to be shown on maps of this scale.

(8) *Glyceria fluitans* is included in 'small grasses' except in Figs. 11.2 and 11.4.

(9) Where a stream is braided, the different channels are marked (1), (2), etc.

LIST OF SYMBOLS

Acorus calamus	*Myriophyllum spicatum*
Agrostis stolonifera and other small grasses	*Nuphar lutea*
Alisma plantago-aquatica	*Oenanthe crocata*
Apium nodiflorum	*Oenanthe fluviatilis*
Berula erecta	*Phalaris arundinacea*
Butomus umbellatus	*Phragmites communis*
Callitriche spp.	*Polygonum amphibium*
Carex acutiformis	*Potamogeton crispus*
Carex spp.	*Potamogeton lucens*
(Catabrosa aquatica, see Agrostis)	*Potamogeton natans*
Ceratophyllum demersum	*Potamogeton pectinatus*
Elodea canadensis	*Potamogeton perfoliatus*
Eleocharis acicularis	*Ranunculus aquatilis*
Eleocharis palustris	*Ranunculus fluitans*
Eleogiton fluitans	*Ranunculus omiophyllus*
Epilobium hirsutum	*Ranunculus peltatus*
Equisetum palustre	*Ranunculus* spp.
Eriophorum angustifolium	*Rorippa amphibia*

List of symbols

Ⴟ *(Glyceria fluitans,* see *Agrostis)*

Ⴘ *Glyceria maxima*

Groenlandia densa

Hippuris vulgaris

Ⴘ *Iris pseudacorus*

┼ *Juncas articulatus*

Juncus bulbosus

Juncus effusus (and other spp.)

Lemna minor agg.

Lemna polyrhiza

Lemna trisulca

Mentha aquatica

Menyanthes trifoliata

Mimulus guttatus

Myosotis scorpioides

Myriophyllum alterniflorum

Rorippa nasturtium-aquaticum agg.

Rumex hydrolapathum

Sagittaria sagittifolia

Schoenplectus lacustris

Scirpus spp.

Sparganium emersum

Sparganium erectum

Symphtum officinale

Typha latifolia

Utricularia vulgaris

Veronica anagallis-aquatica agg.

Veronica beccabunga

Zannichellia palustris

Mosses

Benthic algae

Blanket weed (trailing algae)

Enteromorpha sp.

Other species

Mountain streams: general

In mountain streams mosses are frequent, except in the fiercest flow, but most tributaries are without angiosperms, these only occurring when the tributary is protected from severe spate flows. The species present vary from those typical of bogs, when the sediment present is acid peat, to the usual fringing herb and *Callitriche* communities, when the sediment is nutrient-rich. Tributaries nearly or quite in the lowlands below, which have very small catchments and negligible scour, can bear tall emergents. Plants are sparser or absent in reaches with the greatest force of water, whether this force is from normal flow or from spates. The force is due to the fall from hill top to stream being particularly great, or to the stream slope being particularly steep, or to the rainfall being both high and irregular.

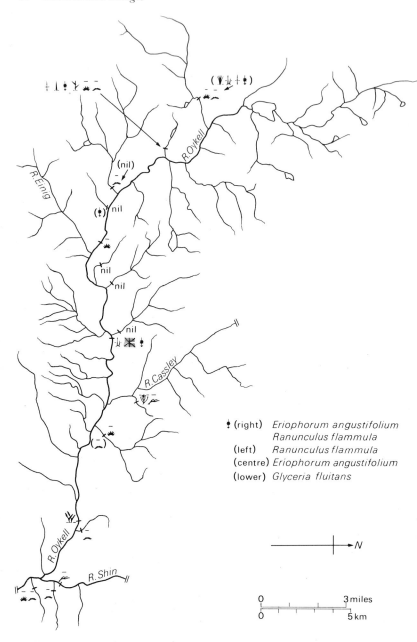

Fig. 11.2. Very mountainous river on Resistant rock. R. Oykell (N. Scotland); 1974.

In these mountain streams submerged plants enter somewhat lower than the fringing herbs, the position depending on the water depth and force. The leaf size of *Ranunculus* increases downstream. If luxuriant vegetation develops in middle or lower reaches, the dominant plants are probably *Ranunculus*. Nutrient status increases downstream. (This is apart from the presence of any tributaries from bogs.) Species of low trophic status (e.g. *Myriophyllum alterniflorum*) occur in upper reaches only, and are effectively absent from fertile rocks (e.g. limestone). The more eutrophic species are normally confined to lower reaches (e.g. *Sparganium emersum*). The comparative length of 'upper' and 'lower' zones varies with substrate and flow type. Downstream eutrophication is greater the longer the river, other factors being equal. Fringing herbs and *Callitriche* decrease downstream, being found in shallow, fairly stable waters. Tall monocotyledons at the edges increase downstream, as banks with deep nutrient-rich soil increase.

Plants of mountain streams show downstream variation with changes in water force, water volume, trophic status and substrate stability.

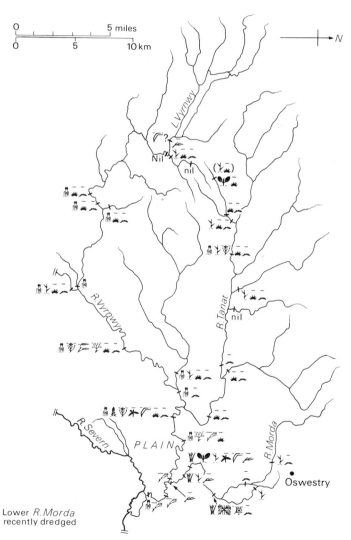

Fig. 11.3. Mountain river on Resistant rock. R. Vyrnwy (Severn); 1973.

Lower *R. Morda* recently dredged

181

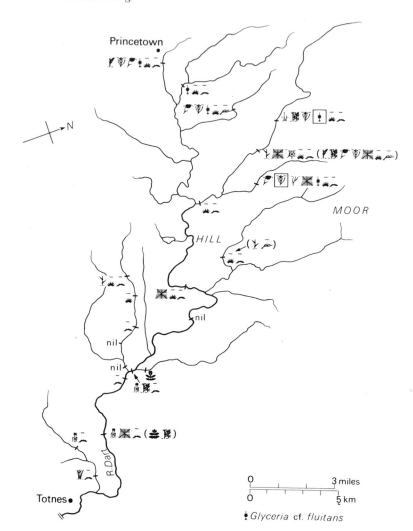

Fig. 11.4. Stream on Resistant rock rising on moorland, and becoming mountainous lower. R. Dart (Devon); most records 1973.

♣ *Glyceria* cf. *fluitans*

Mountain streams: individual descriptions

The very mountainous stream in Fig. 11.2 has strong normal flows and frequent, very fierce spates. These keep the channel nearly bare, even mosses being sparse. Plants can grow near the mouth, where water force is least, and in small tributaries from bogs receiving little run-off. The general absence of plants is due to the fierce flow, and not to low nutrient status.

A typical mountainous stream on Resistant rock is shown in Fig. 11.3. (This is typical if the reservoir is ignored.) Small tributaries which are flatter or have less fall from hill top to stream bed usually have fringing herbs. Short grasses may be washed into the channel from the banks, but then on the channel bed the plants are usually poorly anchored and

182

temporary. *Phalaris arundinacea* grows well on banks, gravel bars, etc. where it is intermittently flooded. Somewhat downstream, a short- or medium-leaved *Ranunculus aquatilis* enters. Lower, where the water volume is larger and the flow less strong, this is replaced by the long-leaved *Ranunculus fluitans*. The only species showing eutrophic influence are confined to the downstream flatter parts which have the most fine sediment and the least water force. Small tributaries in the foothills are dominated by tall emergents because there is no swift storm flow. The mountain stream in Fig. 11.4 rises on a high-level plain with moorland vegetation. The drop in land level from the moor to the stream is very small so fierce flow is absent, and oligotrophic plants (Chapters 1, 8 and 13) grow well. When the river leaves the plain the ground drops steeply, and the stream is a typical mountainous one. Lowland tributaries have tall emergents, as before.

Upland streams: general

Upland streams have a lower water force than mountain ones and thus have more fine sediment and a higher trophic status. Many upland streams are on sandstone, and so necessarily have more fine sediment than streams on Resistant rocks. Oligotrophic species are absent, even on Resistant rocks. Fringing herbs potentially occur in all tributaries. Upper reaches do not have fierce spates as they do in the mountains, nor do they dry in summer as often happens in the lowlands. Submerged plants are frequent. Spates are still present, though, and these can sweep away even plentiful vegetation.

Upland streams: individual descriptions

A typical upland sandstone stream is shown in Fig. 11.5. It has a well-developed and species-rich fringing herb vegetation, and *Callitriche* and *Ranunculus* grow high into the hills. Where plants are sparse or absent this is more because of shade than because of fierce spates. Downstream, silt increases, flow decreases and eutrophic species enter. In Fig. 11.5 this transition is sudden, as the stream enters a flood plain (and there is pollution from Taunton).

The stream in Fig. 11.6 is a less common type, being upland above and lowland below. There is much less vegetation. The main stream is polluted (see Chapter 22). Deep storm flows, silting and the consequent unstable substrates reach far into the tributaries. Part is shaded. The fall from hill top to stream is less (often <200′; <60 m) and this allows silting and *Sparganium emersum* to occur in upper reaches. Good vegetation occurs in two stable habitats: fringing herbs, etc. in stable small hard-rock tributaries with good anchorage, gently sloping banks and low silting; and a eutrophic channel flora in the flood plain in stable slow silted reaches not liable to much disturbance during storms.

Other upland streams with spates are shown in parts of Figs. 11.16, 11.17 and 11.18, and one on clay, with lesser storm flows, in Fig. 11.13. This last is described below with the other clay streams.

Fig. 11.5. Upland river on hard sandstone, with a flood plain. R. Tone (Somerset); most records 1973.

🐝 *Ranunculus penicillatus* or a similar hybrid
❋ *Scrophularia aquatica*

Lowland chalk streams: general

Chalk streams are lowland, without spates. They show little fluctuation in depth and little silting. Fringing herbs typically occur throughout, in species-rich populations. In shallow parts they grow on the bed as well as at the sides. The non-eutrophic species (see Chapters 1, 8 and 12) occur throughout the stream, while semi-eutrophic species are common

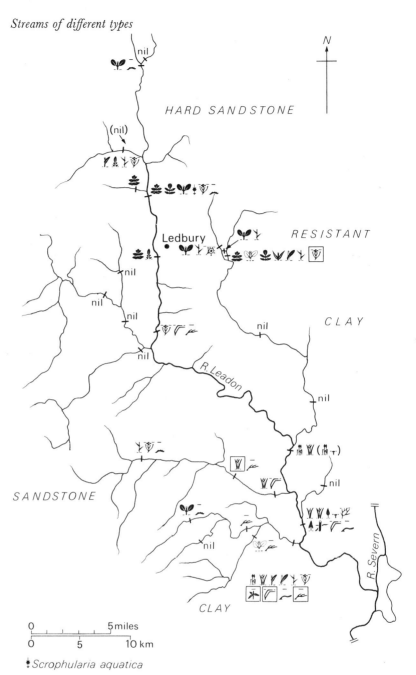

Fig. 11.6. Upland river rising on hard sandstone (and locally on Resistant rock), flowing on to clay below. R. Leadon (Severn); most records 1973. (Upper tributaries on the right are on Resistant rock.)

‡*Scrophularia aquatica*

in lower reaches and may be sporadic above. In small brooks, *Ranunculus* spp. are short-leaved (e.g. short-leaved *Ranunculus peltatus*, *Ranunculus peltatus* × *Ranunculus trichophyllus*). As the water volume increases, medium-leaved forms take over (typically *Ranunculus calcareus*). This is the same progression of growth forms as in the hill streams. If short-leaved forms are absent, small brooks are without *Ranunculus*, for medium-leaved forms cannot tolerate either the very shallow water or

Fig. 11.7. Lowland chalk river. R. Itchen (Hampshire); most records 1973.

the intermittent drying that short-leaved forms can. Submerged plants grow farther up than in other stream types. This is mainly because of the perennial springs at the sources, but also because the submerged *Callitriche* and *Ranunculus* can survive on damp mud. (Gravelly stretches of winter-bournes dry out quickly, and bear fringing herbs or, if dry for longer, no aquatics.)

Plants of chalk streams show downstream variation with changes in trophic status, water depth and substrate stability.

Lowland chalk streams: individual descriptions

Fig. 11.7 is of a typical chalk stream, which was described earlier in [5]. Fig. 11.8 shows a chalk stream recently severely damaged by a decrease in its water flow. The damage is discussed in Chapter 20. Fringing herbs are sparse and submerged plants almost absent by the end of the drought period. There is little downstream variation.

Mixed mountain and lowland, limestone and non-limestone streams: individual descriptions

An unusual combination of mountain and lowland streams, with limestone and non-limestone influences, is shown in Fig. 11.9. The upper tributaries are mountain limestone ones, with little vegetation, but

186

(a)

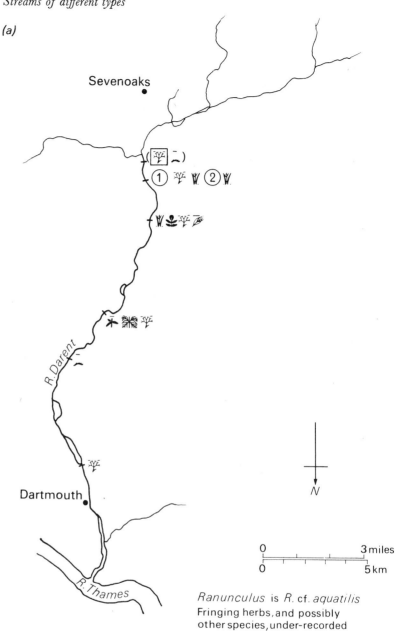

Fig. 11.8. Lowland chalk
river. R. Darent (Thames);
(a) 1969.

Ranunculus is *R.* cf. *aquatilis*
Fringing herbs, and possibly
other species, under-recorded

when these enter the flat Vale of Pickering the water force decreases and
is no longer the controlling factor. The chemical status is both limestone
and eutrophic. *Ranunculus* dominates, with some *Potamogeton pectinatus*.
The streams, though, are still spatey, fluctuating in depth, and the
fringing herbs which are so abundant in lowland limestone streams are
very sparse because of these fluctuations.

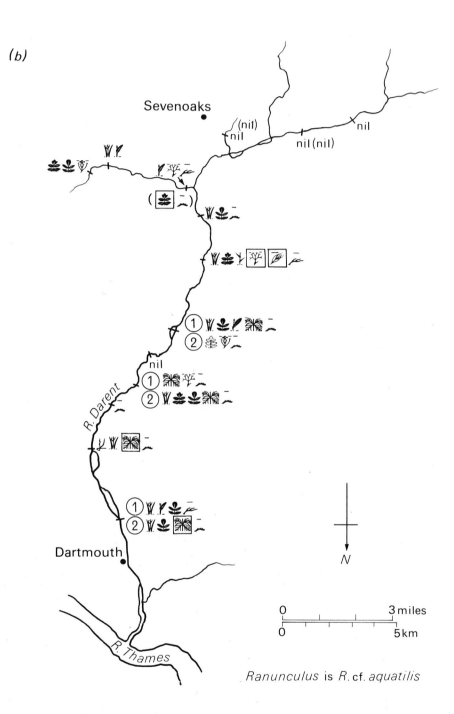

Fig. 11.8 (*cont.*). Lowland chalk river. R. Darent (Thames); (*b*) 1972.

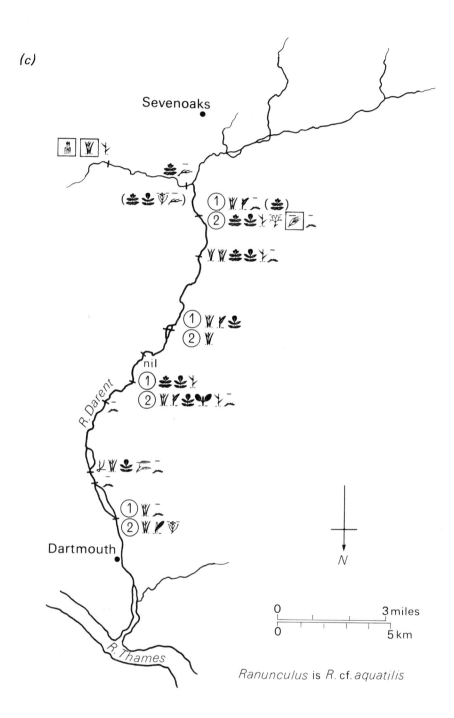

Fig. 11.8 (*cont.*). Lowland chalk river. R. Darent (Thames); (*c*) 1974.

Ranunculus is *R.* cf. *aquatilis*

189

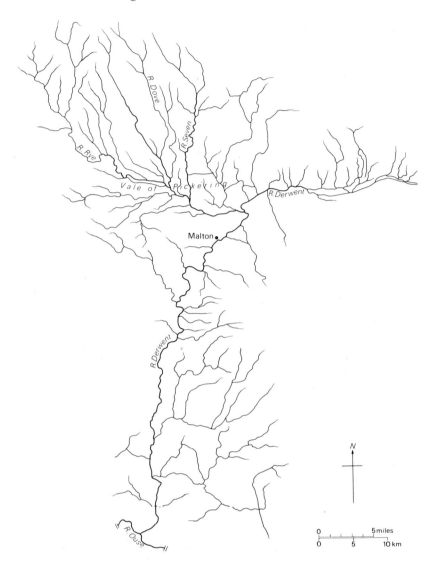

Fig. 11.9. River rising on mountain limestone, flowing into an alluvial plain, then passing downstream on to lowland clay (etc.). R. Derwent, (Yorkshire); most records 1973.

Lowland sandstone streams

Lowland sandstone streams are rather like chalk ones, but species diversity is less, fringing herbs are less frequently found on the channel bed, and submerged plants extend less far upstream (because the upper parts often dry in summer). There is downstream eutrophication, though this is not shown very well on the short streams in Figs. 11.10 and 11.11. Fig. 11.10 has the faster flow.

Lowland and upland clay streams: general

In clay streams the upper tributaries are ditches which are not normally

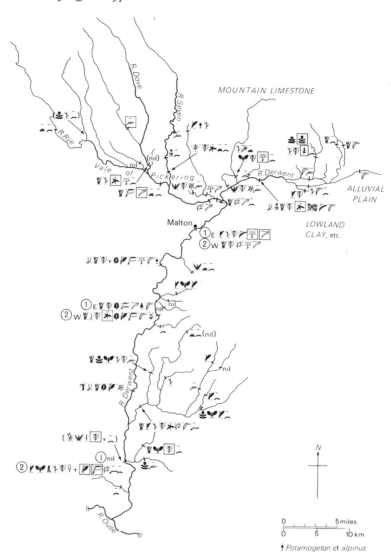

Fig. 11.9 (*cont.*).

‡ *Potamogeton* cf. *alpinus*

scoured much, and tall emergents are common. Next, where discharges are larger and there is some scour, fringing herbs and tall emergents are sparse. These, possibly with *Callitriche*, continue well into the zone of perennial flow downstream. Clay streams have more silt, a clay bed is more eroded, and their flow regime is more unstable than other lowland streams. These factors lead to poor anchorage, much storm loss, and consequently little vegetation in the middle reaches where scour is greatest. Narrower streams are often shaded by hedges. Fringing herbs are sparse in middle reaches and are uncommon downstream because of the silting and flooding. Eutrophic and semi-eutrophic species enter

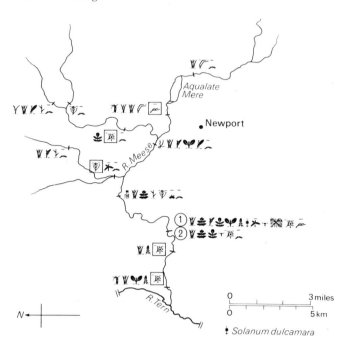

Fig. 11.10. Lowland river on soft sandstone. R. Meese (Severn); most records 1972.

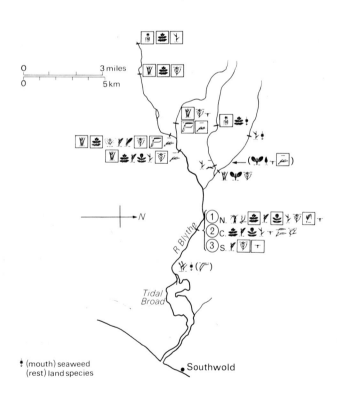

Fig. 11.11. Lowland river on soft sandstone. R. Blythe (East Suffolk and Norfolk); most records 1972.

192

farther up than in other stream types, being found as soon as the sub-strate is sufficiently stable and silty for the survival of the species concerned.

In the middle reaches, flow type is often variable. Some (rare) zones have swift flow, coarse substrates and few plants. Other areas have moderate to fast flow, fairly firm gravelly beds and little silting, and these bear the semi-eutrophic species of faster flows (e.g. *Zannichellia palustris*). These species are unable to grow in upper reaches because there is too little water, or in lower ones because there is too much silting. The slower middle and the lower reaches bear as eutrophic a plant community as can exist in natural conditions in this country. The plants often grow on deep silt.

Plants of clay streams show downstream variation with changes in water flow and depth, substrate type and stability, and trophic status.

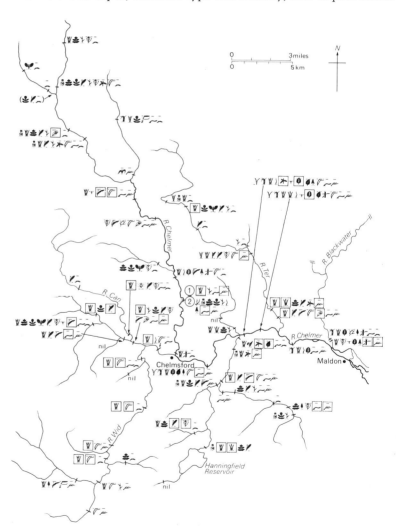

Fig. 11.12. Lowland clay river. R. Chelmer (Essex); 1972.

Fig. 11.13. Upland clay river. R. Chew (Bristol Avon); most records 1972.

Ranunculus is *R. peltatus* or *R. trichophyllus*

Lowland and upland clay streams: individual descriptions

The stream in Fig. 11.12 is a typical lowland clay stream. That in Fig. 11.13 is an upland stream, not entirely on clay but, as regards the plants, effectively on clay. The flow is greater than in the lowland stream and so silting is less. This means that fringing herbs can grow farther downstream, and *Ranunculus* can grow in places hilly enough to provide a firm anchorage, a moderate flow and little silting. Eutrophic plants still enter well upstream. The stream is unfortunately not long enough to show a full downstream eutrophic vegetation, though this progression can be seen in Fig. 11.6 where the lower part is on clay.

The stream in Fig. 11.14 also rises in hilly clay. In the upper reaches flows are swift, and short-leaved *Ranunculus* grows well. The stream is still small when it enters the flat alluvial plain and it is very soon similar to the drains that rise within the plain. This contrasts with the streams in Figs. 11.3 and 11.9 when they are in plains. In these the volume of swift water is much greater and the vegetation remains that of moving water. The only difference between the stream in Fig. 11.14 and the channels of the plain is that the stream vegetation is slightly more eutrophic, presumably because more silt is present. The Cheddar

Streams of different types

Gorge stream, on hard limestone, is a tributary. In the Gorge the stream bears a chalk stream vegetation, though this is much affected by human interference. It has a short-leaved *Ranunculus* in the shallower swifter zones, and a medium-leaved species in the deeper slower areas. When the stream flows into the plain, *Ranunculus* is still present, though in an unusual form, growing as a short carpet across the channel bed.

Conclusions

It can be seen, therefore, that vegetation does change downstream. This is attributable mainly to changes in:

flow regime;
trophic status and silting;
water depth; and
substrate stability.

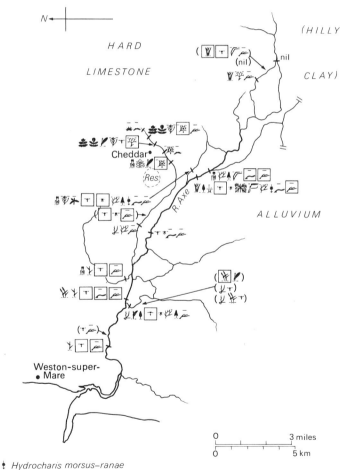

Fig. 11.14. River rising on slightly hilly clay, but soon flowing into an alluvial plain. One tributary on hard hilly limestone. R. Axe (Somerset); most records 1972.

‡ *Hydrocharis morsus–ranae*
Potamogeton spp.
Ranunculus in Cheddar *R. Yeo* is *R. calcareus* (deeper)

195

Fig. 11.15. Lowland river influenced by chalk, fertile sandstone and clay. R. Avon (Avon and Dorset); most records 1972.

R.Bourne, perennial flow starts below Idmiston
♣ in brooks, land species
Ranunculus: *R.Wylye*, *R.peltatus* changing to *R.calcareus*; *R.Nadder*, *R.trichophyllus*
and a medium–leaved form; *R.Ebble*, *R.* cf. *aquatilis*; *R.Avon*, *R.* cf. *calcareus*

These trends apply to all streams, but the actual effect on the vegetation depends on the other habitat factors influencing each stream. If rivers flow over several rock types, for instance, the vegetation shows a combination of downstream variation and geological variation (see below; and Chapters 12 and 13).

196

CHANGES IN BEDROCK

General principles

When bedrock changes, flow regime, substrate type, silting and trophic status may all change too. Changes in flow regime are due to changes in topography, and the effects on vegetation are described in Chapters 2, 3, 5, 6, 12, 13, 14 and 20. When a stream crosses a geological boundary, the dissolved substances and suspended particles in the water that are derived from the original rock type will be carried over the boundary towards the mouth of the stream. The relative amount of these found beyond the boundary on the new rock type decreases as material enters from this new rock, the greatest change occurring when a tributary entirely on the new rock flows into the stream. Consequently the influence of the original rock type diminishes downstream.

The effect on the plants depends on how much the rock types differ in their influence on the plants, and on how the rock types are distributed within the catchment. There is, for instance, no change on passing from clay bedrock to thick boulder clay over a different rock type, or again no change on passing from one Resistant rock to another (schist, hard shale, gneiss, granite, etc.), unless topography alters also. There is most change in the lowlands when passing between chalk and clay, and in the hills when passing between sandstone and Resistant rocks.

The trophic status of a site depends on the whole catchment, and a second rock type has more effect the more of the catchment it covers. It also has more effect the closer it is to the source of the stream, the least effect being seen when it is sited away from the stream at the downstream end of the catchment. When streams cross to a new rock type, they at first bear the same vegetation as the original rock type because the sediment and water are both derived from this. The original type of vegetation extends farther downstream the larger the river is, since large streams carry more sediment, etc. The change in a river starts either after several miles on the new rock, or when the first tributary of any size enters from the new rock, if this happens first. After this the influence of the new rock increases, both with the length traversed and with the number and size of tributaries entering (Figs. 11.15–11.18).

Soft rock (lowland) catchments

A chalk–clay river never changes completely to the vegetation appropriate to the downstream rock, perhaps because the British examples of such rivers are too short (Fig. 11.15). The vegetations of chalk and sandstone streams are more similar to each other than either is to clay streams, and so there is less effect on passing from chalk to sandstone or vice versa. Nevertheless a stream rising on a fertile sandstone (e.g. greensand) and then flowing on chalk contains some eutrophic (i.e sandstone) plants throughout its length. Small amounts of a less fertile sandstone placed at the side of the catchment, however, have little or no effect. In the reverse case, where a stream rises on chalk and flows on to

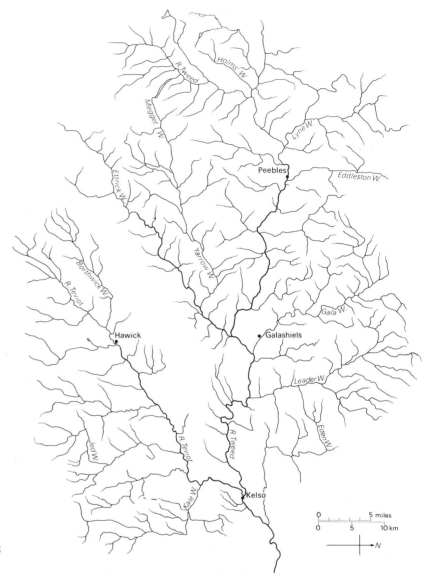

Fig. 11.16. Mountain river influenced by Resistant rock and hard sandstone. The bedrock is Resistant except for streams marked as being on hard sandstone. R. Tweed; most records 1973.

sandstone, the effect of the chalk quickly disappears.

Hard rock (mountain and upland) catchments

In the mountains two further factors influence the effects of changes in bedrock. First, a fierce flow can override all other habitat factors. It does not matter what plants can potentially grow on each rock type if the water force prevents all of them from growing. This often happens with catchments of hard limestone and of Resistant rocks. (Where water force decreases enough for plants to grow, if limestone is present then *Ranunculus* is probably more likely to be present also, and the

Fig. 11.16 *(cont.).*

Ranunculus: in upper parts *R.aquatilis*; in lower reaches *R.fluitans* (or hybrid);
and in middle and lower parts *R. calcareus*
Glyceria cf. *fluitans* is frequent in smaller streams and recorded under '℣'

Ranunculus is more likely to be *Ranunculus calcareus*.) The second factor is
that streams have a low trophic status on Resistant rocks. In a chalk–
clay interaction both rocks have a strong trophic influence and mixed
effects result, but when Resistant rock streams interact with others the
only result is to dilute the influence of the second rock. Where streams
of roughly equal size from Resistant rock and sandstone join, the
downstream vegetation is similar to that of the more fertile tributary.
Where the sandstone tributary is very small compared with that from

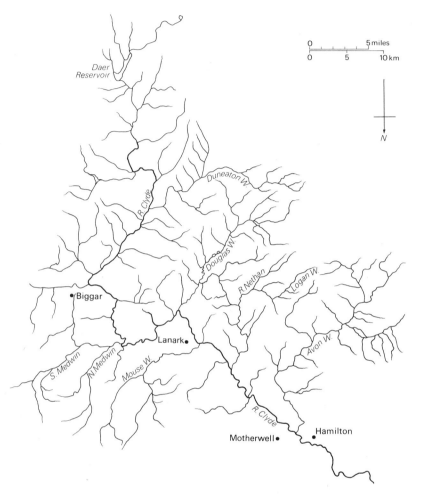

Fig. 11.17. Mountain river influenced by Resistant rock, hard limestone and calcareous sandstone. R. Clyde; most records 1972. Mouse W., North Medwin and South Medwin are mostly on calcareous sandstone or hard limestone. Douglas W. is influenced by hard limestone. R. Avon and Rotten Calder, though they have considerable hard limestone, show little influence of this because of the hilly and moorland topography.

Resistant rock, though, the downstream influence of the sandstone may be negligible, particularly if the flow is swift and spatey with little silt deposited (Fig. 11.16). (The plant distribution in R. Tweed is much more fully described in [14] and [15], where further controlling factors are considered.) However, a small area of fertile rock may have far-reaching effects below a boundary. Fig. 11.17 shows one such example where vegetation intermediate in trophic status between the short fertile tributaries and the long infertile ones occurs for many miles downstream of the confluence. Fig. 11.18 shows the overriding influence of sandstone, when present in a large proportion of the catchment. Fringing herbs can appear very soon after a stream passes from Resistant rock to sandstone (Figs. 11.16 and 11.18) because the banks on which they grow (in the larger streams) are mainly on sandstone soil once the boundary is crossed.

‡ *Potamogeton* × *sparganifolius* (mostly)
Ranunculus: upstream, *R. aquatilis;* downstream, *R. fluitans* (usually)
Glyceria cf. *fluitans* occurs in various small and species-rich streams,
 and is recorded under '⅄'

Fig. 11.17 *(cont.)*.

DOWNSTREAM PROPAGATION

Viable fragments, fruits and winter buds of river plants are washed
downstream by the current. Many do not become lodged at all, while
others lodge so far downstream that the habitat is too different for good
growth and the plant dies (e.g. *Ranunculus* lodging in the slow flow and
deep silt of a stream in a flood plain). A few, however, may lodge on
obstructions and then grow and become established (see, e.g. Chapter
10). Plant parts which cannot become permanently established may
lodge and grow for a few weeks or months. Some *Ranunculus* clumps in
slow reaches develop like this and clumps of fringing herbs downstream
are often formed from shoot fragments washed from above. Although
the clumps do not grow large and are soon washed away, new fragments arrive from upstream and form new clumps. Even upstream

Fig. 11.18. Upland river
influenced by Resistant rock
and hard sandstone. R. Lugg
(Wye); most records 1972.

*Ranunculus: R.cf. aquatilis above; R. fluitans in lower R. Lugg;
and R. calcareus in R. Arrow and part of lower R. Lugg*

stands of fringing herbs may start from such fragments. *Rorippa nasturtium-aquaticum* agg. populations are particularly frequently started from fragments, and watercress beds upstream are important in providing *Rorippa nasturtium-aquaticum* agg. populations in the main river [12].

Lakes sometimes occur along a river. They do not normally alter the general vegetation, except where they act as settling tanks for streams with much sediment pollutions, etc. They do, however, usually have a different and richer flora, and sometimes lake species extend somewhat into the river below the lake, usually as temporary populations established from fragments [47].

Plants which usually or sometimes form temporary populations are much affected by downstream propagation (e.g. *Callitriche* spp., *Elodea canadensis, Lemna minor* agg.). In the R. Tweed river system Dr N. T. H. Holmes has recorded a distribution of *Potamogeton* spp. which suggests downstream propagation from whatever site the species first entered the river [14, 15]. In contrast, Dr F. H. Dawson has found that in Dorset chalk streams different physiological strains of *Ranunculus calcareus* occur in a downstream progression – which indicates that *Ranunculus calcareus* is not propagating here by fragments.

After severe losses of plants, e.g. from fierce spates or dredging,

202

recolonisation is partly from plants upstream and partly from underground parts left in place. Propagation from washed-down plant parts is made easier in these circumstances, since the habitat is initially empty and so the places most suitable for plant growth are available for colonisation. Also, if the affected zone is short, it is likely to have a similar environment to the zone immediately upstream of it, and so fragments washed down are likely to be of species which will grow well in the bare habitat.

I2

Vegetation of streams on soft rocks

Streams on soft rocks can be, and often are, choked with plants. They can be classified by flow regime and nutrient status. Chalk streams have clear water, fairly swift and stable flow, and firm gravelly substrates. They are species-rich, with a lot of Ranunculus *and fringing herbs. Soft sandstone streams are siltier, with less species diversity and often less vegetation. They have more* Callitriche, *eutrophic species extend further upstream, and fringing herbs, and often* Ranunculus, *decrease. Clay streams are the commonest type. They tend to have less flow, a less stable substrate, more silt and more turbid water. Upper reaches are often shaded, or choked by emergents. Eutrophic (clay) species grow throughout the zone of perennial flow, and chalk species are sparse. If catchments contain two or more rock types, the vegetation is influenced by each in those parts where the stream has flowed over the different rock types.*

The soft rocks of the lowland south and east of Britain comprise the soft limestones of chalk a⎯ ⎯i oolite, the clays, and the soft sandstones of Mesozoic, Tertiary or, if they are present in large quantity, Quaternary age. (The Triassic sandstones of hills in the north-west are here classed as hard rocks.) Watercourses influenced mainly by alluvium are described in Chapter 14. There are no severe storm flows in streams on soft rocks, so potentially they can contain choking vegetation. Fig. 12.1 shows the relationship between the main stream types, which is discussed more fully on p. 223. Much basic ecological information about these streams is also given in Chapter 11 and in Fig. 10.2.

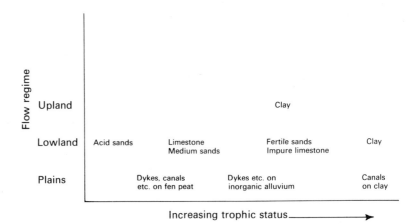

Fig. 12.1. Stream types on soft rocks.

CHALK STREAMS

Chalk is a very pure fine-grained limestone. Many streams which are classified for geographical or fishery purposes as chalk streams differ in their plants from the streams solely on chalk, and are botanically classed as streams of mixed catchments (e.g. R. Great Stour in Kent, R. Frome in Dorset). A chalk stream can, however, have small amounts of sandstone, drift or alluvium in the catchment, or some clay-with-flints on the hills, without affecting the plants.

Fig. 12.2. A chalk landscape.

Chalk usually forms rolling hills (Downs) up to *c*. 800′ (250 m) (Fig. 12.2). Much of the brook water comes from springs, and so the flow is fairly swift. Little silt is washed in from the land and much of the silt in the upper reaches is organic in origin. The silt within plant clumps, and the silt shoals downstream, are less than in other streams on soft rocks. The bed is usually firm and gravelly, with a little deposited sediment. The water is unusually clear, seldom being coloured or turbid except in storms. Because of its origin in springs the water temperature is more stable than that of other stream types. The relatively high winter temperatures promote the growth of *Rorippa nasturtium-aquaticum* agg. and allow the species to be grown commercially in winter. The relatively high spring temperatures encourage the early growth of *Ranunculus*, etc. However, the stable temperature has little overall effect on the vegetation. In drying brooks, the high summer water temperatures in pools do not harm the plants and so do not alter the character of the vegetation.

Width–slope patterns for individual stream types (Figs. 12.3, 12.5, 12.7, 13.3, 13.5, 13.7 and 13.9) cannot be directly compared with those described in Chapter 6. Large rivers, especially in the lowlands, usually have more than one rock type in the catchment, so the patterns for the separate rock types apply mainly to small streams. The width–slope pattern for an individual species cannot be larger than that for the stream type (Fig. 12.3). Small chalk brooks are often *c.* 2 m wide with a slope steeper than 1:200; larger brooks are often *c.* 4 m with a slope of *c.* 1:400; and rivers may reach *c.* 20 m wide with a slope of 1:1000. There are, of course, more small brooks than large rivers (Fig. 12.3*a*). Even the rivers, though, are seldom over 1–1.5 m deep and have a stable firm bed which allows good anchorage, and under such conditions plants are able to grow right across the channel. Relatively more of the large than the small channels are full of plants, and there are only a few sites in which angiosperms are sparse or absent (Fig. 12.3*a*). The characteristic species are shown in Table 12.1. Additional habitat notes are given below.

TABLE 12.1 *Chalk stream species*

(*a*) *Channel species*
    Undamaged channels contain higher frequencies
Occurring in at least 40% of sites recorded:

    Ranunculus spp.   *Rorippa nasturtium-aquaticum* agg.

Occurring in at least 20% of sites recorded:

    Apium nodiflorum   *Myosotis scorpioides*

    Callitriche spp.   *Sparganium erectum*

    Mentha aquatica   *Veronica anagallis-aquatica* agg.

Occurring in at least 10% of sites recorded:

    Berula erecta   *Veronica beccabunga*

    Lemna minor agg.

Other species which, though sparse, are characteristic of some part of the range:

    Elodea canadensis   *Oenanthe fluviatilis*

    Glyceria maxima   *Phalaris arundinacea*

    Groenlandia densa   *Schoenoplectus lacustris*

    Lemna trisulca   *Sparganium emersum*

    Mimulus guttatus   *Zannichellia palustris*

    Myriophyllum spicatum   Mosses

(*b*) *Bank species*
Occurring in at least 40% of sites recorded:

TABLE 12.1 (*cont.*)

 *Epilobium hirsutum*   *Urtica dioica*

Occurring in at least 20% of sites recorded:

 *Phalaris arundinacea*

Occurring in at least 10% of sites recorded:

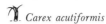 *Apium nodiflorum* + *Berula erecta*   *Filipendula ulmaria*

Carex acutiformis   *Rorippa nasturtium-aquaticum* agg.

Sparganium erectum

Notes on Tables 12.1–12.5, 13.1–13.5, 14.1–14.2
The data in these tables come from surveys of watercourse sites (mostly from bridges) carried out between 1969 and 1974. Some 4500 records were considered for the analyses, only sites on catchments of mixed geology or with severe pollution being discarded and all other sites (upstream and downstream, mildly polluted, shaded, recently dredged, etc.) being included. This number of sites, however, is too small to ensure that any other survey will produce the same data – though the main pattern should be similar – and the tables are intended as summaries only. Ecological subdivisions are discussed in the text, particularly in Chapters 8 and 11–14. Because sites that are shaded, recently dredged, etc. are included, species frequency is less than in well-vegetated rivers (see figs. in Chapter 11).

Some categories of plants have been omitted or grouped in the tables. Mosses are considered as a group only; *Enteromorpha* sp. is the only alga included; and the small grasses (*Agrostis stolonifera* and, less commonly, *Catabrosa aquatica* and *Glyceria* spp.) are omitted as the earlier records did not distinguish between temporary plants on fallen clods and permanent populations. On the banks, the only species recorded were those which also occur in the channel and those which are characteristic of (drier) marshes.

The species are listed in frequency categories, of species occurring in at least 40%, 20–39%, and 10–19% of the sites recorded of that geological type. If plants are sparse in a stream type, a further category of 5–9% is added.

Chalk streams which are permanently dry near the source bear land plants there. Winter-wet brooks bear sparse land plants and aquatics, the proportion of aquatics increasing with the length of the flooded period. The channel is seldom covered with aquatics if it is dry for more than late summer. Apart from land plants, the plants of the drier channels are usually:

Mentha aquatica   *Urtica dioica*

Myosotis scorpioides   *Veronica anagallis-aquatica* agg.

Phalaris arundinacea   *Veronica beccabunga*

Rather lower, the following enter:

Apium nodiflorum   *Rorippa nasturtium-aquaticum* agg.

In the zone flooded for most of the year, the following can come in:

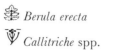

Berula erecta   *(Lemna trisulca*, rare*)*

Callitriche spp.   *(Mimulus guttatus*, infrequent*)*

Lemna minor agg.   *Ranunculus* spp., short-leaved

By this stage, *Urtica dioica* is only on the bank, and *Phalaris arundinacea* is mainly so. The term 'fringing herb' is conveniently used to describe, collectively, the short semi-emerged dicotyledons *Apium nodiflorum*, *Berula erecta*, *Mentha aquatica*, *Mimulus guttatus*, *Myosatis scorpioides*, *Rorippa nasturtium-aquaticum* agg., *Veronica anagallis-aquatica* agg. and *Veronica beccabunga*. These form clumps, usually based on the sides even in small channels, but also extending into the middle. *Apium nodiflorum* and *Berula erecta* may form submerged carpets over a good deal of the bed, and *Rorippa nasturtium-aquaticum* agg. may form emerged carpets in flooded parts with stable low flows. *Callitriche* and short-leaved *Ranunculus* grow in the centre. The *Ranunculus* species is commonly *Ranunculus peltatus*, though *Ranunculus aquatilis*, *Ranunculus trichophyllus* and intermediates may also occur. All can tolerate drying in late summer, and often develop aerial leaves in shallow water or on damp ground. Such brooks are typically covered with vegetation unless they are cut or otherwise managed.

With perennial flow the following can be found:

Sparganium erectum   Mosses

(Hippuris vulgaris, uncommon*)*

Brooks with perennial flow from the source may contain any of the above species. The typical pattern of plants in the channel is shown in Fig. 10.2 *(e)*. Fringing herbs with occasional *Sparganium erectum* occur at the sides, the fringing herbs being mixed together, or in small, often contiguous patches, and not growing very large or luxuriant. Any of the species may extend into the channel in shallow water. Submerged carpets of *Apium nodiflorum* and *Berula erecta* are common and emerged ones of *Rorippa nasturtium-aquaticum* agg. occur occasionally. *Callitriche* and *Ranunculus* are more prominent than in shallower brooks, and with increasing depth and flow *Callitriche* becomes concentrated at the sides.

In middle reaches (Fig. 10.2*f*), *Ranunculus* is the main channel dominant, and is here in medium-leaved forms which cannot tolerate drying. *Ranunculus calcareus* is the commonest, but *Ranunculus penicillatus*, *Ranunculus aquatilis*, *Ranunculus peltatus* and intermediates also occur. *Callitriche* is usually present but inconspicuous. The fringing herbs become increasingly concentrated at the edges as scour and depth increase in the centre. *Apium nodiflorum* and *Berula erecta* still form large submerged carpets in shallower channels (usually those less than 40 cm deep, though carpets are occasionally in water *c.* 75 cm deep), but *Rorippa nasturtium-aquaticum* agg. seldom forms carpets because the plants are so easily washed away. *Lemna trisulca* and *Mimulus guttatus* are usually absent.

At the margins there may be a band (up to *c.* 1 m wide) of floating fragments of fringing herbs and *Callitriche* (Fig. 10.2*f*), and there may also be a narrow band of tall monocotyledons on the bank, usually *Glyceria maxima* and *Sparganium erectum*, though *Carex acutiformis* and (in small patches) *Phalaris arundinacea* are also characteristic. Bands of tall monocotyledons and of fringing herbs are usually mutually exclusive, though sparse fringing herbs may be found on the channel side of a fringing herb band, and patches of tall monocotyledons may occur behind fringing herbs.

In the downstream or siltier parts of these middle reaches, species not especially characteristic of chalk streams first appear. *Elodea canadensis* is usually found in species-rich sites, and may grow just as small shoots within other clumps. It is luxuriant only in the more eutrophic places such as silt shoals.

In the lower reaches, where the channel is deeper, fringing herbs are excluded from the bed and *Callitriche* usually occurs in shallow or sheltered places. The water is still clear and fairly swift, the bed remains firm and stable (suitable for plants with shallow root wefts) and plants can usually grow right across the channel. *Ranunculus* is the usual dominant, often *Ranunculus calcareus*, sometimes *Ranunculus penicillatus* or other species. Their leaves are typically longer than in the middle reaches.

The distance from the water to the top of the bank is less than in other stream types, often only *c.* 0.5 m. The banks are steep with little silt, and the fluctuations in water level are greater. Because of this, fringing herbs are seldom prominent, though small patches, often temporary, are usually present where they can anchor. The narrow band of tall monocotyledons which appeared upstream is now likely to be continuous unless the zone is grazed.

Nutrient status increases in the lower reaches, allowing the entry of semi-eutrophic and eutrophic species.

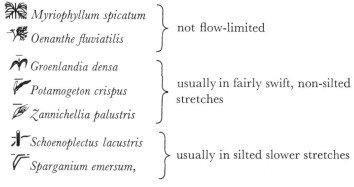

Myriophyllum spicatum	not flow-limited
Oenanthe fluviatilis	
Groenlandia densa	usually in fairly swift, non-silted stretches
Potamogeton crispus	
Zannichellia palustris	
Schoenoplectus lacustris	usually in silted slower stretches
Sparganium emersum,	

One or more of these occur in *c.* 15% of all sites recorded. *Groenlandia densa* does also rarely occur in upper parts with only slight indication of a non-chalk influence, but apart from this the presence of any of these species in upper or middle reaches, or of many or large amounts of them in lower reaches, indicates eutrophication, probably from man's activities. (They can also occur in silted flood plains, where there is, for example, a more eutrophic rock in the catchment that has been overlooked.)

The difference between upper, middle and lower reaches is one of stream size and vegetation physiognomy, and is controlled by the topography and water regime rather than by the actual distance from the source. The typical downstream patterning of a large chalk stream in good condition is shown in Fig. 11.7. Upper reaches which are winter-wet and summer-dry usually have 1–2 aquatic species per site. Upper brooks with perennial flow have *c.* 4 species per site and un-damaged lower reaches at least 7 species. A small number of sites have 10 or more species in normal conditions. A healthy chalk stream looks roughly full (25% or more) of plants throughout the zone of perennial flow.

The width–slope patterns for the more characteristic plants are show in Fig. 12.3(*b*)–(*g*), demonstrating the wide habitat ranges and the downstream variation.

Apium nodiflorum + *Berula erecta*, *Callitriche obtusangula* + *stagnalis* and *Ranunculus* spp. grow well in all sizes of chalk stream. Luxuriant stands of *Callitriche* have a smaller pattern, avoiding the extremes of flow. Short-lived *Ranunculus* species occur upstream of medium-leaved forms, as is demonstrated here for *Ranunculus peltatus* and *Ranunculus calcareus*, the two most frequent chalk species. The central parts of the habitat range can bear both species. *Ranunculus aquatilis*, *Ranunculus penicillatus*, *Ranunculus trichophyllus*, hybrids and intermediate forms occur less frequently.

The other fringing herbs, Fig. 12.3(*c*), are restricted to brooks with less extreme width–slope characters. *Myosotis scorpioides* and *Veronica* spp. are more confined to the fringes than *Apium nodiflorum* and *Berula erecta*, though they also may occur on the bed of brooks, and *Mentha aquatica* is almost confined to the fringes. Emerged fringing herbs collect silt and grow more luxuriantly in it, so because silting is little on the chalk, shoots of fringing plants are often smaller than on clay. The low silting and the firm bed to the channel are the main reasons why fringing herbs grow better in channels here than they do in other streams. *Sparganium erectum* is somewhat restricted (Fig. 13.3*e*), probably because silting is low and flow is swift.

The slow-flow eutrophic species (Fig. 12.3*f*) are confined to wide flat streams with the most silt. Those more characteristic of faster flow (Fig. 12.3*g*) occur in somewhat narrower and steeper streams, though they do not grow as far upstream as on sandstone and clay. Small swift chalk streams are unsuitable for eutrophic species because of their trophic status, substrate and flow.

Downstream patterning is described in Chapter 11 and is illustrated in Fig. 11.7, which also shows typical plant communities in different sites.

Chalk outcrops are found in the south and east of England. The most westerly chalk streams are tributaries of R. Frome, Dorset. They extend eastwards to Kent, and north through the Berkshire Downs and Chilterns to Hertfordshire and Cambridgeshire. The next main outcrop is in

Plate 1. A tree swamp, North
Carolina. The water is dark
(peat-stained), fairly deep and
very slow-moving.

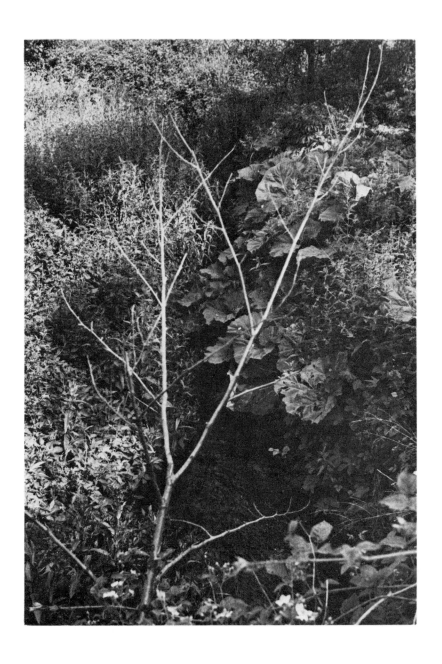

Plate 2. A brook shaded by
tall herbs on the banks.
(Bristol Avon.)

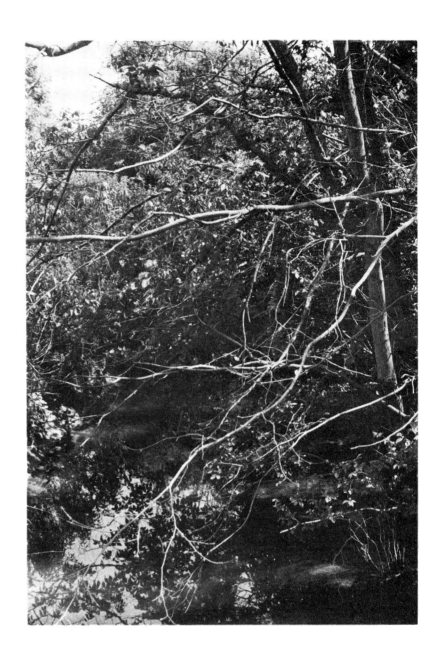

Plate 3. A brook shaded by
hedge. (R. Ivel, Great Ouse.)

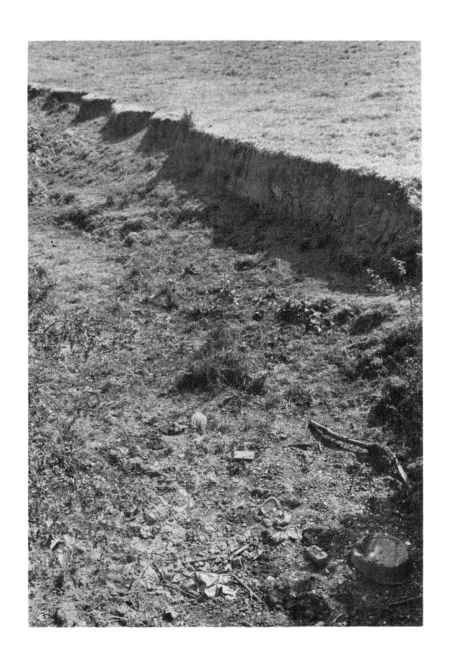

Plate 4. A drying stream bed.
(R. Quin, Lee.)

Plate 5. Plant distribution in relation to speed of flow, Sutherland. Plants occur only in the flood plain, some distance beyond the end of the slope, where both normal and spate flows are less fierce.

Plate 6. *Potamogeton pectinatus* in R. (Severn) Avon at a polluted site. Compare this picture of the typical British habit of this species with Plate 15 which shows the typical North American habit.

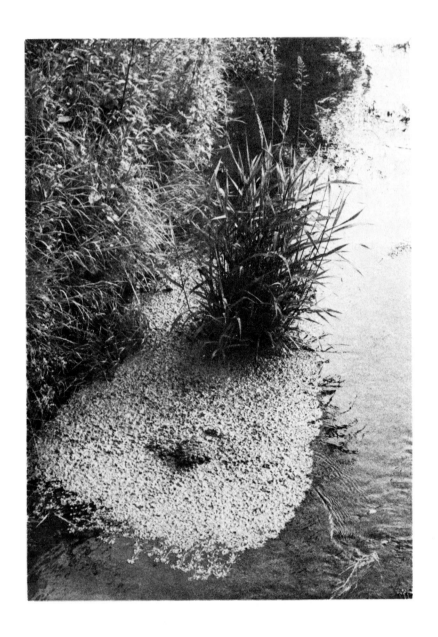

Plate 7. *Phalaris arundinacea*
clump providing protection
for *Callitriche*. (Sutton Benger
Brook, Bristol Avon.)

Plate 8. Fragments of fringing
herbs and other submerged
and emerged plants caught
on obstructions, and growing
well (in the absence of storm
flow). (R. (Bristol) Avon.)

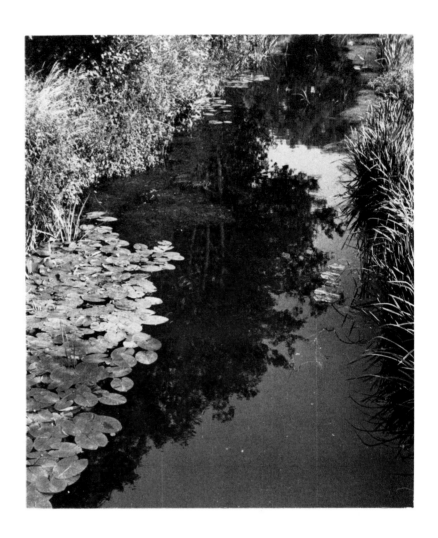

Plate 9. Upstream of a bridge
in a reach with very slow
flow. *Agrostis stolonifera,
Potamogeton pectinatus,
Sparganium emersum, Sparganium
erectum* and *Veronica
beccabunga* can be seen. (R.
Marden, Bristol Avon.)

Plate 10. Downstream of the
reach shown in Plate 9, below
the bridge. Here flow is fast
but becoming slower into the
picture. *Ranunculus* is
abundant and *Sparganium
erectum* can be seen. (R.
Marden, Bristol Avon.)

Plate 11. A species-rich canal.
(N. Coventry.)

Plate 12. A brook near its
source (the source being a
flooded ditch), Quebec.
Vegetation is intermediate
between that of a ditch and a
brook. *Alisma plantago-aquatica*,
Eleocharis sp., *Polygonum
lapathifolium* agg., *Sagittaria
latifolia* and *Typha latifolia*
are present.

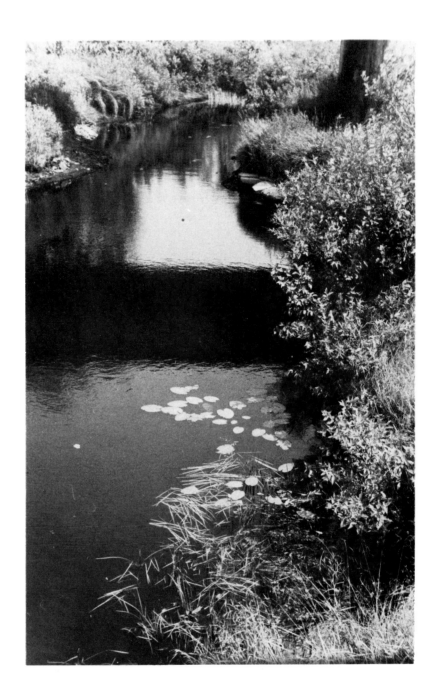

Plate 13. A stream which is unusually species-rich for Quebec, with slow flow. The substrate is silty. *Eleocharis* sp., *Nuphar variegatum*, *Polygonum lapathifolium* agg., *Potamogeton epihydrus* and *Sparganium chlorocarpum* are present.

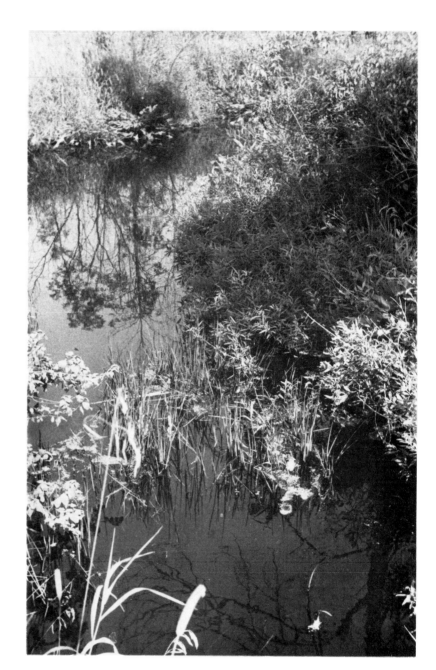

Plate 14. A stream near its source in a tree swamp, Augusta Creek, Michigan. The summer water level is as stable as it is in Britain (contrast Fig. 15.4). The water is peat-stained and dark, but not too dark for plant growth. *Ceratophyllum demersum* (submerged and not visible), *Lemna minor*, *Peltandra virginica*, *Phalaris arundinacea*, *Solanum dulcamara* and *Sparganium americanum* are present.

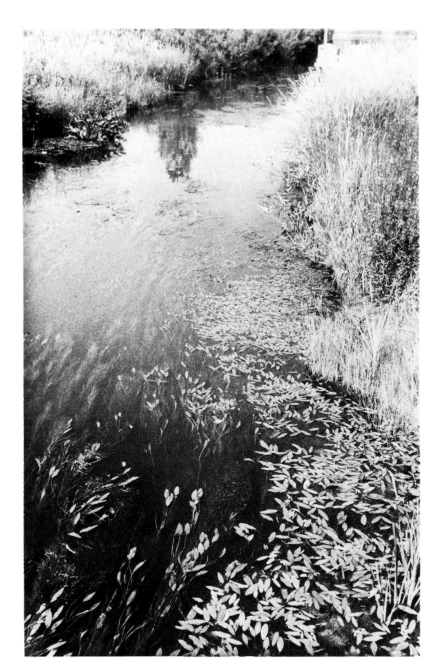

Plate 15. A stream choked with plants, Michigan. This is very unusual in North America. The summer water level is as stable as in Britain and flow slow to moderate. The vegetation comprises *Eleocharis erythropoda* agg., *Heteranthera dubia, Phalaris arundinacea, Pontederia cordata, Potamogeton* cf. *gramineus, Potamogeton pectinatus, Sparganium americanum* and *Typha angustifolia. Potamogeton pectinatus* is in its typical North American form. Compare this with Plate 6 showing the typical British habit.

Plate 16. A dyke with *Phragmites communis*. This would create a flood hazard if the water level rose substantially. (Fenland.)

Plate 17. A recently dredged dyke, which would cause little obstruction to flow if the water level rose. (Fenland.)

Plate 18. A chalk stream in
its normal state, May 1975.
There is dense *Ranunculus*,
species-rich vegetation, good
flow and a gravel substrate.
(R. Ivel, Great Ouse.)

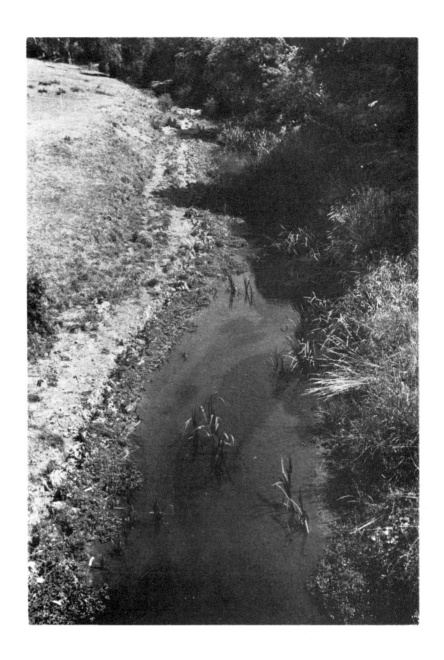

Plate 19. The same chalk
stream site as in Plate 18
severely damaged by drought,
June 1976. There is sparse
and small *Ranunculus* and
Sparganium erectum, but little
else present, little flow and
thick silt on the bed. (R.
Ivel, Great Ouse.)

Plate 20. Cutting *Ranunculus* by scythe. (R. Rhee, Great Ouse.)

Plate 21. A weed-cutting boat. (Photograph kindly supplied by Rolba Ltd, East Grinstead, Sussex.)

Plate 22. A 3.0 m Bradshaw
weed-cutting bucket mounted
on a JCB 807 excavator.
(Photograph kindly supplied
by G. H. Bradshaw & Son
(Contractors) Ltd,
Stibbington, Peterborough.)

Plate 23. Alder (*Alnus glutinosa*) roots growing down a stream bank and stabilising the channel. (Photograph kindly supplied by Drs A. Krause & W. Lohmeyer.)

Plate 24. A stream shaded, and its banks stabilised, by alder (*Alnus glutinosa*), and requiring very little maintenance. (Photograph kindly supplied by Drs A. Krause & W. Lohmeyer.)

Plate 25. Another stream which is shaded and its banks stabilised by alder (*Alnus glutinosa*) and therefore requires very little maintenance. (Photograph kindly supplied by Drs A. Krause & W. Lohmeyer.)

Plate 26. Ungrazed banks
(*left*) with tall herbs and
grazed banks (*right*) with short
grasses but showing damage
from trampling by horses.
(R. Rhee, Great Ouse.)

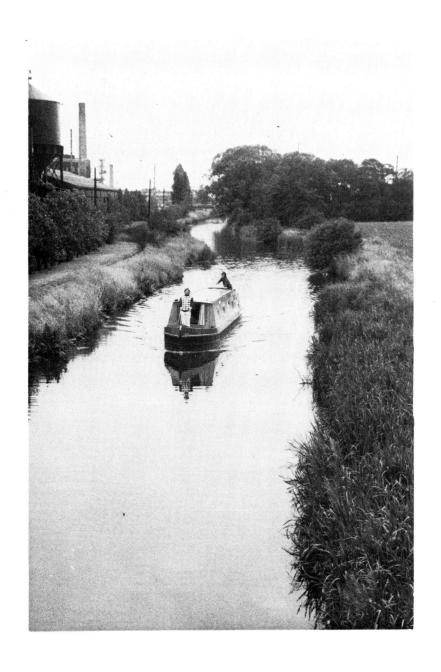

Plate 27. A canal which has
considerable boat traffic and
supports only some emergent
vegetation. (Near
Wolverhampton.)

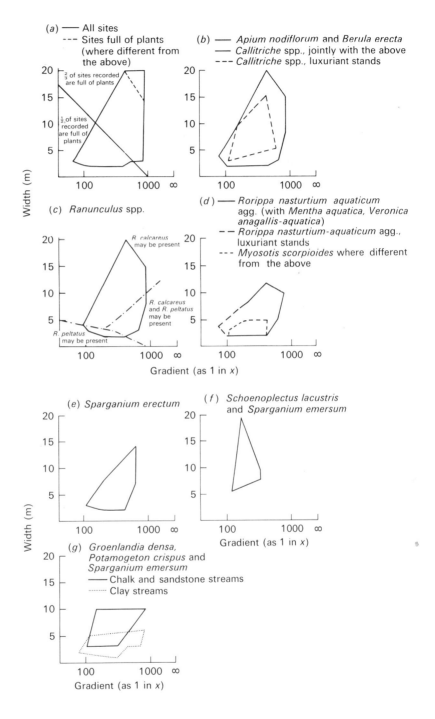

Fig. 12.3. Width–slope patterns in chalk streams. Here, and in the comparable figures in this chapter and Chapter 13, large rivers are usually excluded, as they so often flow over more than one rock type. Consequently the species sites occurring to the right of the patterns in the figures in Chapter 6 are usually absent. The patterns for each rock type should be compared with pattern (*a*) on each figure, as this is the pattern for the total sites included for that rock type. Gradient and width are represented as in Fig. 6.1., etc.

Lincolnshire, where most streams of the Wolds are on mixed catchments, and the most northerly outcrop is in Yorkshire, R. Hull being the main chalk river there. There is little variation between the plants found in streams in different geographical areas. However, *Apium nodiflorum* is a common channel species in Dorset, while in Hampshire and Berkshire, *Berula erecta* is the more abundant. *Ranunculus calcareus* is more frequent in the south-west. R. Hull, Yorkshire, has lower hills than the others, and a large part of the catchment is covered by alluvium: only small parts have a pure chalk vegetation.

Chalk Downs sometimes rise sharply from flatter ground on other rock types (e.g. the Sussex Downs with greensand and clay in the vale below). The chalk streams are then very short, often summer-dry, and narrow, making it difficult or impossible to diagnose chalk vegetation. Wetter brooks usually bear the characteristic fringing herbs and have the characteristic sparkling water.

OOLITE STREAMS

Oolite is the other common soft British limestone, though oolite streams are less frequent than chalk ones. The total area is less, much of it is covered by Boulder clay outcrops, so that the catchments are mixed limestone clay and often only short brooks are influenced solely by oolite.

Oolite is less pure than chalk, and more semi-eutrophic and eutrophic species grow in the channels. Some brooks have the same vegetation as chalk ones of a similar size. Usually, however, there is less:

 *Callitriche* spp.   *Ranunculus* spp.

and more

Myriophyllum spicatum   *Sparganium erectum*

Nuphar lutea   *Enteromorpha* sp.

Sparganium emersum

There is usually less vegetation in the channels and a higher proportion of sites are almost without macrophytes.

The main outcrops are the Cotswolds, Northamptonshire Wolds, and the ridges projecting north from the latter into Lincolnshire. (There are smaller outcrops joining these areas and running south into Somerset. The streams of the North Yorks Moors on Coralline oolite are classed as hard limestone and described in Chapter 13.) The other soft limestones are of only local importance, e.g. bands in the Corallian, so it is very rare to find streams of any size exclusively on these limestones. In mixed catchments the vegetation is the same as for oolite.

Fig. 12.4. A soft sandstone landscape.

SANDSTONE STREAMS

Sandstone hills can be as high as chalk ones, but the ground is usually lower with the streams flowing more slowly (Fig. 12.4). There are often at least intermittent patches of Boulder Clay, so it is not easy to determine the vegetation of exclusively sandstone streams. The vegetation described here as characteristic could be slightly influenced by clay.

Sandstone streams are partly fed by underground aquifers, so their flow type is nearer that of soft limestone than is that of other rock types. However, sandstone has more very narrow brooks, and brooks are often

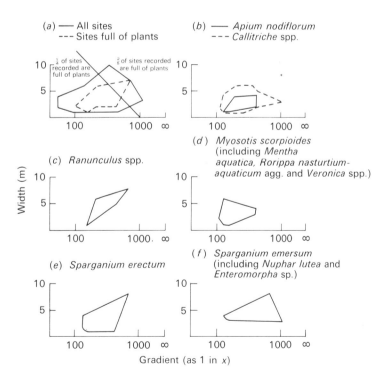

Fig. 12.5. Width–slope patterns in sandstone streams. For details see Fig. 12.3. For *Groenlandia densa, Potamogeton crispus* and *Zannichellia palustris* see Fig. 12.3 (*g*).

213

closer together than they are on chalk. Because of the fewer springs and greater run-off, more brooks rise on steeper slopes than is the case on chalk (Fig. 12.5a), but more of the stream length is on flatter ground and, as it happens, Britain has no large rivers on soft sandstone. There is more loose sand and silt in the streams, and the downstream increase in silting is greater. Consequently the potential damage from scour is also greater. When the firm bed is on sandstone itself, this is more easily eroded than the firm bed of a chalk stream. The water tends to be more turbid but still with little colour, and though water temperature is more variable this is probably of little significance to the plants (see p. 205).

The smallest sandstone brooks are often 1–2 m wide with a slope

TABLE 12.2 *Soft sandstone species*[a]

(a) *Channel species*
    Undamaged channels contain higher frequencies
Occurring in at least 40% of sites recorded:

 𝖂 *Callitriche* spp.

Occurring in at least 20% of sites recorded:

 𝖸 *Myosotis scorpioides*　　　 𝖂 *Sparganium erectum*

 𝖶 *Sparganium emersum*

Occurring in at least 10% of sites recorded:

 🌿 *Apium nodiflorum*　　　 🌼 *Ranunculus* spp.

 𝖸 *Epilobium hirsutum*

Other species which, though sparse, are characteristic:

 🌱 *Alisma plantago-aquatica*　　　 🍃 *Nuphar lutea*

 ✳ *Elodea canadensis*　　　 ⌐ *Potamogeton natans*

 🌿 *Mentha aquatica*　　　 🌿 *Veronica beccabunga*

(b) *Bank species*
Occurring in at least 40% of sites recorded:

 𝖸 *Epilobium hirsutum*　　　 *Urtica dioica*

Occurring in at least 10% of sites recorded:

 Filipendula ulmaria　　　 𝖂 *Sparganium erectum*

 𝖽 *Phalaris arundinacea*

---

[a] See notes to Table 12.1.

steeper than 1 : 100, and the largest truly sandstone streams are seldom over 8 m wide with slopes of *c.* 1 : 750. There are relatively fewer sites which are full of plants, and relatively more which are effectively without angiosperms. Also, the width–slope pattern for sites choked with plants is smaller than that for all sites taken in total, showing that luxuriant vegetation does not occur in more extreme habitats. Because of the less stable bed, less stable flow and more turbid water, downstream reaches with plants growing right across the channel are less common than on chalk (Fig. 12.5*a*), and the habitat is generally less suitable for plant growth. The characteristic species are shown in Table 12.2. The typically chalk species are less frequent, though *Callitriche* has increased and the eutrophic species are commoner. The loss in chalk vegetation is, however, greater than the gain in eutrophic vegetation.

Fringing herbs: These are less frequent and in small populations that in chalk streams. *Apium nodiflorum*, *Berula erecta* and *Rorippa nasturtium-aquaticum* agg. no longer form extensive carpets over the channel. All show greater decreases downstream than they do on chalk.

Non-emergents: *Ranunculus* decreases, and the increase in *Callitriche* cannot compensate for the decline in the other carpet-forming species. Luxuriant *Callitriche* is characteristic of sandstone. The *Ranunculus* spp. are more diverse, and rarely include *Ranunculus calcareus*. *Potamogeton natans* is characteristic, though infrequent.

Characteristic of small upper brooks:

   Epilobium hirsutum   *Sparganium erectum*

   Phalaris arundinacea   Other tall herbs

Entering somewhat lower, with more water and scour:

   Apium nodiflorum   *Veronica beccabunga*

Entering in the zone of perennial flow:

   Mentha aquatica   *Ranunculus* spp.

   Myosotis scorpioides   *Veronica anagallis-aquatica* agg.

Downstream:

Eutrophic species occur mainly in lower reaches, but can extend upstream to the limits of perennial flow. In general they are commoner and occur further upstream than on chalk (because silting occurs further upstream). Mosses, on the other hand, are much scarcer because there is a small proportion of stable bed. This is a different downstream progression to that on chalk, attributable to the differences in substrate and flow.

Species diversity is usually lower than on chalk. On the banks there is an increase in:

   Epilobium hirsutum   *Urtica dioica*

   Filipendula ulmaria

and the fringing herbs, etc. (see above) decrease. The width–slope patterns (Figs. 12.5*b–f* and 12.3*g*) show that less of the potentially colonisable area is occupied by the chalk species and more by the eutrophic ones.

In the width–slope pattern, fringing herbs and *Ranunculus* occupy less of their potential areas than on the chalk. This is also true for *Callitriche*, even though this is more frequent on sandstone. *Sparganium erectum* has a large pattern. The eutrophic species of slow flows have a wider habitat range, extending into steeper and narrower brooks. Downstream patterning is described in Chapter 11, and Figs. 11.10 and 11.11 show typical plant communities in different sites.

The larger outcrops of soft sandstone are the Tertiary sands of southern England, the greensands rather more to the north and east, the Crags of East Anglia, and the Mesozoic sandstones of the Cheshire–Shropshire region. Other outcrops are smaller, or have more clay in the stream catchments. Sandstone streams vary more in their vegetation than do chalk streams. Where the land is nearly flat (hills rarely up to 200′ (60 m)) and flows are slow, *Ranunculus* is likely to be absent. The acid sands of the New Forest bear oligotrophic species upstream, e.g.

 *Callitriche hamulata*   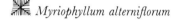 *Myriophyllum alterniflorum*

Juncus bulbosus

Downstream eutrophication is rapid because of the silting, and the vegetation soon becomes eutrophic. Some greensand has a high trophic status and the vegetation is then more nearly that of clay streams.

CLAY STREAMS

Clay is the commonest rock of lowland Britain, and large rivers as well as small brooks can occur on exclusively clay catchments (Fig. 12.6). The large rivers have, however, been omitted from the analyses here

Fig. 12.6. A clay landscape.

in order to keep these comparable with the chalk and sandstone data. Clay has a strong influence on river vegetation, and clay communities occur also when small outcrops of other bedrock are in the catchment, and when thick Boulder Clay overlies other rock types. This demonstrates well that, within the lowlands, flow pattern is less important for plants than rock types. Streams on thick Boulder clay over chalk receive water from springs in the chalk, but the substrate is derived from Boulder Clay and it is this which determines the trophic status. Clay streams have an unstable flow pattern (see Chapter 5). The non-porous land surface leads to the brooks being close together and flows being very dependent on run-off. There is more storm scour than on chalk or sandstone. Much silt enters the streams and is deposited in the lowlands, giving, together with an erodible firm bed, an unstable substrate to the channel.

In clay country the land is usually low, but in some parts hills reach *c.* 800′ (250 m). Because of the more spatey flow and the unstable silt, hilly clay streams have a different character, and are here classed as upland rather than lowland (Fig. 12.1) – in contrast to chalk and sandstone streams, which are all classed as lowland. Whether upland or lowland, the small brooks are usually on steep slopes, as they are dependent on run-off. In the upper reaches the channel bed is often the solid clay, and this is more easily eroded than other firm beds. Even gravel beds on clay are less stable than those on chalk. Even in these upper reaches deposited silt is more important, and sand less important than on the other rocks, and deep silt commonly occurs in lower reaches. The water tends to be more turbid than in the other streams, and is often somewhat grey, particularly in winter. Water temperature fluctuates more, but again this is only of minor importance (see p. 205).

Small clay brooks are often 1–2 m wide with a slope steeper than 1:80, while the largest clay rivers are over 20 m wide with slopes flatter than 1:1000. Comparing clay with sandstone, the proportion of sites full of plants is about the same and, again, full sites do not occur in the whole width–slope range (Fig. 12.7*a*). The full sites are in the most silted habitats. However, there are even more effectively empty sites, though this is partly due to the increased tree shading, and even fewer sites with plants right across the channel. The latter are frequent only in slow silted streams with water less than 1 m deep, which provide a substrate that is relatively stable (for species growing well in deep silt) and allow enough light to reach the channel bed.

The characteristic species are shown in Table 12.3. In general the chalk species are much less, and though the eutrophic species have increased, the gain is on average less than the loss. The downstream decrease in fringing herbs and *Callitriche* is faster than on sandstone, contrasting markedly with the small decrease on chalk. The fringing herbs are effectively confined to the edges, and in shallow water they form the luxuriant, silt-accumulating, nutrient-rich and short-lived clumps described in Chapters 3, 5 and 10. They do not form the stable carpets across the channel which are so characteristic of chalk streams. The eutrophic species extend far upstream, and show greater frequency and diversity than those on clay or sandstone. In large rivers plants are usually confined to the sides (see Chapters 5 and 10). In lower reaches tall monocotyledons are frequent. These tend to form wide and intermittent bands, but in quiet open places the fringe may be continuous. (Chalk streams typically have narrow continuous bands, and sandstone ones show an intermediate pattern.)

TABLE 12.3   *Clay species*[a]

(a)   Channel species
      Undamaged channels contain higher frequencies
Occurring in at least 40% of sites recorded:

   Sparganium erectum

Occurring in at least 20% of sites recorded:

   Apium nodiflorum   *Sparganium emersum*

Occurring in at least 10% of sites recorded:

   Callitriche spp.   *Nuphar lutea*

   Epilobium hirsutum   *Phalaris arundinacea*

   Lemna minor agg.   *Rorippa nasturtium-aquaticum* agg.

Other species which, though sparser, are characteristic of part of the range:

   Alisma plantago-aquatica   *Sagittaria sagittifolia*

   Butomus umbellatus   *Schoenoplectus lacustris*

   Glyceria maxima   *Typha latifolia*

   Potamogeton crispus   *Zannichellia palustris*

   Potamogeton perfoliatus   *Enteromorpha* sp.

   Rorippa amphibia

(b)   Bank species
Occurring in at least 40% of sites recorded:

   Epilobium hirsutum   *Urtica dioica*

Occurring in at least 20% of sites recorded:

   Phalaris arundinacea

[a] See notes to Table 12.1.

Clay streams

Notes to Table 12.3

As more clay stream sites are small than are large, the characteristic species of the larger clay streams, *Sagittaria sagittifolia*, etc., are not among the most frequent. Upper sites, with little sediment and a relatively low nutrient status, contain most of the *Apium nodiflorum*, *Callitriche* spp., etc. (see text). The characteristic species of chalk and soft sandstones, in contrast, occur on small as well as large streams.

Characteristic of small upper brooks:

Epilobium hirsutum   Land species

Sparganium erectum

Entering somewhat lower, with more water and scour, and less choking by plants:

Apium nodiflorum   *Veronica beccabunga*

Phalaris arundinacea

Entering just above or below the zone of perennial flow:

Callitriche spp.

Entering in the zone of perennial flow:

In swifter flow: *Potamogeton crispus*

                     Zannichellia palustris

If lime is present: *Oenanthe fluviatilis*

In slower flow: *Nuphar lutea*   *Sparganium emersum,*
                                             etc.

                     Sagittaria sagittifolia   *Enteromorpha* sp.

                     Schoenoplectus lacustris

Large rivers characteristically bear mainly deep-rooted eutrophic species:

Elodea canadensis (shallow-rooted)   *Schoenoplectus lacustris*

Lemna minor agg. (free-floating)   *Sparganium emersum*

Nuphar lutea   *Sparganium erectum*

Sagittaria sagittifolia   *Enteromorpha* sp. (shallow-anchored)

In less eutrophic parts there may also be:

Alisma plantago-aquatica   *Potamogeton perfoliatus*

Polygonum amphibium   Fringing herbs

219

With some organic or industrial pollution there may also be:

Potamogeton pectinatus

Sparganium erectum is important throughout, in contrast to chalk streams. It can choke the channel in flatter narrower brooks and can dominate, and indeed create, silt banks downstream.

Elodea canadensis is more frequent, and more frequently luxuriant, than on chalk or sandstone.

Potamogeton natans may occur in upper reaches which receive little silt and little pollution.

Species diversity is as high as on chalk, some downstream sites having 10 or more species. The most frequent number of angiosperms present (including the large number of species-poor small streams), in sites with river plants, is 3, similar to that of sandstone. This number, however, is artificially lowered by pollution. Because towns are most often sited on clay, more clay streams suffer from pollution than do chalk and sandstone ones, and an important effect of pollution is to decrease species diversity (see Chapter 22).

On the banks, there is, in comparison to chalk, an increase in:

Epilobium hirsutum   *Urtica dioica*
Filipendula ulmaria

The fringing herbs, etc., characteristic of chalk (p. 208) decrease, and so do the tall grasses and *Sparganium erectum* which are frequent on sandstone banks.

The width–slope patterns (Figs. 12.7*b–f*, 12.3*g*) show the restricted habitat distribution of the chalk species, and the increased distribution of the eutrophic ones.

Apium nodiflorum has as restricted a distribution as on sandstone, but *Rorippa nasturtium-aquaticum* agg. and *Veronica* spp. have a wider range, *Veronica* spp. extending into steeper streams. *Ranunculus* is more restricted than on sandstone and its most frequent species are *Ranunculus penicillatus* and *Ranunculus trichophyllus*. These are characteristic of unpolluted streams. The *Callitriche* pattern is like the sandstone one, and more restricted than that on chalk. The populations are, however, of much sparser plants than before, and they are less frequent in lower reaches. *Sparganium erectum* is very widely distributed. The eutrophic species of faster flow grow further upstream than they do on chalk or sandstone (Fig. 12.3*f*). The species of slow flow occur wherever silt can accumulate, and show a very different pattern to that on chalk. *Enteromorpha* has a large pattern for the first time. Like other submergents it avoids small brooks likely to dry in summer.

Downstream patterning is further described in Chapter 11 and Figs. 11.13–11.15.

The upland streams have much less silting compared with the lowland streams and there is an increase of:

🌿*Apium nodiflorum*   🌿*Rorippa nasturtium-aquaticum* agg.

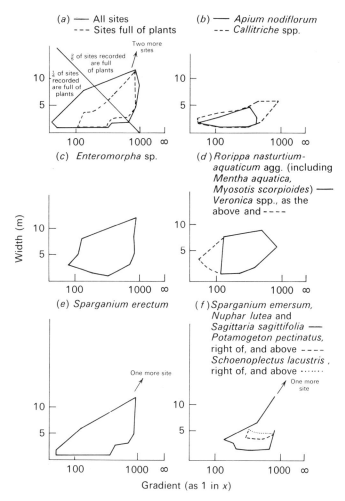

Fig. 12.7. Width–slope patterns in clay streams. For details see Fig. 12.3. For *Groenlandia densa, Potamogeton crispus* and *Zannichellia palustris* see Fig. 12.3 (*g*).

Callitriche spp.   Mosses

Ranunculus spp.   (Other fringing herbs, to a lesser extent)

These are more frequent than in lowland streams and also extend farther downstream. The eutrophic species may be less frequent than in the lowlands.

Clay is found in those parts of south and east England which are not on limestone or sandstone or in alluvial plains.

STREAMS OF MIXED CATCHMENTS

Large lowland rivers derived from more than one rock type are usually slow-floating and silted, and frequently somewhat polluted. They are very similar to large clay rivers in their general vegetation, though a few more mesotrophic species may be present. Where smaller rivers are

221

concerned, however, mixed catchments result in river vegetation which reflects the geology of the catchments and the proportion of each rock type. A chalk stream rising on fertile sandstone, e.g. R. Avon and R. Wylye, Wiltshire (see Fig. 11.15) shows some eutrophic influence throughout. An equally small amount of acid sandstone at the source or of any sandstone downstream and at the sides, has little effect. When sandstone and chalk occupy roughly equal portions in the catchment, the river vegetation is intermediate between that of chalk and that of sandstone. When clay occurs in a sandstone catchment, or vice versa, it has a noticeable influence on the plants.

Many streams are mixed limestone–clay. This clay may be solid or drift (Quaternary, Boulder Clay, etc.), but there are only a few differences in the plants. When Boulder Clay overlies limestone, and when solid clay occurs upstream of limestone, *Apium nodiflorum* and perhaps *Callitriche* spp. are likely to be more frequent than where limestone is upstream of clay. Comparing solid with drift clay, *Ranunculus* spp. are apt to be more frequent on the solid clay. When passing from mainly limestone to mainly clay catchments, chalk species decrease and clay ones increase, but the habitats in which alterations occur differ for different species.

Chalk species decreasing towards clay:

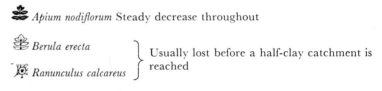

Apium nodiflorum Steady decrease throughout

Berula erecta
Ranunculus calcareus } Usually lost before a half-clay catchment is reached

Most frequent in mixed catchments:

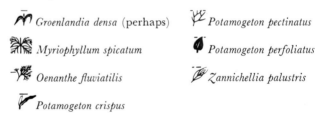

Groenlandia densa (perhaps)   *Potamogeton pectinatus*

Myriophyllum spicatum   *Potamogeton perfoliatus*

Oenanthe fluviatilis   *Zannichellia palustris*

Potamogeton crispus

Several of these species occur in chalk streams with eutrophic influence, and in clay ones with little deposited silt.

Clay species increasing away from chalk:

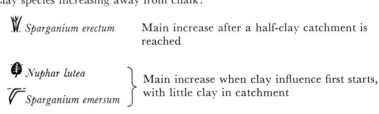

Sparganium erectum   Main increase after a half-clay catchment is reached

Nuphar lutea
Sparganium emersum } Main increase when clay influence first starts, with little clay in catchment

When Boulder Clay intermittently overlies chalk, the river plants are sparse in swifter flows because of the silty unstable substrate, but are (part-) limestone in type. The *Ranunculus* spp., however, do not usually include *Ranunculus calcareus*. When the rock boundary is crossed, the distance on the new rock that elapses before the plants change varies with the size of stream and the tributaries entering (see Chapter 11).

Limestone upstream, clay downstream. Two examples:

1. Clay species entering 2 miles and a town downstream of the rock boundary.
2. Still chalk vegetation 2 miles downstream of boundary. After 8 miles and the entry of both chalk and clay tributaries, stream with semi-clay vegetation in slower silted reaches and semi-chalk vegetation in swifter gravel reaches.

Clay upstream of limestone. Two examples:

1. Chalk species entering 2 miles downstream of the rock boundary.
2. Larger stream. After 5 miles and the entry of a chalk stream, chalk, clay and intermediate species present. After 9 miles and two more chalk tributaries a mainly chalk vegetation, but clay species still present.

Clay brooks with intermittent bands of Corallian limestone have limestone species for several miles below the band if at least half the catchment is limestone, and these are luxuriant where the catchment is three-quarters limestone.

STREAM CLASSIFICATION

The watercourses on soft rocks and alluvial plains, except for streams on mixed catchments, are classified in Fig. 12.1. The vertical axis shows flow regime. There are minor differences in flow pattern between lowland chalk, sandstone, and clay streams, but these are small compared with their differences from upland and mountain streams. The hill heights on upland clay are no greater than those on chalk both reaching around 800′ (250 m). However, the greater run-off, more spatey and swifter flow, and the less silted and unstable substrate in upland compared with lowland clay streams means there is a real difference in the vegetation, and such a difference is not found within the chalk streams. The horizontal axis in Fig. 12.1 represents trophic status (described in Chapter 8). Chalk streams are mesotrophic, with remarkably little eutrophic influence even in the lower reaches. Acid sand may be oligotrophic, but most sandstone streams are intermediate in nutrient status between those of chalk and of clay.

13

Vegetation of streams on hard rocks

Streams on hard rocks are often spatey in highland areas, and usually contain only a little vegetation. Like the streams on soft rock they can be classified by flow regime and nutrient status. Streams on Resistant rocks where there is bog peat in the substrate bear nutrient-poor species. The upper reaches of mountain streams normally have fierce flow and thus there are few river plants. Scour-tolerant species increase downstream, and Ranunculus *is usually the only plant which can be luxuriant. Coal Measures streams are often polluted and species frequency is low. Limestone streams have fringing herbs in the upper reaches and, when water force is low,* Ranunculus *can be luxuriant in lower reaches. Species diversity is low. Hard sandstone streams are much more silty, and so although plants are few where there is strong flow, vegetation can be luxuriant where scouring is less and silt can accumulate. Fringing herbs,* Callitriche *and* Ranunculus *occur in the upper reaches, and* Ranunculus *and more eutrophic species in the lower reaches.*

The rocks termed here as hard are those of north and west Britain. They comprise the hard limestones, mainly Carboniferous ones, the hard sandstones, mainly Old Red Sandstone (but including hilly Mesozoic sandstones, etc.), Coal Measures, and the Resistant rocks. Resistant rocks include the old (Pre-Cambrian and Palaeozoic) shales, slate, gneisses and schists, together with later igneous rocks (basalt, etc.). They are all resistant to erosion and solution and, if topography is similar, bear similar vegetation.

The relationship between the different stream types is shown in Fig. 13.1. This resembles the classification on soft rocks (Fig. 12.1) except that because the range of topography is greater, from plain to mountain, there is more separation with flow regime. (Fig. 13.1 is more fully described below.)

The rainfall and its seasonal distribution determine the amount of water reaching the streams as run-off. The force of this water is influenced by the height of the near-by hills, the fall from these hills to the stream bed and the slope of the stream bed. Of these three, the fall from hill to stream is usually the most important for vegetation type. An example is the North Yorks Moors, which are only *c*. 1000' (300 m) high, but have vegetation typical of mountainous areas in their streams because the fall to the valleys is that characteristic of the mountains (over 600'; 200 m). Local conditions may alter this: for example, a stream running down a very steep mountain-side has the same absence of plants as a flatter stream in a steep gully.

224

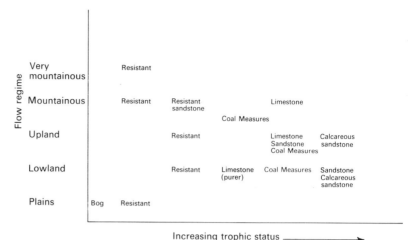

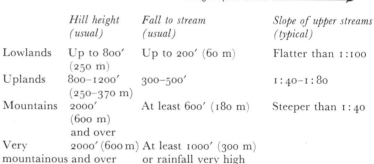

Fig. 13.1. Stream types on hard rocks.

	Hill height (usual)	Fall to stream (usual)	Slope of upper streams (typical)
Lowlands	Up to 800' (250 m)	Up to 200' (60 m)	Flatter than 1:100
Uplands	800–1200' (250–370 m)	300–500'	1:40–1:80
Mountains	2000' (600 m) and over	At least 600' (180 m)	Steeper than 1:40
Very mountainous	2000' (600 m) and over	At least 1000' (300 m) or rainfall very high	

Where water force is low the vegetation may be dense (as it is on the soft lowland rocks). In mountain streams spates can cause severe damage, and these and the swift normal flows usually prevent dense vegetation developing. There is an intermediate situation in the uplands, where vegetation may grow thickly (as in the lowlands) but may also be severely damaged by spates (as in the mountains). Water force is greatest in the sub-group of very mountainous streams, and here river plants are usually absent. As plants are usually much sparser and rarer on hard than on soft rocks, the width–slope patterns quoted in this chapter are less reliable than those in other chapters. Much basic ecological information about these streams is also given in Chapter 11 and in Fig. 10.2.

STREAMS ON RESISTANT ROCKS

Streams on lowlands, plains and bogs

Bog streams occur in blanket bogs in almost flat watersheds (e.g. parts of the Scottish Highlands) and are on, or contain, acid bog peat (Fig. 13.2). They are characterised by having very little force of water. They tend to have little water movement except after rain, to be shallow, and to have water stained by the peat of the land near by. The typical

Fig. 13.2. A stream in blanket bog.

species of these streams are also found in the bogs. For example:

Eriophorum angustifolium   *Potamogeton polygonifolius*

*Menyanthes trifoliata*   *Ranunculus flammula*

(*Sphagnum* spp. occur in similar channels which have still water except after rain.)

On the moorland plains (e.g. Dartmoor), acid peat sediment still reaches the channel, but mineral particles occur as well and the water is less brown. Because the land is less flat, flow is swifter, though the force of water after rain is still low. (Rainfall does not affect the nature of this river community, as streams of this type can be found where there is an annual rainfall of anything between 35″ (87.5 cm) and 80″ (200 cm) per year.) In these stream types sedimentation leads to a decrease, not an increase, in trophic status, because the sediment here includes particles of solid peat (the substrate of lowest nutrient status among those with fine particles). The vegetation (Table 13.1) contains oligotrophic, though not bog species (except in streams intermediate between these and bog streams). For example:

Eleogiton fluitans   *Juncus bulbosus*
Eleocharis acicularis   *Nymphaea alba*

The ubiquitous *Sparganium erectum* is found also. Two species are also present (*Nuphar lutea* and *Sparganium emersum*), which on soft rocks are eutrophic in distribution.

Species diversity is a high as on chalk and clay, some sites containing 10–12 species, and 2–5 species being common. Species diversity in British streams is therefore not influenced by trophic status. (The bog streams are not species-rich, but as they are too shallow for most aquatics

226

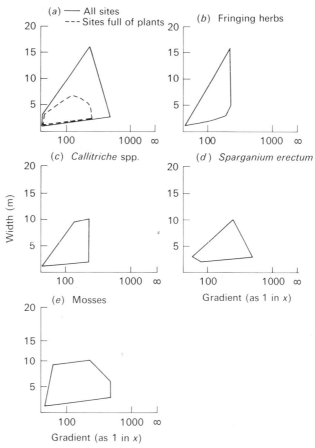

Fig. 13.3. Width–slope patterns of moorland plains streams on Resistant rocks. For details see Fig. 12.3.

their lack of species cannot be attributed to nutrient supply.) Almost half the sites recorded were full of plants, a proportion nearly equal to that for chalk streams and greater than that for all other channel types with flowing water. The smallest tributaries tend to contain fewer species, and submerged species avoid streams liable to dry in summer. The width–slope patterns (Fig. 13.3) show wide distributions for *Callitriche*, *Sparganium erectum* and mosses. The fringing herbs avoid the sites of greatest water force (i.e. the steeper, wider streams) and the other species have even more restricted habitat ranges.

As is typical of hard rocks, the banks have frequent:

Juncus effusus

and, compared to the soft rocks, much less:

Epilobium hirsutum   *Urtica dioica*

Sparganium erectum

This particular hard rock habitat also has frequent:

Phragmites communis

227

The moorland plains streams seldom extend far downstream, as topography usually changes and the river plants change to those species characteristic of upland or mountain streams (e.g. Chapter 11, Fig. 11.4). They occur chiefly in south-west England and south-west Scotland.

In the more truly lowland country, gentle hills mean somewhat swifter water flow and, more importantly, no acid peat. The sediment, therefore, is inorganic, and, as on soft rocks, silt increases trophic status. The streams usually occur as small tributaries in or near the flood plain of a larger river and resemble soft rock streams of a similar size, though they are more likely to be kept clear of plants by water flow.

With little water: *Phalaris arundinacea, Urtica dioica,* and other tall herbs

With some scour: Sparse fringing herbs, with perhaps *Callitriche* or *Sparganium erectum* entering lower

With much scour: No plants

Streams in the uplands and mountains

The streams on Resistant rocks form a continuous series from the bog tributaries to the very mountainous streams. It is, however, convenient to divide them into those with and without frequent destructive spates. The hill streams, from upland to very mountainous, show decreasing vegetation and increasing water force. Some examples are shown in Fig. 13.4 and Figs. 11.2 and 11.3, which also demonstrate downstream variation. The typical species are shown in Table 13.1.

Fig. 13.4. A Resistant rock landscape (mountainous).

TABLE 13.1 *Resistant rocks speceis*[a]

1. *Moor and plains streams (mainly south-west England, south-west Scotland)*
(*a*) Channel species
   Occurring in at least 40% of sites recorded:

   🌱*Callitriche* spp.   🌿 Mosses

   Occurring in at least 20% of sites recorded:

   Eleocharis acicularis   *Potamogeton natans*

   Eleogiton fluitans   *Ranunculus omiophyllus*

   Juncus bulbosus   *Sparganium erectum*

   Occurring in at least 10% of sites recorded:

   Myosotis scorpioides   *Oenanthe crocata*

   Other species which, though sparser, are characteristic:

   Myriophyllum alterniflorum   *Phragmites communis*

   Nuphar lutea   *Sparganium emersum*

   Nymphaea alba

(*b*) Bank species
   Occurring in at least 40% of sites recorded:

   Juncus effusus

   Occurring in at least 10% of sites recorded:

   Eupatorium cannabinum   *Phragmites communis*

   Filipendula ulmaria   *Urtica dioica*

2. *Streams of the north of Scotland (rare elsewhere)*
(*a*) Channel species
   Occurring in at least 40% of sites recorded:

   🌿 Mosses

   Occurring in at least 10% of sites recorded:

   Juncus articulatus

   Occurring in at least 5% of sites recorded:

   Eleocharis palustris

   Other species which, though sparser, are characteristic of some part of the range:

TABLE 13.1 (*cont.*)

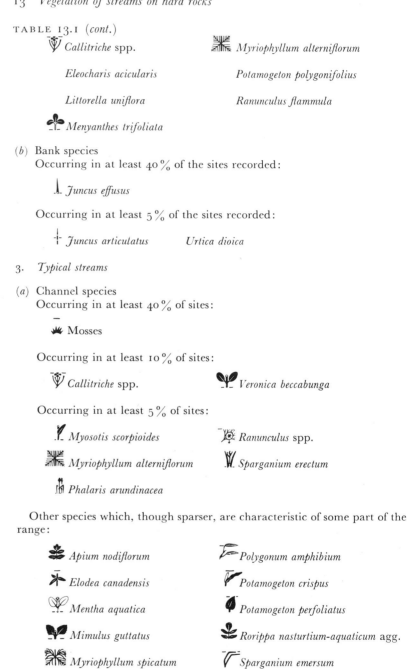

Callitriche spp.   *Myriophyllum alterniflorum*

Eleocharis acicularis   *Potamogeton polygonifolius*

Littorella uniflora   *Ranunculus flammula*

Menyanthes trifoliata

(*b*) Bank species
    Occurring in at least 40% of the sites recorded:

Juncus effusus

Occurring in at least 5% of the sites recorded:

Juncus articulatus   *Urtica dioica*

3.   *Typical streams*

(*a*) Channel species
    Occurring in at least 40% of sites:

 Mosses

Occurring in at least 10% of sites:

Callitriche spp.   *Veronica beccabunga*

Occurring in at least 5% of sites:

Myosotis scorpioides   *Ranunculus* spp.

Myriophyllum alterniflorum   *Sparganium erectum*

Phalaris arundinacea

Other species which, though sparser, are characteristic of some part of the range:

Apium nodiflorum   *Polygonum amphibium*

Elodea canadensis   *Potamogeton crispus*

Mentha aquatica   *Potamogeton perfoliatus*

Mimulus guttatus   *Rorippa nasturtium-aquaticum* agg.

Myriophyllum spicatum   *Sparganium emersum*

(*b*) Bank species
    Occurring in at least 20% of sites recorded:

Filipendula ulmaria   *Phalaris arundinacea*

Juncus effusus   *Urtica dioica*

In the upper reaches of very mountainous streams the water force is excessive, even boulders being moved in spates, so these reaches are usually without angiosperms, mosses or prominent algae. In small streams of lower force, but in otherwise mountain habitats, *Juncus articulatus* is often present towards the sides, and mosses on the larger stones.

Further downstream on very mountainous rivers, where bog peat may enter the channel, there may sometimes be oligotrophic species present. For example:

Eleocharis acicularis   *Littorella uniflora*

✝ *Juncus articulatus*   *Ranunculus flammula*

The upper reaches of mountainous streams usually have sparse mosses (though these may be dense in the occasional stream with less water force, and in groves where trees can stabilise the channel) and very sparse fringing herbs. The upper reaches of upland streams, and the somewhat lower reaches of mountain ones, have more diverse species, and more frequent plants (Table 13.1). Mosses as a group are usually common in such habitats, often more so than angiosperms.

Callitriche spp. (including *Callitriche hamulata*) are fairly frequent, but most populations, and most well-grown stands, are in the smaller streams with shallower water and more stable substrates.

Mimulus guttatus is a fringing herb important on hard rocks, though infrequent on soft ones.

Myriophyllum alterniflorum is an oligotrophic species which is locally frequent, though less common in Wales.

Phalaris arundinacea tends to be on stone or gravel banks flooded in high discharges, dry in low ones.

Veronica beccabunga is much the most frequent fringing herb on hard rocks, except that in south-west England it is replaced by *Apium nodiflorum* (the most frequent fringing herb species on soft rocks).

Downstream, water force is less, and water volume greater. Short-leaved species of

Ranunculus may enter here and may even cover the channel.

Sparganium erectum and

Elodea canadensis also enter here.

Further downstream, species of the penultimate group (i.e. mosses, *Callitriche*, *Mimulus guttatus*, *Myriophyllum alterniflorum*, *Phalaris arundinacea* and *Veronica beccabunga* tend to decrease, and the incoming species are those requiring a larger water volume, less water force, or a higher nutrient status. For example:

Myriophyllum spicatum   *Ranunculus fluitans*

Potamogeton perfoliatus   *Sparganium emersum*

Further eutrophication brings in more eutrophic species.

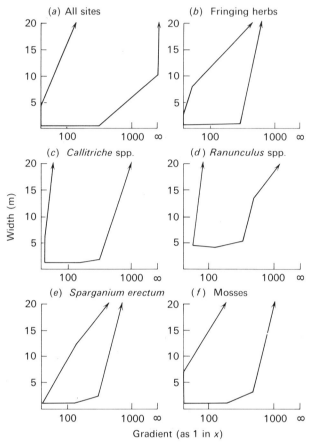

Fig. 13.5. Width–slope patterns of hill streams on Resistant rocks. For details see Fig. 12.3.

Species diversity is low. A few sites have 5–7 species, but most sites with angiosperms have only 1–2 species of these. With increased water force, fewer plants occur. The number of species present in over 5% of sites recorded varies from 12 (on the andesites of northern England and southern Scotland), to none in the very mountainous streams. The proportions of sites full of plants, and of those effectively empty, vary in the same way. The width–slope patterns in Fig. 13.5 show larger habitat ranges than might have been expected and wider ranges than on soft rocks. This is because local conditions may decrease water force and permit plants to grow.

Mosses have a large width–slope pattern, their habitat of large stones being common throughout upland and mountainous streams. *Callitriche* has a quite large pattern, as it grows both on stable medium-grained substrates and temporary silt or sand banks. *Myriophyllum alterniflorum* is an upstream species. Fringing herbs vary somewhat with rock type or topography.

On the banks of mountain streams, the most important plants are:

Filipendula ulmaris   *Phalaris arundinacea*

Juncus effusus   *Urtica dioica*

232

Locally important in the west of England are:

 *Apium nodiflorum*   *Eupatorium cannabinum*

(*Epilobium hirsutum*, which is so frequent on soft rocks, is unimportant in the hills.)

Some controlling factors of plant distribution are illustrated well in R. Tweed (Figs. 11.17 and 11.18). Megget Water, Yarrow Water and R. Tweed rise in hills over 2000′ (600 m). The first two have falls from hill to stream of over 1000′ (300 m) in the upper reaches. Because of this great fall, water force is great, and angiosperms are almost absent and mosses sparse. A few tributaries without spate flow can bear dense fringing herbs and *Callitriche* on silt. R. Tweed has a lesser fall (nearer 800′; 250 m), and most sites recorded were on a channel slope flatter than 1:100. The force of water is thus less and fringing herbs and *Myriophyllum alterniflorum* are present, though sparse. (See also [14] and [15] for a fuller and more detailed study of R. Tweed.)

The Cheviot hills, lower in the catchment, are also over 2000′ (600 m), but in contrast to those mentioned above their rainfall is less, 40–50″ (100–125 cm) instead of 50–60″ (125–150 cm), and, more importantly, Oxnams Water, Kale Water, etc. have falls to the stream of only 200–400′ (60–120 m). There is less moor than farther upstream, so perhaps a rather richer sediment. The main difference, however, is in the force of the water, which is less because the fall (and perhaps partly because the rainfall) is less. The streams bear a good vegetation of:

Callitriche spp.   *Ranunculus* spp. (short-leaved)

Phalaris arundinacea   Fringing herbs

Resistant rocks outcrop over most of the highland and Southern Uplands of Scotland, Wales, South-west England, the Lake District and part of the Pennines.

COAL MEASURES STREAMS

Coal Measures streams are usually polluted from the coal mines and associated industries. In the unpolluted areas the vegetation is sparse, and so is not easy to classify. Hill types range from lowland to nearly mountainous, though falls from hill to channel seldom exceed those of the upland category. Lowland tributaries can contain abundant vegetation:

With intermittent water: Tall emerged aquatics
With perennial flow: *Callitriche* spp., *Lemna minor* agg., *Potamogeton* spp., *Sparganium erectum*

In the hills, mosses are much less frequent than on Resistant rocks (Table 13.2, Fig. 13.6). *Callitriche* spp. are most frequent in smaller streams, as usual. *Sparganium erectum* is common. *Potamogeton pectinatus* is

TABLE 13.2 *Coal measures streams*[a]

(a) *Channel species*

Undamaged channels contain higher frequencies

    Occurring in at least 20% of sites recorded:

 Mosses

    Occurring in at least 5% of sites recorded:

      Callitriche spp.   *Sparganium erectum*

      Potamogeton pectinatus   *Veronica beccabunga*

    Other species which, though sparser, are characteristic:

      Alisma plantago-aquatica   *Enteromorpha* sp.

(b) *Bank species*

    Occurring in at least 40% of sites recorded:

      Urtica dioica

    Occurring in at least 20% of sites recorded:

      Epilobium hirsutum   *Eupatorium cannabinum*

[a]See notes to Table 12.1.

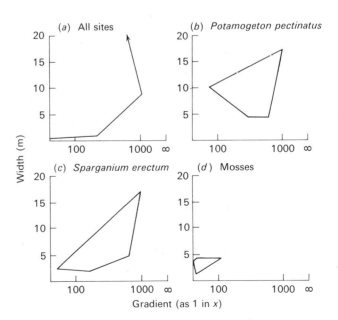

Fig. 13.6. Width–slope patterns of Coal Measures streams. For details see Fig. 12.3.

fairly frequent, except in streams with little water. This last species is characteristic of polluted sites, but it is not clear if it grows well in completely unpolluted parts of Coal Measures streams. The species frequency is low. Over half the sites recorded are effectively empty, though about 10% of the sites are full of plants. Bank plants are infrequent also, but of this group *Epilobium hirsutum* and *Urtica dioica* are the most frequent, and *Juncus effusus* is not uncommon.

The main outcrops of Coal Measures are in the Pennines, South Wales and Southern Scotland.

LIMESTONE STREAMS

Fig. 13.7. A hard limestone landscape (mountainous).

The most widespread hard limestone is Carboniferous limestone (Fig. 13.7). It is a less pure limestone than chalk and most streams have a little non-limestone in the catchment (e.g. Millstone Grit, Boulder Clay). These streams have a somewhat higher trophic status for plants than does chalk and are closer to oolite streams (Chapter 12; Table 13.3). A lowland stream can be almost indistinguishable from a chalk one (compare Figs. 11.7 and 11.14), and upland ones resemble those on upland sandstone (see below). In the upland streams fringing herbs are luxuriant, but because of the spates they are usually confined to the sides of the channel, growing in a carpet across the centre only where the water force is low. Several submergents may be frequent also.

TABLE 13.3 *Hard limestone species[a]*

(a) *Channel species*
   Upland channels have higher frequencies, mountain ones, lower.
   Occurring in at least 20% of sites recorded:

   🌱 *Veronica beccabunga*   🌿 Mosses

235

TABLE 13.3 *(cont.)*

Occurring in at least 10% of sites recorded:

Ψ *Sparganium erectum*

Occurring in at least 5% of sites recorded:

Ψ *Callitriche* spp.   ⚘ *Rorippa nasturtium-aquaticum* agg.

❀ *Ranunculus* spp.

Other species which, though sparser, are characteristic:

⚘ *Apium nodiflorum*   ⚶ *Myriophyllum spicatum*

⋔ *Groenlandia densa*   ⚘ *Zannichellia palustris*

❦ *Mimulus guttatus*

(b) Bank species

Occurring in at least 40% of sites recorded:

Urtica dioica

Occurring in at least 20% of sites recorded:

Filipendula ulmaria   ⚘ *Phalaris arundinacea*

Occurring in at least 10% of sites recorded:

⌊ *Juncus effusus*   ⚘ *Rorippa nasturtium-aquaticum* agg.

[a]See notes to Table 12.1.

Rorippa nasturtium-aquaticum agg. is the fringing herb which is the most likely to form carpets, to spread very rapidly and to be washed away in spates.

Callitriche spp. and *Ranunculus* spp. enter further downstream than in the lowlands, only where water volume increases and the flow is somewhat less fierce.

Groenlandia densa is characteristic, though infrequent. (It also occurs on hard limestone near Munich [48].) On soft rocks *Groenlandia densa* and *Potamogeton crispus* have a similar habitat range (see Chapter 12), but on hard rocks *Groenlandia densa* is found on limestone while *Potamogeton crispus* is most frequent on sandstone and Resistant rocks.

In the R. Clyde system there are streams at *c.* 1000′ (300 m) with a low fall from hill to stream of only *c.* 200′ (60 m) so that flow is slow and thick silt may accumulate. The channels may be choked with, for example:

Potamogeton alpinus   ⋁ *Sparganium emersum*

⊐ *Potamogeton natans*   Ψ *Sparganium erectum*

(*Ranunculus* enters where flow is swifter)

Mountain-type vegetation occurs in lower hills here more than on any other rock type. In the North Yorks Moors it is found with hills only 1000′ (300 m) to 1500′ (450 m) high, but the more important determining factor is the fall from hill to stream, which at *c*. 800′ (250 m) is well into the mountainous category. Semi-mountainous streams can bear dense *Ranunculus* (including *Ranunculus calcareus*), some fringing herbs and some semi-eutrophic species (such as *Myriophyllum spicatum*) (Table 13.3). The same community, but with more fringing herbs, also develops in lower hills, where although storm flows are very deep their force is low (Fig. 13.8). Mountain streams, in contrast, are almost barren, even more so than those on Resistant rocks.

Downstream, trophic status increases, and in large rivers a semi-eutrophic vegetation may develop in flatter reaches where water force is lower, with, for example:

Groenlandia densa *Zannichellia palustris*

Myriophyllum spicatum

A river with mountain limestone tributaries entering a flat alluvial plain was described in Chapter 11 (Fig. 11.9). In the plain the streams, which are still small, have dominant *Ranunculus*, though the fluctuations in water level prevent fringing herbs from growing well.

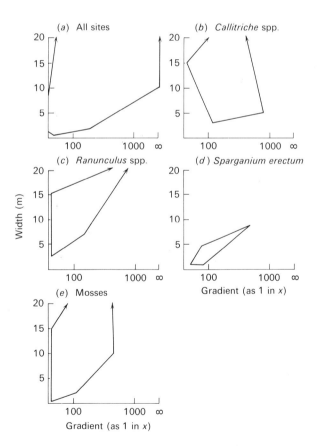

Fig. 13.8. Width–slope patterns of hard limestone streams. For details see Fig. 12.3.

Hard limestone streams are species-poor, rarely having more than 3–4 angiosperm species at the more spatey sites and most places with angiosperms having only one species. Half the sites recorded are effectively empty, and only 12% of them are full of plants. The width–slope pattern of all streams is large (Fig. 13.8), and sites full of plants occupy a much larger area than usual (Fig. 6.5). Some outcrops are in areas with low water force, and here the fertile rock and stable substrate encourage plant growth. The patterns for *Callitriche* and *Ranunculus* are fairly large. On the banks, *Juncus effusus* is less important than on the Resistant rocks and *Urtica dioica* more important. This reflects the fact that limestone is much the more fertile rock.

The variation of vegetation with force of water is greater than on Resistant rocks or Coal Measures. This is probably because limestone is a more fertile rock. Fierce flow, however, overrides any other factor, so the streams range from empty, through mesotrophic limestone, to a more nearly eutrophic habitat. The variation with geographical location is also considerable, but is mainly due to variations in water force and volume. For example, in the more southerly areas it so happens that water force is lower, and so vegetation is dense in both lowland and upland stream types, with *Ranunculus* being dominant. Most of the streams in north-east England, by contrast, are mountainous and empty, while those in north-west England have less water force and sparse vegetation including fringing herbs and *Groenlandia densa*, as well as *Ranunculus*, etc. The Durness limestone bears only small streams, the chief species being emergents which include:

Caltha palustris	┼ *Juncus articulatus*
Filipendula ulmaria	*Mimulus guttatus*

The main outcrops of hard limestone are the Carboniferous limestone of the Pennines and northern England. The most southerly region is the Mendip Hills, and the most northerly, the Durness limestone, outcrops in the far north-west of Scotland.

SANDSTONE STREAMS

Old Red Sandstone is the most widespread hard sandstone, though other types are locally important (Fig. 13.9). There is more silt and sand entering these streams than those on other hard rocks, and consequently there is a greater quantity of sediment deposited in sheltered areas and when flow slackens. Appreciable silting also extends farther upstream, and in fact sandstone streams with a fall of 600′ (200 m) from hill to stream bed are likely to have an upland vegetation, though on other rock types they would have a mountain one. These streams are therefore more eutrophic. Because sandstone also tends to form lower hills, the rivers contain relatively more reaches where silting can occur than do those on other hard rock landscapes.

Fig. 13.9. A hard sandstone landscape (upland).

In relation to limestone, sandstone streams may either be more nutrient-poor (when there is swift flow and no silting) or more nutrient-rich (when there is slower flow and silting). Mosses are less common in sandstone than other hill streams (Table 13.4), presumably because sandstone beds are more easily eroded. Oligotrophic species are rare, if not absent.

TABLE 13.4   *Hard sandstone species[a]*

1.   Except far north of Scotland
(*a*) Channel species
Upland channels have higher frequencies, mountain ones, lower.
  Occurring in at least 10 % of sites recorded:

Callitriche spp.   *Veronica beccabunga*

Sparganium erec'   Mosses

  Occurring in at least 5 % of sites recorded:

Phalaris arundinacea   *Rorippa nasturtium-aquaticum* agg.

Ranunculus spp.

  Other species which, though sparser, are characteristic of part of the range:

Mimulus guttatus   *Potamogeton crispus*

Myriophyllum alterniflorum   *Potamogeton natans*

Myriophyllum spicatum   *Sparganium emersum*

(*b*) Bank species
  Occurring in at least 20 % of sites recorded:

Epilobium hirsutum   *Phalaris arundinacea*

Filipendula ulmaria   *Urtica dioica*

239

TABLE 13.4 (*cont.*)

Occurring in at least 10 % of sites recorded:

↓ *Juncus effusus*

2. *Caithness, etc.*
(*a*) Channel species
Occurring in at least 20 % of sites recorded:

⊥ *Juncus articulatus*   ⸜ Mosses

⋓ *Sparganium erectum*

Occurring in at least 10 % of sites recorded:

⊻ *Callitriche* spp.   ❧ *Mimulus guttatus*

† *Equisetum* spp.   ⸖ *Myosotis scorpioides*

⸙ *Mentha aquatica*

(*b*) Bank species
Occurring in at least 40 % of sites recorded:

Filipendula ulmaria   ⅃ *Juncus effusus*

Occurring in at least 20 % of sites recorded:

Caltha palustris   ⸗ *Phalaris arundinacea*

⅄ *Iris pseudacorus*

⸖ *Myosotis scorpioides*   *Urtica dioica*

Occurring in at least 10 % of sites recorded:

Ranunculus flammula   ⋓ *Sparganium erectum*

Upland brooks typically have a rich fringing herb vegetation, and, if spates are low, these can grow across the bed. In contrast to chalk streams, however, these carpets are usually semi-emerged in shallow water. As on soft sandstone, *Callitriche* is more frequent than *Ranunculus*, though both are important. As water volume and trophic status increase downstream, *Ranunculus* increases and eutrophic species become prominent.

Myosotis scorpioides and *Veronica beccabunga* are the most important fringing herbs.

Callitriche spp. enter far upstream.

Ranunculus spp. enter below *Callitriche*, and usually occur only if other submergents are present. Short-leaved forms occur above long-leaved ones. This can dominate downstream.

Potamogeton spp., most often *Potamogeton crispus*, and *Sparganium emersum* enter downstream, and usually occur only if other submergents are present too.

Nuphar lutea, Potamogeton pectinatus and *Schoenoplectus lacustris* can grow well in lower reaches if flow is quiet and silting considerable.

The effect of the fall from hill to stream is shown well by a comparison of R. Teme and R. Lugg, two adjoining rivers in the Welsh Marches. R. Teme has the larger catchment, and the fall in the lower reaches is *c.* 400′ (120 m). *Ranunculus* is the dominant species. R. Lugg (Fig. 11.18) has a quieter, more silty habitat in the lower reaches, with a fall of only *c.* 200′ (60 m). Here eutrophic species are the most important, with *Ranunculus* only sparse and confined to the swifter parts.

R. Tone (Fig. 11.5) is a typical upland stream, except that there is an unusually large proportion of its length in the flood plain and, as it is sited in south-west England, *Apium nodiflorum* is the most frequent fringing herb.

The upland streams are slightly more species-rich than those on other hard rocks, with some sites having 6 or more species and the most frequent number (in sites where there are any plants) being 2. Ten species are found in at least 5% of the sites recorded. The proportions of full and empty sites are similar to those on limestone. The width–slope patterns (Fig. 13.10) include more of the flatter sites than those of other hard rocks.

Mosses have a wide habitat range, even though they are so sparse. *Callitriche* also has a large pattern, as on soft sandstone and chalk. *Sparganium erectum* is rather more restricted, presumably because of swifter flow. Fringing herbs decrease in sites liable to deep flooding. *Potamogeton crispus* and *Sparganium emersum* avoid the more extreme conditions. On the banks the more lowland plants (*Epilobium hirsutum, Urtica dioica*) are prominent, while the more mountainous species (*Juncus effusus*) is still frequent. Fringing herbs are frequent, as they are on chalk.

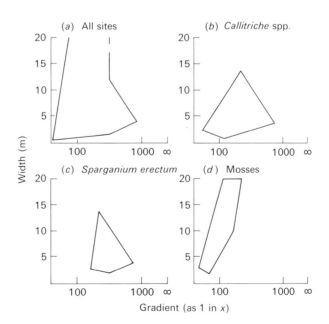

Fig. 13.10. Width–slope patterns of hard sandstone. For details see Fig. 12.3.

TABLE 13.5   *Calcareous sandstone species[a]*

(a) Channel species
Occurring in at least 40% of sites recorded:

🌿 Mosses

Occurring in at least 20% of sites recorded:

🌿*Veronica beccabunga*

Occurring in at least 10% of sites recorded:

🌿*Myosotis scorpioides*   🌿*Sparganium erectum*

Occurring in at least 5% of sites recorded:

🌿*Potamogeton natans*   🌿*Rorippa nasturtium-aquaticum* agg.

🌿*Ranunculus* spp.   🌿*Sparganium emersum*

Other species which, though sparser, are characteristic:

🌿*Callitriche* spp.   🌿*Polygonum amphibium*

🌿*Elodea canadensis*   🌿*Schoenoplectus lacustris*

🌿*Phalaris arundinacea*   🌿*Veronica anagallis-aquatica*

(b) Bank species
Occurring in at least 40% of sites recorded:

Urtica dioica

Occurring in at least 20% of sites recorded:

🌿*Epilobium hirsutum*   🌿*Juncus effusus*

Filipendula ulmaria   🌿*Phalaris arundinacea*

Occurring in at least 10% of sites recorded:

🌿*Sparganium erectum*

[a]See notes to Table 12.1.

Where Old Red Sandstone forms a nearly lowland topography, slow flow and silting increase, and there are also increases in:

🌿*Callitriche* spp.   🌿*Potamogeton natans*

🌿*Elodea canadensis*   🌿*Sparganium emersum*

The low-lying parts of Caithness should be considered separately from other areas. These have low water force but because the water is stained very dark, plants cannot grow in deep water. Vegetation is sparse, and in rivers plants are restricted almost entirely to the shallow edges. The species present (Table 13.4) show that they are influenced both by the sandstone and the effects associated with a combination of very mountainous conditions (see above) and low scour.

Mountain streams are few. Their substrate is somewhat erodible and any sediment is unstable, so plants cannot anchor easily. No sites recorded were full of plants and over 80% were empty. Mosses are frequent where large stones can occur (in a small width–slope pattern, Fig. 13.9), but other plants are rare, though they do include oligotrophic species (*Myriophyllum alterniflorum*). No typical site bears submerged angiosperms.

The fall from hill to stream is again the most important factor determining the plants of the mountain streams. For instance, one river (R. Rhee, Severn) has abnormally little vegetation considering that its hills are only 1700′ (520 m) high and that the land is fertile grassland. The fall, however, is *c.* 1000′ (600 m), which is well into the mountainous category.

Hard sandstones outcrop mainly in west England, the Welsh Marches, northern England and much of central and eastern Scotland.

CALCAREOUS AND FELL SANDSTONE STREAMS

Compared with the other hard sandstones these streams contain more lime, and usually rise in lower hills. The vegetation is more eutrophic (Table 13.5) and is, of course, denser in streams not liable to spate.

Fringing herbs comprise nearly half of the more frequent species and *Rorippa nasturtium-aquaticum* agg. is particularly common.

Callitriche spp. are unusually low for sandstone.

Potamogeton natans is frequent, as is usual on sandstone.

Ranunculus spp. are frequent, but tend to be in flatter parts (Fig. 13.8).

Sparganium emersum is the most frequent eutrophic species, as is usual on sandstone.

Sparganium erectum is frequent, but tends to be in flatter parts. Mosses are frequent, but tend to be in flatter parts.

On the banks, fringing herbs are much less common than on pure sandstone.

STREAMS ON MIXED CATCHMENTS

Many large rivers have more than one rock type in their catchment, and as is the case with soft rocks, mixed rock types lead to mixed vegetation types. However, as hard rocks have fewer species and less total vegetation than soft rocks, the effects on the plants are less, and are less easy to detect. In the hills, Boulder Clay is usually in scattered patches, too small, in comparison to the stream discharge, to affect the plants. Where, however, thick Boulder Clay covers most of the catchment, as in part of Anglesey, the vegetation is that of clay (see Chapter 12). The coarser drifts, e.g. moraines, have little effect on stream vegetation. When streams pass over different types of Resistant rocks, topography may vary, and if so, plants vary also. Granite and basalt, for instance, form flatter country, and their streams contain more:

Callitriche spp.   *Glyceria maxima*

Streams which have flowed over both sandstone and Resistant rocks are necessarily usually large, and perhaps because of this have more *Ranunculus* and less fringing herbs than streams exclusively on one of these rock types.

Streams frequently rise on Resistant rocks and later flow on to sandstone. When this happens, some species decrease with decreasing sandstone, for example:

Callitriche spp.   *Sparganium erectum*

Potamogeton crispus

In mixed catchments, some species (excluding those whose distribution is related to stream size) are more frequent than they are on either rock type separately. Examples of such species are:

Elodea canadensis   *Sparganium emersum*

Potamogeton natans

On the banks, an increase in the proportion of Resistant rocks produces an increase in:

Juncus effusus   *Phalaris arundinacea*

and an increase in sandstone an increase in:

Epilobium hirsutum   *Urtica dioica*

There are fewer streams which rise on sandstone and then cross on to Resistant rocks. In these, with decreasing proportions of sandstone, *Rorippa nasturtium-aquaticum* agg. decreases. *Ranunculus* increases downstream, as it does on either rock type separately, but probably because of the increasing water volume and substrate stability rather than because of the mixed catchment. Sandstone and Resistant rock streams may flow into each other, and the vegetation downstream of their confluence depends on the total proportion of the sandstone and on the flow regime, as described in Chapter 11. If over half the catchment is on sandstone then the vegetation is usually indistinguishable from that of sandstone streams. Not many streams have catchments of both calcareous sandstone and Resistant rocks, but those that do have a more diverse flora than streams entirely on calcareous sandstone, with more:

Elodea canadensis   *Sparganium erectum*

Myosotis scorpioides   *Enteromorpha* sp.

Ranunculus spp.

and with less mosses. However, at least part of this difference is due to the larger size of the stream channels on mixed catchments. In streams on mixed limestone and Resistant rock, mosses increase with the proportion of Resistant rocks. The other differences observed (on mixed catchments there is less *Rorippa nasturtium-aquaticum* agg. and more *Ranunculus* and perhaps *Sparganium emersum*) may well be due solely to the increase in channel size.

The general effects of mixing rock types in different proportions are described more fully in Chapter 11.

STREAMS IN FLOOD PLAINS

Hill rivers have great fluctuations in depth in their flood plains. If silting is considerable then the substrate will be unstable, at least near the hills where water movement is greater, and so the plants present must be those species able to anchor securely to the bank. For example:

Polygonum amphibium   *Phragmites communis*

On deep stable silt the vegetation is likely to be lowland eutrophic in character (see Chapter 12, p. 221). On a gravel bed with a swifter flow, likely species are:

Elodea canadensis   *Sparganium emersum*

Ranunculus fluitans

STREAM CLASSIFICATION

The topography of streams on hard rocks varies greatly, and so their vegetation is strongly influenced by flow regimes (Fig. 13.1). For example, in many mountain streams the fierce flow keeps plants almost absent regardless of the type of rock in the catchment. As water force decreases, though, geological differences become increasingly important influences on vegetation.

One such influence that varies with rock type is silting. Silting increases trophic status. Consequently, as the water force decreases on each rock type and silt can be deposited, the streams become of higher trophic status (see Fig. 13.1). There are two exceptions to this general pattern. The first arises because the flatter landscapes on Resistant rock vary from acid bogs to fertile arable country, and as the sediment entering the stream from bogs is more oligotrophic than that from arable land, here it is the soil and land use that influence stream vegetation. The reasons for the other exception are less clear-cut. The Cheddar Gorge stream is on limestone with a lowland chalk vegetation, but, unlike the much longer upland limestone streams of north-west England, has no eutrophic species. However, this could well be simply because the Cheddar stream is too short to show downstream eutrophication. The colour of the water is also determined by rock type and varies from clear to very dark brown (peat-stained). It is, however, only in northern Scotland that large streams are consistently sufficiently dark for this to have a regional effect on the plants.

14

Vegetation of channels with little flow

Dykes have little water movement, no scour and a fine substrate. Tall emerged monocotyledons dominate in shallow channels, and deeper or recently dredged ones can have a very diverse and variable vegetation dominated by submerged and floating plants. Undisturbed canals have a similar vegetation to dykes. Much management, whether by herbicides, boat traffic or other aspects of land or channel use, leads to species-poor watercourses.

Low-lying alluvial plains must have man-made drainage channels if the land is to bear good pasture or arable. These drainage channels have different names in different parts of Britain, but here we will adopt the names used in the Fenland, the largest such plain in the country. There the small channels between fields are called 'dykes', and the larger ones which collect dyke water and carry it to the rivers are 'drains'. (Channels used for navigation are called 'lodes', but their vegetation can be combined with that of drains.) Other common names are rhynes, in the west of England and South Wales (pronounced, and sometimes spelt, reens), and in the south of England, sewers. There is no effective scour from flow in these channels. Water may move freely by gravity, but in the larger plains there is usually a complex system of flood gates or pumps to ensure drainage. After heavy rain the water level may rise considerably but the water moves too slowly to do more than float off non-anchored plants.

The other main type of flat man-made channel is the canal, used for

Fig. 14.1. Alluvial plain with dykes, and marsh nearer the river.

247

navigation. Canals usually lie above ground water level and so must have non-porous beds and sides, usually of clay, to prevent the water draining away. Their water levels vary less than those of drains, as they are not flood-relief systems. They receive little run-off, and excess water can drain from them. All these man-made channels have negligible flow, and require constant maintenance and management. Canals have low banks, because water level never fluctuates much, while dykes and drains may have high or low banks.

DYKES AND DRAINS

Dykes occur throughout Britain (Fig. 14.1; Table 14.1). The smaller and shallower ones often dry in summer. These typically bear tall emerged monocotyledons, either only one species or several, each dominating for short stretches. (Fig. 14.2).

Typha latifolia

Phragmites communis

Carex acutiformis

Fig. 14.2

Glyceria maxima is most frequent in dykes which are hardly more than shallow depressions, and outside East England.

Phragmites communis is the commonest in the Fenland, Broadland and Romney Marsh, and in slightly deeper dykes than *Glyceria maxima*.

Phalaris arundinacea is rather less frequent than *Glyceria maxima* or *Phragmites communis* and is usually dominant only in dykes which are dry for most of the summer.

Epilobium hirsutum, Urtica dioica and other non-aquatics are found in dykes which are dry for most of the year.

Carex spp., *Scirpus* spp., *Sparganium erectum* and *Typha* spp. are infrequent and less likely to dominate long stretches of dyke.

Dykes which occasionally dry for short periods may bear a few submerged or floating plants, for example:

Callitriche spp.   *Ranunculus* spp. (short-leaved)

Lemna spp.

248

TABLE 14.1   *Dyke and drain species*[a]

---

1. The Fenland
(mainly arable farmland)
Undamaged channels have higher frequencies
(i) Peat
(*a*) Channel species
    Occurring in at least 20 % of sites recorded:

    Glyceria maxima   *Phragmites communis*

    Lemna minor agg.   *Sparganium erectum*

    Occurring in at least 10 % of sites recorded:

    Alisma plantago-aquatica   *Ceratophyllum demersum*

    Callitriche spp.   *Nuphar lutea*

    Carex acutiformis   *Enteromorpha* sp.

    Other species which, though sparser, are characteristic:

    Elodea canadensis   *Sagittaria sagittifolia*

    Myriophyllum spicatum   *Sparganium emersum*

    Polygonum amphibium   *Typha* spp.

    Potamogeton natans

    and, in species-rich smaller dykes:

    Hottonia palustris   *Potamogeton* spp. (grass-leaved)

    Myriophyllum verticillatum   *Ranunculus* spp. (short-leaved)

(*b*) Bank species
    Occurring in at least 20 % of sites recorded:

    Glyceria maxima   *Urtica dioica*

    Phragmites communis

    Occurring in at least 10 % of sites recorded:

    Epilobium hirsutum   *Juncus effusus*

[a] See notes to Table 12.1.

TABLE 14.1 (*cont.*)

(ii) Silt
(*a*) Channel species
    Occurring in at least 40 % of sites recorded:

        Phragmites communis

    Occurring in at least 20 % of sites recorded:

        Agrostis stolonifera   *Glyceria maxima*

        Callitriche spp.

    Occurring in at least 10 % of sites recorded:

        Lemna minor agg.   *Ranunculus* spp.

        Potamogeton pectinatus   *Enteromorpha* sp.

    Other species which, though sparser, are characteristic:

        Alisma plantago-aquatica   *Potamogeton perfoliatus*

        Apium nodiflorum   *Rorippa nasturtium-aquaticum*
                                    agg.

        Ceratophyllum demersum   *Sagittaria sagittifolia*

        Myriophyllum spicatum   *Sparganium emersum*

        Nuphar lutea

(*b*) Bank species
    Occurring in at least 40 % of sites recorded:

        Urtica dioica

    Occurring in at least 20 % of sites recorded:

        Phragmites communis

    Occurring in at least 10 % of sites recorded:

        Glyceria maxima   *Phalaris arundinacea*

(iii) Clay (at edge of plain)
(*a*) Channel species
    Occurring in at least 20 % of sites recorded:

        Apium nodiflorum   *Rorippa nasturtium-aquaticum*
                                    agg.

        Callitriche spp.   *Sparganium erectum*

        Glyceria maxima   *Veronica beccabunga*

TABLE 14.1 (*cont.*)

🌾 *Phalaris arundinacea*　　　　　　　　⌒ *Enteromorpha* sp.

🌿 *Phragmites communis*

Occurring in at least 10 % of sites recorded:

🌾 *Agrostis stolonifera*　　　　　　　🌿 *Veronica anagallis-aquatica*

🌱 *Alisma plantago-aquatica*

Other species which, though sparse, are characteristic:

🌱 *Myosotis scorpioides*　　　　　　　🌿 *Polygonum amphibium*

🌿 *Nuphar lutea*

2. *Somerset Levels*
(mainly pasture)
(i) Peat
(*a*) Channel species
　　Occurring in at least 40 % of sites recorded:

🌿 *Lemna minor* agg.

Occurring in at least 20 % of sites recorded:

🌿 *Glyceria maxima*　　　　　　　🌿 *Sparganium erectum*

🌿 *Lemna trisculca*

Occurring in at least 10 % of sites recorded:

🌾 *Agrostis stolonifera*　　　　　　　🌿 *Lemna polyrhiza*

🌱 *Alisma plantago-aquatica*　　　　　🌾 *Phalaris arundinacea*

🌿 *Ceratophyllum demersum*　　　　　🌿 *Sagittaria sagittifolia*

Hydrocharis morsus-ranae (abundant stands)

(*b*) Bank species

Occurring in at least 20 % of sites recorded:

🌿 *Glyceria maxima*　　　　　　　　　　*Urtica dioica*

Occurring in at least 10 % of sites recorded:

🌾 *Phalaris arundinacea*

TABLE 14.1 *(cont.)*

(ii) Silt
(*a*) Channel species
Occurring in at least 40% of sites recorded:

ᴛ *Lemna minor* agg.

Occurring in at least 20% of sites recorded:

Ⴉ *Ceratophyllum demersum*   *Sagittaria sagittifolia*

Nuphar lutea   *Sparganium emersum*

Phalaris arundinacea   *Sparganium erectum*

Potamogeton pectinatus   *Enteromorpha* sp.

Occurring in at least 10% of sites recorded:

Callitriche spp.   *Lemna polyrhiza*

Elodea canadensis   *Potamogeton natans*

Hydrocharis morsus-ranae (sparse
stands)

Other species which, though sparser, are characteristic:

Alisma plantago-aquatica   *Schoenoplectus lacustris*

Phragmites communis

(*b*) Bank species
Occurring in at least 40% of sites recorded:

Phalaris arundinacea

Occurring in at least 20% of sites recorded:

Phragmites communis

Occurring in at least 10% of sites recorded:

Epilobium hirsutum   *Urtica dioica*

Glyceria maxima

3.   *Romney Marsh*
(i) Silty, mainly grassland
(*a*) Channel species
Occurring in at least 40% of sites recorded:

ᴛ *Lemna minor* agg.   *Phragmites communis*

TABLE 14.1 (*cont.*)

Occurring in at least 20% of sites recorded:

🌱 *Agrostis stolonifera*   🌿 *Glyceria maxima*

🌱 *Alisma plantago-aquatica*   〰 *Enteromorpha* sp.

(10% level unsatisfactory)

Other species which, though sparse, are characteristic:

🌱 *Butomus umbellatus*   ⚘ *Lemna trisulca*

🌿 *Callitriche* spp.   *Nymphoides peltata*

🌿 *Ceratophyllum demersum*   🌿 *Potamogeton pectinatus*

Hydrocharis morsus-range   🌾 *Sparganium erectum*

(*b*) Bank species
Occurring in at least 40% of sites recorded:

🌾 *Phragmites communis*   *Urtica dioica*

Occurring in at least 20% of sites recorded:

🌿 *Epilobium hirsutum*

(10% level unsatisfactory)

(ii) Sandy, mainly arable
(*a*) Channel species
Occurring in at least 40% of sites recorded:

⚘ *Lemna minor* agg.   〰 *Enteromorpha* sp.

🌾 *Phragmites communis*

Occurring in at least 20% of sites recorded:

🌿 *Ceratophyllum demersum*   🌿 *Potamogeton pectinatus*

⚘ *Lemna trisculca*   🌿 *Scirpus maritimus*

(10% level unsatisfactory)

Other species which, though sparse, are characteristic:

🌱 *Alisma plantago-aquatica*

(*b*) Bank species
Occurring in at least 40% of sites recorded:

🌾 *Phragmites communis*   *Urtica dioica*

(10% level unsatisfactory)

TABLE 14.1 (*cont.*)

4. *Pevensey Levels and nearby*
(silty, mainly grassland)
(*a*) Channel species
    Occurring in at least 40% of sites recorded:

    Lemna minor agg.   *Phragmites communis*

Occurring in at least 20% of sites recorded:

    Glyceria maxima   *Lemna polyrhiza*

    Hydrocharis morsus-ranae   *Sparganium erectum*

    Lemna minor agg.

Occurring in at least 10% of sites recorded:

    Agrostis stolonifera   *Polygonum amphibium*

    Callitriche spp.   *Sagittaria sagittifolia*

    Elodea canadensis

Species which, though sparser, are characteristic:

    Ceratophyllum demersum   *Potamogeton natans*

5. *Norfolk, clay areas*
    (mainly grassland, some species locally very frequent – noted as local)
(*a*) Channel species
    Occurring in at least 40% of sites recorded:
    Phragmites communis

Occurring in at least 20% of sites recorded:

    Callitriche spp.   *Myriophyllum spicatum*

    Lemna minor agg.

Occurring in at least 10% of sites recorded:

    Apium nodiflorum   *Rorippa nasturtium-aquaticum* agg.

    Hippuris vulgaris (local)   *Scirpus maritimus* (local)

    Myriophyllum verticillatum (local)   *Sparganium erectum*

    Phalaris arundinacea   *Enteromorpha* sp. (local)

    Potamogeton pectinatus

6. *Other areas*
Prominent species in dykes of:

Dykes and drains

S. Wales (silt, grassland): *Alisma plantago-aquatica*, *Callitriche* spp., *Glyceria maxima*, *Hydrocharis morsus-ranae*, *Lemna minor* agg., *Potamogeton pectinatus*, *Sparganium erectum*

Essex marshes (clay, arable and grass): *Alisma plantago-aquatica*, *Apium nodiflorum*, *Callitriche* spp., *Lemna minor* agg., *Phalaris arundinacea*, *Phragmites communis*, *Rorippa nasturtium-aquaticum*, *Scirpus maritimus*, *Sparganium erectum*, *Enteromorpha* sp.

Yorkshire plains (clay, silt, etc.): *Callitriche* spp., *Epilobium hirsutum*, *Lemna minor* agg., *Rorippa nasturtium-aquaticum* agg., *Sparganium emersum*, *Veronica anagallis-aquatica*, *Veronica beccabunga*

Strine plain, Salop (silt): *Agrostis stolonifera*, *Glyceria maxima*, *Juncus effusus*, *Phalaris arundinacea*

Dykes and drains which are normally flooded and are unpolluted show much variation in vegetation as a result of different habitat factors. These factors and their effects are described below.

Depth

In small channels up to 0.5 (–1) m deep, tall emerged monocotyledons are not confined to the sides by scour and flow as they are in flowing streams, and are able to dominate. Thus, if the channels are to be kept clear, human interference is necessary to restrict these plants to the side fringes. So when these channels bear submerged and floating plants, as most do, the occurrence of these different types of vegetation is dependent on management. On steep banks of deep drains it is impossible for them to spread into the channel, so killing them with herbicides (as is often done) is quite unnecessary (Fig. 14.3; also see Chapters 19 and 21).

The submerged and floating dyke species include both plants more frequent in shallow water (e.g. *Callitriche* spp.) and ones more frequent in deep water (e.g. *Nuphar lutea*). If the water level is stable, a species-rich flora of submerged plants can be present in only *c.* 25 cm of water. Usually, however, water level fluctuates, and species-rich communities occur in water normally at least 0.5 m deep. If a dyke does dry in summer, but conditions are otherwise suitable for a species-rich community, the plants present will be short emergents such as:

Alisma spp.   *Rumex* spp.

Juncus spp.   Short grasses

Ranunculus scleratus

Width

Channel width as such has little effect on the plants, though a wide channel is also more likely to be deep. If fringing herbs are present they are more likely to become luxuriant in narrow channels. They are emerged, and so anchored only at the banks. Loosely anchored patches are blown away in strong winds and large patches are more likely to

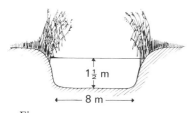

Fig. 14.3

255

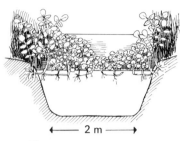

Fig. 14.4

remain in place if anchored to both sides of the dyke (Fig. 14.4). *Rorippa amphibia* and *Rorippa nasturtium-aquaticum* agg. are the most frequent species.

Turbidity

Turbid water decreases the light reaching the bed, and so prevents submerged plants from growing in deep channels. Turbidity is more important in drains than streams because drains are usually steep-sided and, unlike most streams, do not then have any shallow water at the sides of the channel where light can penetrate turbid water sufficiently for plants to be able to grow.

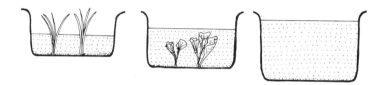

Fig. 14.5

In dykes, lack of water movement means that upward growth of plants is easy. Thus if a plant can become established on the bed, it can grow into brighter light and then do well. Turbidity, therefore, does not have any gradation of effect: either a species cannot grow on the bed, or it can potentially grow luxuriantly in the dyke (Fig. 14.5). Species do, however, differ in the amount of light which they require (see Chapter 7). Those tolerating turbid water best usually have leaves above the water which can receive full light, e.g. emergents at the edges and, in the centre:

  Nuphar lutea   *Sparganium emersum*

  Sagittaria sagittifolia

The species-richness of emerged and floating plants in turbid dykes is independent of the turbidity, provided toxic pollution is absent.

The water in dykes and drains may seep in through the soil of the alluvial plain, in which case it will usually be clear when it emerges into the channel, or it may come down in streams from the higher land around. One cause of the high turbidity in the Pevensey Levels, for example, is the very turbid water of the clay streams feeding them. Water in peat plains, such as part of the Fenland, is usually clear, but may, however, be stained brown, and if so has the same light-blocking effect on the plants as turbid water. In still water, especially when the larger plants are killed by herbicides, phytoplankton can grow well and these make the water turbid. Another effect of killing edge plants by herbicides is that it makes the banks less stable, and the fine particles washed into the channel as a result may also add to the turbidity. Boats often stir up silt on the channel bed, and thus create high turbidity. Turbid and clear-water dykes are sometimes found together. This can happen when turbid river water has access to part of a dyke system, or when herbicides have been used on just part of a dyke. The turbid dykes then usually have both less total vegetation and fewer species present than those with clear water (also see Chapter 22).

Substrate type and trophic status

Dykes and drains have fine-textured substrates, their beds being the sub-soil of the alluvial plain together with any deposited sediment. The subsoil may be sand, mineral silt or organic matter, sand being the least frequent. The Fenland and the Somerset Levels, for instance, are partly on peat and partly on silt. The ground level of the channel is raised by sediment from the land being washed in during storms and lowered by intermittent dredging, but because scour is absent the substrate can be consolidated by root wefts in a way which is impossible in flowing waters (see Chapter 3).

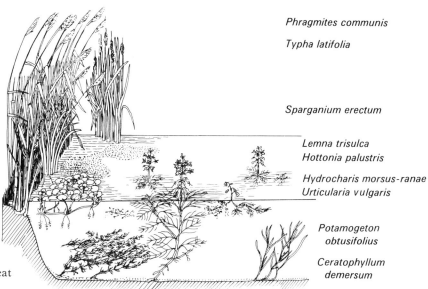

Phragmites communis

Typha latifolia

Sparganium erectum

Lemna trisulca
Hottonia palustris

Hydrocharis morsus-ranae
Urticularia vulgaris

Potamogeton obtusifolius

Ceratophyllum demersum

Fig. 14.6. A mesotrophic peat dyke.

In general peat dykes tend to be mesotrophic as do silt ones which do not receive eutrophic or polluted water (Fig. 14.6). Mesotrophic dykes are characterised by:

Lemna trisulca   *Hydrocharis morsus-ranae*

Both these however, may be absent, and *Hydrocharis morsus-ranae* is increased by dredging. Other typical species include:

Ceratophyllum demersum   *Potamogeton* spp. (grass-leaved)

Hottonia palustris   *Stratiotes aloides*

Lemna polyrhiza   *Utricularia vulgaris*

257

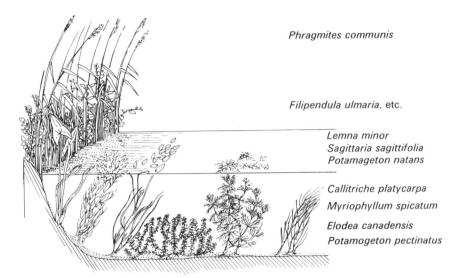

Phragmites communis

Filipendula ulmaria, etc.

Lemna minor
Sagittaria sagittifolia
Potamageton natans

Callitriche platycarpa
Myriophyllum spicatum

Elodea canadensis
Potamogeton pectinatus

Fig. 14.7. A semi-eutrophic dyke.

The dykes of intermediate nutrient status, on e.g. clay, sand, slightly eutrophicated silt or eutrophicated peat, tend, if they are species-rich, to have (Fig. 14.7), for example:

Callitriche spp.   *Myriophyllum spicatum*

Ceratophyllum demersum   *Myriophyllum verticillatum*

Elodea canadensis (luxuriant growth)

Lemna minor agg.   *Potamogeton natans*

Lemna polyrhiza   *Potamogeton pectinatus*

                                      Sagittaria sagittifolia

                                      Enteromorpha sp.

The most eutrophic dykes and drains are on silt, often with some source of eutrophication (Fig. 14.8). Typical species include:

Lemna minor agg.   *Sagittaria sagittifolia*

Nuphar lutea   *Sparganium emersum*

Potamogeton perfoliatus   *Enteromorpha* sp.

The larger species, however, prefer fairly deep channels.

These groups are neither consistent nor exclusive, but they do illustrate the general trends as seen in, for example, the Somerset Levels dykes on peat, silt and eutrophicated silt.

Polamogeton pectinatus has a different distribution in dykes from that discussed in earlier chapters for streams. Here it is in a different form, with shorter leaves and shoots, and is usually very much finer and more

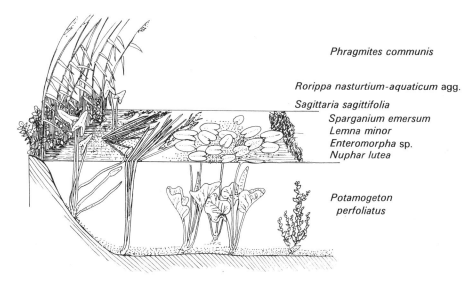

Phragmites communis

Rorippa nasturtium-aquaticum agg.

Sagittaria sagittifolia
Sparganium emersum
Lemna minor
Enteromorpha sp.
Nuphar lutea

*Potamogeton
perfoliatus*

Fig. 14.8. A eutrophic silt dyke.

delicate. Its occurrence is sporadic and is most frequent after dredging and in brackish water. In Romney Marsh, which still received sea water flooding early this century, the delicate form of *Potamogeton pectinatus* is frequent. In East Anglia, in dykes near the coast which receive sea water flooding every decade or two, *Potamogeton pectinatus* is often dominant, in a short but coarse form.

Geographical differences

Here, only the larger alluvial areas are being considered. More differences would arise if small flood plains with only one or two dykes were included in the comparison.

A few species are almost confined to one part of the country, for example:

Azolla filiculoides	S. and S. W. England
Nymphoides peltata	E. England
Stratiotes aloides	East Anglia

Other species have habitat preferences, which affect their distribution within an alluvial plain. For example:

Scirpus maritimus	Rare or absent far inland, frequent towards the coast, e.g. Essex, Norfolk. In Romney Marsh, perhaps commoner on sand than on clay.
Fringing herbs	Mainly at the edge of the plains, where dykes receive some water from the incoming streams. Sporadic occurrences elsewhere sometimes occur on locally higher ground.

259

Management

All dykes and drains are, or were, managed, and the frequency and depth of dredging, the frequency and season of cutting, and the type, season and frequency of herbicide application all affect the vegetation. The effects are discussed in Chapter 21.

Species-richness

Dykes may be choked with plants or empty, depending on the drainage pattern and management. Drains in use, however, must be kept clear as they are then able to carry more water. The plants are kept down partly by direct management and partly by making the channels deep and steep-sided so that the habitat is unsuitable for river plants. A variety of means have been used for direct management, both mechanical and chemical. It is generally observed that mechanical methods do not decrease species diversity, unless habitats unsuitable for plants are created (e.g. deep and turbid drains). Chemical methods, on the other hand, eliminate many species, and were used in some parts of the country for a long time before modern herbicides were available. Pollution, and severe eutrophication and turbidity, also eliminate many species (see also Chapter 22), but herbicides appear to be the most common cause of species-poor dykes and there is a correlation between the proportion of these and the variation in herbicide use in different areas.

The plants most likely to be present in species-poor dykes are:

Agrostis stolonifera   *Lemna minor* agg.

Callitriche spp. (less)   *Enteromorpha* sp.

(Also, as described above, shallow dykes left unmanaged will become species-poor as a result of domination by tall monocotyledons such as *Phragmites communis*.)

The species-rich dykes are those which are kept fairly clear of tall plants by mechanical means, are not polluted or much eutrophicated, have a usual minimum water depth of at least 20 cm, and are not shaded from the banks. Such dykes are likely to contain many, often 10–12, submerged and floating species. Species characteristic of, but rare in, dykes tend to occur together, rather than to be scattered evenly throughout the dyke sites recorded. This is attributable to human interference making the species-poor dykes unsuitable for the less vigorous plants. It is the species-rich dykes which are described in the following section on patterns and communities.

Patterns and communities

In dykes, and to a lesser extent in the larger drains, the plant communities vary in both space and time. There are seasonal differences in the time at which species reach their maximum size, e.g.

260

ⓥ *Callitriche* spp.	Maximum in spring
Hottonia palustris	Maximum in late spring
🌿 *Myriophyllum* spp.	Maximum in summer
⚘ *Rorippa amphibia*	Maximum in late summer

The differences in communities within the growing season are more prominent in dykes than in streams, as there are fewer other checks to growth in the dykes.

Because of the absence of flow, plants can occur at all levels within the water, and also above it (Fig. 10.2, Figs. 4.5–14.7). In a channel full of plants the upper layer will, therefore, shade the lower. The emergents are of course the tallest plants, and their effect depends on the intensity of the shade they cast (see Chapter 7). The effect of floating plants varies somewhat with the species, for example:

> *Enteromorpha* spp. and *Lemna* spp. can be thick enough to prevent other plants gowing below.
>
> *Callitriche* spp. are submerged as well as floating. The floating leaves form an incomplete cover, but the submerged ones, mainly in the upper part of the water, add to the shade so that only shade-tolerant species grow below.
>
> *Potamogeton natans* also forms an incomplete cover, but its submerged parts cast little shade.

There is also a characteristic group which grows in the upper part of the water. For example:

✳ *Lemna trisulca*	🌿 *Myriophyllum spicatum*
Hottonia palustris	❋ *Ranunculus* spp.

(*Ranunculus* spp. belong in this group because they are usually delicate and unable to remain rooted because of dying from the base. Detached shoots therefore float to the surface and grow there. See Chapter 3.)

Finally there are the plants occupying mostly the bottom of the water. These are often shade-tolerant. They include, for example:

ⓨ *Ceratophyllum* spp.	*Potamogeton* spp. (grass-leaved)
✦ *Elodea canadensis*	

In what appears to be the natural conditions for such communities, many species grow well but none becomes the sole and dominant species – except in shallow dykes with tall emerged monocotyledons. The most prominent species may vary from year to year at the same place. This may be from several causes. For example, one species may find the weather unsuitable and be replaced by more tolerant plants (e.g. *Myriophyllum verticillatum* killed in a hot summer and replaced for a few years by *Elodea canadensis* and *Fontinalis antipyretica*); there may be changes after cutting (e.g. *Lemna* spp. replacing *Callitriche* spp.); and there are usually changes after dredging (described in Chapter 21), for example:

> *Hottonia palustris* replacing *Callitriche* spp.
> *Ceratophyllum demersum* replacing *Myriophyllum verticillatum*
> *Potamogeton natans* replacing *Myriophyllum spicatum*

In streams, plant patterns within a reach are usually repetitive and are composed of the same few species, both the patterns and the species usually being controlled by simple habitat factors. In a short stretch of dyke, however, there can be a series of dominants within an apparently uniform habitat. For example, within 75 m were found (Fig. 14.9):

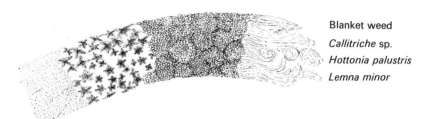

Blanket weed
Callitriche sp.
Hottonia palustris
Lemna minor

Fig. 14.9

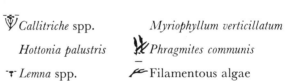

Callitriche spp.	*Myriophyllum verticillatum*
Hottonia palustris	*Phragmites communis*
Lemna spp.	*Filamentous algae*

This variation of community is an important feature of dykes. In dykes, also, species occur together which in streams have different habitat preferences. For example:

Alternative habitat:

Lemna trisulca	Chalk brooks
Nuphar lutea	Slow-flowing silted eutrophic rivers, e.g. on clay
Potamogeton natans	Particularly in sandstone streams
Ranunculus aquatilis	Most frequent in sandstone or swift hill streams

CANALS

Canals are similar in size to drains, but can be allowed to bear more vegetation as they are not needed for the drainage of flood water (Fig. 10.2; Table 14.2; Plate 11). If boat traffic is dense, however, plants are sparse or absent (see Chapter 21), and in some canals vegetation is reduced by serious pollution (see Chapter 22). Vegetation, as always,

TABLE 14.2   *Canal species*[a]

(Sites with severe pollution or dense boats are excluded.)

(*a*) Channel species

Occurring in at least 40 % of sites recorded:

⫞ *Lemna minor* agg.   ⫟ *Sparganium erectum*

Occurring in at least 20 % of sites recorded:

⫠ *Ceratophyllum demersum*   ⫟ *Rumex hydrolapathum*

⫠ *Glyceria maxima*   ⫟ *Sparganium emersum*

⫟ *Nuphar lutea*

⫟ *Potamogeton pectinatus* (including mildly polluted sites)

Occurring in at least 10 % of sites recorded:

⫟ *Carex acutiformis*   ⫟ *Sagittaria sagittifolia*

⫟ *Elodea canadensis*   ⫟ *Enteromorpha* sp.

(*b*) Bank species

Occurring in at least 20 % of sites recorded:

⫟ *Epilobium hirsutum*   ⫠ *Glyceria maxima*

Filipendula ulmaria   *Urtica dioica*

Occurring in at least 10 % of sites recorded:

⫟ *Phalaris arundinacea*   ⫟ *Sparganium erectum*

[a]See notes to Table 12.1.

varies with depth. Ship canals are few, and are too deep and turbid to bear much vegetation, so they are not described here. Disused canals tend to silt up, and are commonly dominated by *Glyceria maxima*. The pleasure boat canals are *c.* (1–) 1.5 m deep, and may have a varied and rich vegetation. As in dykes, water may become turbid from incoming particles, from pollution, or from boats whose movement stirs up silt on the bed. When bottom silt is disturbed, less light reaches the bed, poorly anchored plants are uprooted, and when the silt settles it tends to smother plants in the water. Under such disturbance, submerged plants are lost first and tall deep-rooted emergents remain the longest. Species which are more frequent in clearer water include:

⫠ *Ceratophyllum demersum*   ⫟ *Myriophyllum spicatum*

⫟ *Elodea canadensis*   ⫟ Mosses

There is usually a band of emergents at the sides of canals, generally wider than that found in drains since canals usually have a small ledge of shallow water at their sides on which the plants can grow. Plants of shallow water are, however, restricted in that they cannot grow beyond this ledge (Fig. 10.2). A fringe of tall plants adds to the amenity value of a canal, improves the stability of the banks, and does not hinder navigation because boats usually keep to the middle, except for passing and mooring. Floating and submerged plants are also mainly at the sides, both because of the boats (their propellers damage the plants) and because of the shallower water there.

The more mesotrophic canals are those which are neither on clay nor comprise part of a river system. The most mesotrophic are on peat, and their characteristic species include:

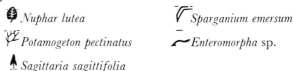

Ceratophyllum demersum	*Lemna trisulca*
Hydrocharis morsus-ranae	*Potamogeton natans*
Lemna polyrhiza	*Stratiotes aloides*

These canals have more species-rich sites than the other groups. The most northerly lowland canals, (Forth–Clyde and Union) have a rather different flora.

The eutrophic canals are those on clay, including eutrophic canalised rivers. The vegetation is like that of clay streams of equivalent size, though more still-water species may be present. The characteristic species include:

Nuphar lutea	*Sparganium emersum*
Potamogeton pectinatus	*Enteromorpha* sp.
Sagittaria sagittifolia	

In canalised rivers the ground level changes and there are intermittent locks, beside each of which is a swifter stretch with a weir. As the boats pass through the lock surplus water flows over the weir. The swifter reach has vegetation like that of a clay stream of equivalent size, and usually has more *Schoenoplectus lacustris* and less *Sagittaria sagittifolia* than the slow reach by the lock.

In the third type of canal, plants are almost or completely absent. The Caledonian Canal in the Scottish Highlands has a stone and gravel bed, deep water, and negligible vegetation.

15

North American streams: habitat and vegetation patterns

Streams in some of the temperate and warm-temperate parts of North America were surveyed (Fig. 15.1). They differ from those of Britain in various habitat factors, and in their plant life. The vegetation is generally less affected by geological differences than it is in Britain, but a controlling influence in areas where snow lies throughout the winter can be the great force of the snow-melt floods in spring.

The ground water level is normally higher than in Britain, and wet ditches and tree swamps are common. In peaty forests in particular the water is often brown-stained. Streams bounded by marsh or forest tend to be more species-rich than those where crops extend right to the water's edge. Pollution is generally less than in Britain, the human population being sparser.

Some species occur in both North America and Britain, but most are confined to one or other. Some groups, such as Potamogeton *spp. and broad-leaved, deep-rooted semi-emergents are better represented in North America, while others, such as* Callitriche, Ranunculus *spp. and fringing herbs, are better represented in Britain. Even within the limited area surveyed there was considerable variation in the species found in different geographical areas.*

In North America, as in Britain, most published research on the ecology of water plants describes lakes rather than streams. Described here are the results of a 6-week survey in regions somewhat comparable to Britain. Because the study was short and comprised only *c.* 1550 site records, the interpretation of the data has necessarily been less sensitive than that in the rest of this book. As this is the only such study, however, it is offered in the hope that it will stimulate more detailed research.

The areas surveyed (Fig. 15.1) all have temperate or warm-temperate climates. Regions with few river plants, such as mountains and the parts south of the Great Lakes with much agricultural silting, were avoided. A short description of Nova Scotia was kindly supplied by Dr M. J. Harvey.

The terms river, stream and brook are synonymous in both North America and Britain. However, smaller and medium-sized streams are commonly called creeks in North America, while in Britain this term applies only to small coastal channels. A slow stream in a swampy area may be call a swamp. A canal may be a man-made navigation channel, as in Britain, or part of an irrigation–drainage system. The word 'ditch' covers both the narrow small channels beside roads, between fields, etc., and the somewhat larger ones more often called drains, dykes, etc.

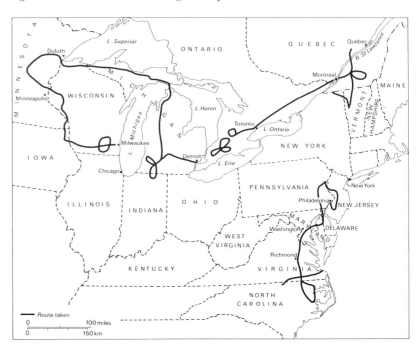

Fig. 15.1. The parts of North America studied for Chapters 15–17.

in Britain. The word river, creek etc. is usually placed after the name of the watercourse, as in Maitland River, Badfish Creek, Beaver Dam Swamp. The comparable British usage is R. Cam, A. (for Afon) Teifi (Wales), but Sixpenny Brook, Bere Stream, Douglas Water (usually Scotland), Culag Burn (Scotland), Sirhowy R. (Wales), and St John's Beck (usually northern England). In Quebec, the usage is R. Noire, R. Nicolet, etc. A rock in North America is a stone, and not a boulder as it is in Britain. For consistency, the British usage is followed in these chapters.

Because of difficulties of identification in such a short survey, plant names here are frequently cited as aggregates (agg.).

Most of the habitat factors that affect river plants are the same in both Britain and North America. Important influences that are found in North America only, and have therefore not been covered earlier in the book, are described below.

GEOLOGY AND TOPOGRAPHY

Most of the area that was surveyed is lowland; plains, gently rolling country or dissected plateaux. Hills were visited only in Quebec (north and south) and Vermont, though steep gorges were studied along large rivers.

Thick glacial drift is common in the north, varying from clays to fertile or infertile sands or gravels. Older clays, soft sandstones and

266

similar rocks occur down the east coast. Limestones and dolomites (both referred to here as limestone) frequently occur beneath the drift, and occasionally outcrop at the surface. The limestone streams are very similar to those of Britain, but the general limestone influence is less since there is so much drift. Resistant rocks outcrop in the Canadian Shield, New England and occasionally elsewhere.

Geology has less effect on vegetation than in Britain, and although limestone vegetation is distinctive, on the available evidence it is less easy to separate the vegetations of the other rock types. Geology does, however, affect river plants through the ground water supply in summer and through topography. The nature of the glacial drift can also influence vegetation through turbidity (also see section below).

FLOW

There is one most important aspect of flow that is not encountered in Britain: snow-melt floods. In the northern regions snow lies through the winter and causes very high discharges when it melts in spring. In the east, New Jersey may have winter-lying snow, but the southern areas visited do not have this. The size of the snow-melt discharge in relation to the lowest monthly flow varies according to the sources of summer flow (springs, rainfall, etc.) and the amount of snow, but can be up to 350 times the minimum even in Wisconsin [49] and is often very high indeed in Quebec. It is lowest where flow is stabilised by much spring water or many impoundments. The effect on river plants depends on the force as well as the discharge (see Chapter 5, etc.) and the force varies within the northern region visited, being usually greatest to the east.

In lowland Quebec rivers often have negligible flow and low discharge in summer; for example, the water may occupy only 5 m of a channel *c*. 30 m wide. Yet despite this the substrates are usually coarse, even with boulders, if the rivers rise in low hills (see Chapters 2, 3 and 5). Because of this variation in discharge, water level, scouring and substrate stability vary greatly through the year. The mountain streams of Quebec have, of course, a continuously swift flow and consequently little vegetation, so the effect of snow-melt is negligible. In Vermont and Ontario fluctuations in both water level and scour are less and consequently there is more vegetation. The streams are even more stable in the Midwest and the vegetation patterns are more similar to those in Britain (where channel plants are usually flooded and edge plants partly emerged). On the east coast, and near limestone springs, the streams tend to be stable.

GROUND WATER LEVEL

The ground water table is usually much higher in North America than it is in Britain (see Chapter 1) and roadside and field ditches nearly always contain aquatic plants. Because most of these ditches dry out in summer the plants are emergents, but unusually deep ditches or those in swampy areas have submergents. In Britain such ditches usually bear land or marsh plants, water plants being almost confined to channels on alluvial plains (e.g. Fenland dykes). The high water table also means that the proportion of brooks which are summer-dry and without river plants is low, despite the great seasonal fluctuations in flow. Most of these summer-dry streams are on somewhat hilly, sandy areas.

In wooded and swampy places, many streams rise in tree swamps (Plate 1). These flow very slowly, but have continuous, fairly deep water, so submerged plants requiring some depth of water may occur even near the source.

STREAM DEPTH

In streams of moderate or fast flow, water depth is similar to that of British streams of equivalent width (see Chapters 2, 4 and 6). However, swampy streams, particularly in Virginia and North Carolina, may be deep even when narrow. Their flow is very slow, and in some ways they resemble dykes. As their water is often dark (see below), vegetation can be much decreased (see Chapter 7).

FORESTS AND TREES

In North America there are both large forests and large areas with no trees on the river banks. In Britain, in contrast, woods are few and small, but trees and shrubs are common on stream banks (except in mountains). In forests and smaller woodlands where trees overhang the bank, the shading effect means that plants cannot grow in the shallow water at the sides of the channel (Fig. 15.2), and if the water is deep, dark or turbid, plants cannot grow in the centre either. Such streams were seen particularly in Ontario and Michigan.

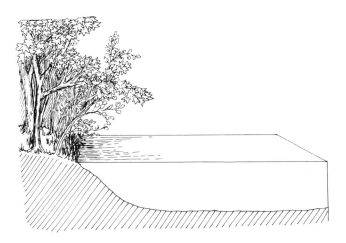

Fig. 15.2. Trees overhanging a deep river and preventing the development of river plants (R. Maitland, Ontario).

In forests, streams usually have as diverse a flora, but less bulk of vegetation, than occurs in similar streams in the open country. Gaps in the tree canopy are frequent and here river plants may be luxuriant (see Chapter 1).

WATER COLOUR AND TURBIDITY

As has already been explained in Chapter 7, dark and turbid water reduces vegetation. Because of the large areas of tree swamps, etc., dark and humus-stained water is common. Such water may either be infertile (as in New England [50] and New Jersey) or fertile (as in Minnesota and Michigan).

Turbidity has many causes. Glacial drift which is easily eroded and fine can, in arable areas where the ground is ploughed, cause much turbidity. Small pockets of this were seen, for example, in Ontario, and it is reportedly widespread south of the Great Lakes. On silted beds, animals can prevent plant growth by stirring up the silt and thus create both turbid water and an unstable bed. Such damage is caused particularly by European carp (e.g. in Wisconsin) and frogs (e.g. locally in Virginia and North Carolina), but can occur with other animals such as turtles. When the disturbance is only mild, deep-rooted emergents are still able to grow at the sides of the channel, but in areas where the populations are dense, plants are completely absent. Industrial and sewage effluents also cause turbidity.

LOGGING RIVERS

Rivers which, as in parts of Quebec, are used for the transport of logs, have little or no vegetation near the water surface or at the channel sides where the logs can damage and sweep away plants.

SAND DAMAGE

Streams are sometimes found on steep sandy slopes, for example by the Mississippi River in Wisconsin. The sandy bed is easily eroded, and the loose sand being washed quickly down the steep channels makes the bed unsuitable for plant establishment. The hills where such streams are found are dry, so many of the smaller brooks dry in summer. Plants may be absent entirely or there may be a few tough species present (e.g. *Eleocharis erythropoda* agg., *Phalaris arundinacea*, short grasses).

BEAVER DAMS

Beaver dams are frequent in Ontario and other sparsely-populated northern areas. The dams each last 10–20 years, creating ponds upstream of themselves of a size varying from *c*. 30 m × 10 m to *c*. 100 × 80 m, depending on the local topography (Fig. 15.3) When the dam and

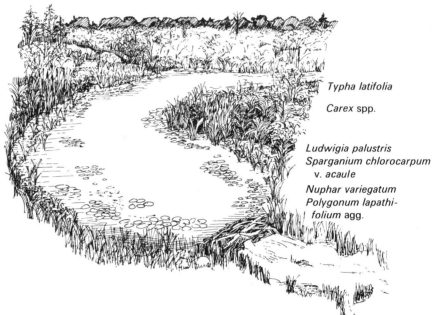

Typha latifolia

Carex spp.

Ludwigia palustris
Sparganium chlorocarpum
   v. *acaule*
Nuphar variegatum
Polygonum lapathi-
  folium agg.

Fig. 15.3. Pool and beaver dam (Ontario).

its accumulated sediment are finally washed away the stream reverts to its former size. The effects of a dam on a stream are thus to cause a short-term increase in channel width and water depth, to increase sedimentation (making the water shallower), and to decrease water movement. The first plant to invade the new habitat is typically *Lemna minor*, perhaps with other duckweeds. Later, if the water is shallow enough, water lilies (*Nymphaea odorata* and *Nuphar variegatum*) and other small river species colonise [51]. In wooded areas heaver dams may kill trees when the area round the original stream channel floods.

LAND USE AND POLLUTION

In general, damage from man's activities seems less in North America than in Britain. In the regions visited the animal life was richer than in Britain, with frogs, insects, fish, birds, etc., all being more prominent. The river plants show as much or more species diversity than in comparable British streams, in spite of snow-melt scouring, greater shading and darker water. This suggests a more species-rich vegetation. The biomass is, however, less, and few streams are sufficiently choked to require weed control. The stream types with least vegetation include mountain streams, Quebec streams with large fluctuations in discharge, swamp forest streams with dark water and steep banks, and summer-dry brooks.

Streams in swamps or marshes, or with a strip of these along their banks tend to be more species-rich than those with crops right up to the banks. Decrease in species-richness is the first result of pollution (see Chapter 22 and, for the USA [52]). Agriculture is also responsible for much turbidity (see above), and herbicides frequently spread to, and damage, at least the emerged river plants.

Effluent pollution is relatively little compared with Britain, as the population is sparse. The effluents from the largest cities were not studied and the only marked pollution seen was in Quebec, where because the streams were unsuitable for much vegetation anyway, there was no pollution damage. Paper mill effluent [53] reduces vegetation but mainly by its turbidity, as in Britain.

Potamogeton pectinatus is the British species most characteristically associated with sewage and industrial effluents (see Chapter 22), and it is infrequent in clean sites. In North America both its habit (Plates 6 and 15) and its habitat differ from those in Britain (see Chapters 12, 13, 17 and 22). Interestingly, the North American habit, with shorter shoots, relatively shorter leaves and less clumped shoots, also occurs in the far north of Scotland.

LIST OF SYMBOLS

When the habits of British and North American species are similar, the same symbol is used for both.

Agrostis stolonifera and other small grasses

Alisma plantago-aquatica

Berula erecta

Callitriche palustris

Carex spp.

Ceratophyllum demersum

Cladium mariscoides

Decodon verticillatus

{ *Eleocharis erythropoda* agg. *Eleocharis* spp.

{ *Elodea canadensis* *Elodea nuttallii*

{ *Equisetum fluviatilis* *Equisetum palustre*

Glyceria canadensis

Groenlandia densa

{ *Nuphar advena* *Nuphar variegatum*

Nymphaea odorata agg.

Peltandra virginica

Phalaris arundinacea

Phragmites communis

Polygonum lapathifolium agg.

Pontederia cordata

Potamogeton amplifolius

Potamogeton crispus

Potamogeton foliosus

{ *Potamogeton gramineus* *Potamogeton illinoiensis* *Potamogeton nodosus*

Potamogeton natans

Potamogeton pectinatus

Heteranthera dubia

Hippuris vulgaris

{ *Juncus effusus*
  Juncus nodosus
  Juncus spp.

Lemna minor

Lemna polyrhiza

{ *Mentha piperita* agg.
  Mentha sp.

Mimulus glabratus agg.

Myosotis scorpioides

Myriophyllum exalbescens

Potamogeton richardsonii

Ranunculus longirostris

Rorippa nasturtium-aquaticum

Sagittaria cuneata

Sagittaria latifolia

{ *Scirpus validus*
  Scirpus spp.

Sparganium americanum

Sparganium chlorocarpum

{ *Sparganium chlorocarpum* v. *acaule*
  Sparganium eurycarpum

Typha angustifolia

Typha latifolia

Utricularia spp.

Vallisneria americana

Zannichellia palustris

Mosses

Characeae

SPECIES DISTRIBUTION

Some species occur in both North America and Britain, but most river plants are confined to one region or the other. Species occurring in both, with similar habits and habitats, include:

Alisma plantago-aquatica

Ceratophyllum demersum

Elodea canadensis
(introduced into Britain)

Equisetum fluviatile

Equisetum palustre

Hippuris vulgaris

Juncus effusus

Phragmites communis

Potamogeton crispus (introduced into North America)

Potamogeton natans

Rorippa nasturtium-aquaticum

Solanum dulcamara

Typha angustifolia

Typha latifolia

Species distribution

 Lemna minor

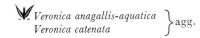 *Veronica anagallis-aquatica*
Veronica catenata } agg.

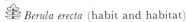 *Lemna (Spirodela) polyrhiza*

 *Zannichellia palustris*

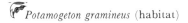 *Myriophyllum spicatum* (introduced into North America)

Phalaris arundinacea

Other species differ more in habit or in habitat between the two regions, for example:

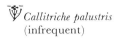 *Berula erecta* (habit and habitat)

 *Potamogeton pectinatus* (habit and habitat)

 Potamogeton gramineus (habitat)

Utricularia vulgaris (habitat)

In some cases a genus may be represented in the two regions by species which are different but which have similar habits and ecological niches. For example:

North America   Britain

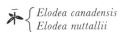 *Callitriche palustris* (infrequent)

 *Callitriche obtusangula*
Callitriche platycarpa
Callitriche stagnalis } (common)

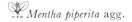 *Eleocharis* spp.

 *Eleocharis palustris*

 { *Elodea canadensis*
Elodea nuttallii

 Elodea canadensis

 Mentha piperita agg.

 *Mentha aquatica*

{ *Myriophyllum exalbescens*
Myriophyllum spicatum (local)

Myriophyllum spicatum

 { *Nuphar advena*
Nuphar variegatum

Nuphar lutea

{ *Ranunculus longirostris* (infrequent)
Ranunculus trichophyllus (rare)

Short-leaved *Ranunculus* spp. including *R. trichophyllus* (common)

 { *Sparganium americanum*
Sparganium eurycarpum

 *Sparganium erectum*

The bracketed species may be difficult to separate taxonomically.

273

Other species pairs show greater differences in habit or habitat, although still showing considerable resemblances. For example:

North America	Britain
Carex stricta	Carex acutiformis (habit)
Cladium mariscoides	Cladium mariscus (habit and habitat)
{ Glyceria canadensis / Glyceria grandis	Glyceria maxima (habit)
Mimulus glabratus	Mimulus guttatus (habit)
{ Nymphaea adorata / Nymphaea tuberosa	Nymphaea alba (habitat)
Sagittaria cuneata	Sagittaria sagittifolia (habit and habitat)
Scirpus validus	Schoenoplectus (Scirpus) lacustris (habit and habitat)
Sparganium chlorocarpum	Sparganium emersum (habit and habitat)

There is also a general resemblance between various groups of *Potamogeton*.

Other common North American species have no effective British counterparts, in similar habitats:

Decodon verticillatus	Pontederia cordata
Heteranthera dubia	Potamogeton foliosus
Leersia oryzoides	Sagittaria engelmanniana
Najas spp.	Sagittaria latifolia
Peltandra virginica	Sagittaria rigida
Polygonum hydropiperoides	Zizania aquatica
Polygonum lapathifolium agg.	Characeae

Similarly, some species which are abundant in Great Britain have no equivalent distribution in North America. For example:

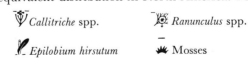

Callitriche spp.	Ranunculus spp.
Epilobium hirsutum	Mosses

DISTRIBUTION OF PLANT GROUPS

The short-rooted fringing herbs so common in Britain (Chapters 12, 13 and 14) are, in North America, almost confined to the rare areas of limestone springs. Similar though more firmly rooted semi-emergents (such as *Polygonum lapathifolium* agg. and *Polygonum hydropiperoides*) occur in a wider range, where scour is not severe or not significant (either because the channel is a ditch, or is a stream near its source, or, in the south, because the land is flat and snow-melt floods are absent).

In Britain, *Alisma* spp. are the only large-leaved, deep-rooted, short edge species dying back in winter. They are never luxuriant and seldom frequent. In North America, however, there is a group of species of similar habit, including:

Alisma plantago-aquatica   (*Sagittaria cuneata*; also grows submerged)

Decodon verticillatus   *Sagittaria engelmanniana*

Peltandra virginica   *Sagittaria latifolia*

Pontederia cordata   *Sagittaria rigida*

These species often form fringing bands along stream banks. They anchor firmly, usually in deep fine-textured substrates, where their overwintering parts remain. They can become established on soft soil and can tolerate considerable winter scour and seasonal fluctuations in water level (Fig. 15.4). They are, therefore, all well adapted to the habitat of the unstable lowland snow-melt streams of North America (see Chapters 3, 5, 6, etc. for the British fringing herb habitat).

Fig. 15.4. River with much seasonal fluctuation in water level, and a band of *Sagittaria latifolia* on dry mud in mid-July (Quebec).

Short rushes and similar monocotyledons are commoner in America, presumably for the same reasons that apply to the former group. *Eleocharis* spp., *Juncus* spp. and *Agrostis stolonifera* are also common.

275

The other main group of edge plants is the tall emerged monocotyledons – deep-rooted and firmly anchored but intolerant of much scour and requiring a fairly fine substrate. *Phalaris* and *Glyceria* have similar ecological niches in North America and Britain (though *Glyceria canadensis* is shorter than *Glyceria maxima*). *Phragmites communis*, is, as in Britain, uncommon on stream banks except near the coast, where it is often dominant. However, in Britain it is the commonest dyke dominant, while in North America its place is taken by *Typha* spp. which are able to tolerate fluctuating water levels. *Carex* spp. *Cladium* and *Scirpus* spp. are well represented in North America, and there are more tall grasses (e.g. *Leersia oryzoides, Zizania aquatica*).

When water level fluctuations and scour are little, as in the swamp forests of Virginia and North Carolina, the three groups of *Polygonum* spp., large-leaved monocotyledons and tall monocotyledons all occur in the same habitat. With increased fluctuations in water level, the *Polygonum* group is lost and the tall monocotyledons move higher up the bank behind the large-leaved fringe. (This was the situation in most of the areas surveyed.)

In the channel, the principal difference between the two regions is that in North America, *Ranunculus* and *Callitriche* are poorly represented and never choking, while in Britain they are well represented and *Ranunculus* is the submergent most commonly causing a flood hazard. In North America this ecological niche is filled by *Potamogeton* spp., but these seldom grow as luxuriantly.

Among the floating-leaved plants, both *Potamogeton* and water lilies are better represented in North America. The water lilies (*Nuphar* and *Nymphaea*) can also tolerate considerable summer drought, growing well on bare mud in late summer with semi-erect leaves (Fig. 17.1). This is uncommon in Britain, because such summer-dry, winter-deep habitats occur but rarely. The American *Sagittaria cuneata* and *Sparganium chlorocarpum* have both submerged and land forms, while the British *Sagittaria sagittifolia* and *Sparganium emersum* are solely aquatic and grow in deeper water. The former are better adapted to a fluctuating water level. However, in the hot dry summer of 1976 these species had more erect and emerged parts and, independently, grew on bare mud or in shallow water, thus resembling much more closely the North American species.

Some plant communities are remarkably similar in both regions. Particularly striking examples are:

1. Small limestone streams with:

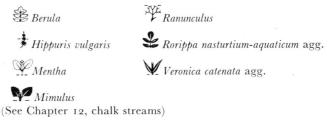

Berula   *Ranunculus*

Hippuris vulgaris   *Rorippa nasturtium-aquaticum* agg.

Mentha   *Veronica catenata* agg.

Mimulus

(See Chapter 12, chalk streams)

2. Medium or large streams with:

🌢 *Nuphar*   🌾 *Scirpus (Schoenoplectus)*

🌱 *Sagittaria*   🌿 *Sparganium*

(See Chapter 12, clay streams)

3. Medium-sized streams with:

🌿 *Callitriche*   🌱 *Rorippa nasturtium-aquaticum* agg.

🌿 *Elodea*   🌾 *Typha latifolia*

🌱 *Myosotis scorpioides*

(Recorded once in Pennsylvania. See Chapter 12, chalk–clay streams with little clay)

4. Ditches dominated by:

🌾 *Carex*   🌾 *Typha angustifolia*

🌿 *Glyceria*   🌾 *Typha latifolia*

🌾 *Phragmites communis*

(See Chapter 14)

As well as these, may other stream types bear some resemblance to British ones, particularly, of course, almost empty channels, and those with prominent *Nuphar* and *Sparganium*, etc., as these look the same wherever they occur (see Plates 9 and 13 and Fig. 10.2). Limestone vegetation is distinctive and easily recognised, but it is rare in North America as so much of the limestone is covered by glacial drift. Coastal and ditch vegetation is likewise recognisable. Elsewhere, vegetation types seem more uniform than in Britain, varying less with rock type and showing less downstream variation. As in Britain, a considerable number of species can potentially occur in any site, those which are actually present depending mainly on the flow regime, substrate and depth. Most submergents are commonest on a stable gravelly substrate with fairly shallow clear water, a noticeable flow in summer and little scour.

Correlations between plant distribution and chemical differences do, however, occur within this general group of streams. The aquatic plants of Minnesota have been classified by water hardness [54]. They include the following:

1. Soft water

🌿 *Callitriche palustris*   🌿 *Potamogeton gramineus* (probably)

Potamogeton alpinus   *Potamogeton pusillus* (probably)

Potamogeton epihydrus   *Sparganium minimum*

2. Somewhat soft water

Equisetum fluviatile   *Potamogeton robbinsii*

Nymphaea odorata   *Sparganium chlorocarpum*

Pontederia cordata

3. Somewhat hard water

Ceratophyllum demersum   *Nymphaea odorata* agg.

Elodea nuttallii   *Sagittaria cuneata*

Heteranthera dubia   *Sagittaria latifolia*

Myriophyllum excelbescens   *Utricularia vulgaris*

Najas flexilis   *Vallisneria americana*

Nuphar variegatum

4. Hard water (but not limestone springs exclusively)

Alisma plantago-aquatica   *Potamogeton pectinatus*

Elodea canadensis   *Potamogeton richardsonii*

Leersia oryzoides   *Sagittaria rigida*

Potamogeton amplifolius   *Scirpus validus*

Potamogeton foliosus   *Sparganium eurycarpum*

Potamogeton illinoiensis   *Zizania aquatica*

Potamogeton nodosus   *Chara* spp.

GEOGRAPHICAL VARIATION

The 5000 miles (8000 km) or so surveyed covered considerable variations in climate. Only a few plants occurred throughout this area. For example:

Lemna minor   *Sagittaria latifolia*

Phragmites communis   *Typha* spp.

Most species, however, are prominent in only part of the range, and in a different area one species may be replaced by an allied one which fills the same ecological niche. For example:

Nuphar advena   and   *Nuphar variegatum*

Nymphaea odorata   and   *Nymphaea tuberosa*

Geographical variation

Potamogeton illinoiensis/nodosus and   Potamogeton alpinus

Sparganium americanum   and   Sparganium eurycarpum

Moving across the country, other species are likely to be replaced in part or all of their habitat by plants of a similar habit. For example:

Glyceria canadensis by other tall monocotyledons

Peltandra virginica by other large-leaved short emergents.

Thirdly, when a certain species is absent, most of the ecological niche may remain unfilled. For example:

Phalaris arundinacea   Sparganium chlorocarpum

If a species is absent, it is rarely because of competition and usually because of such factors as unsuitable flow, pollution or climate. In general, the southernmost streams recorded had the poorest flora. This is partly because of the unsuitable physical habitat (see Chapter 17), but also because the rich northern aquatic flora is becoming attenuated, and only a few species of the aquatic flora of the southern United States are present as far north as this.

16

North American ditches and canals

Ditches are abundant except in the south-east of the area visited. Most of them are dominated by Typha *spp., though other tall monocotyledons may be frequent. Short plants occur after dredging and in wetter dykes.*

DITCHES

In contrast to the situation in Britain, the ditch system in North America is seldom connected to the stream system, and indeed in drier parts the ditches themselves may not even be interconnected. The North American ditches were, of course, dug much more recently and with different aims in mind as regards drainage. They are usually 1–2 m wide, flooded at least in spring, and dry at least in late summer. The substrate is usually mud, with peat in swampy areas, or sand, etc. in dry sandy places. Larger and deeper, summer-flooded ditches are occasionally present. Channels intermediate between ditches and brooks are classed here as streams and described in Chapter 17.

The ditches may be intermittently dredged. In the first summer after dredging there is usually a community of short species with, for example:

Alisma plantago-aquatica	*Polygonum lapathifolium* agg.
Eleocharis erythropoda agg.	*Rorippa nasturtium-aquaticum* (limestone)
Equisetum palustre	*Sagittaria latifolia* (wetter or moving water)

Lemna minor (summer-flooded)

If the bed is dry for most of the summer, short grasses and other land plants occur.

Tall monocotyledons spread quickly and dominate 2–3 years after dredging, as in Britain (Chapters 14 and 21).

The parts of each province and state described in this chapter are shown in Fig. 15.1.

In **Quebec**, ditches are very common, except in the higher hills. The main species are:

Very frequent	*Typha angustifolia*
Frequent	*Typha latifolia* (*Typha* hybrids also present)

280

Ditches

Infrequent   🌿 *Carex* spp.   🌾 *Phragmites communis*

                  Eupatorium perfoliatus (drier)

                                                  🌾 *Scirpus* sp.

In ditches flooded more deeply, or which have been recently dredged, any tall monocotyledons are likely to be confined to the sides, while the beds have:

Most frequent   🌱 *Alisma plantago-aquatica*

Frequent   🌿 *Eleocharis erythropoda* agg.   🌿 *Polygonum lapathifolium* agg.

                  🌿 *Equisetum palustre*   🌿 *Sagittaria latifolia*

                  🌿 *Lemna minor*

Ditches on sandy glacial drift are likely to dry earlier in the year, and to bear mainly short monocotyledons.

In **Vermont**, ditches tend to be narrower and to dry earlier than in Quebec. The main dominants are:

Most frequent   🌿 *Typha latifolia*

Frequent   🌿 *Typha angustifolia*

Less frequent: Drier   🌿 *Cladium mariscoides*   *Lythrum salicaria*

                     🌿 *Equisetum palustre*   🌿 *Phalaris arundinacea*

Wetter   🌾 *Phragmites communis*   🌾 *Scirpus validus*

The **New York** ditches are equally frequent, and the main dominants are still *Typha* spp.

Ontario again has very abundant ditches, though their density decreases somewhat in the west. Dredging is needed particularly in these western arable areas. The main species are:

Most frequent   🌿 *Typha angustifolia*

Frequent   🌿 *Typha latifolia*

Less frequent   🌿 *Carex* spp.   🌿 *Equisetum fluviatile*

Drier   🌿 *Cladium mariscoides*   🌿 *Phalaris arundinacea*

                    🌿 *Glyceria canadensis*   🌿 Grasses, various

                    Lythrum salicaria

Wetter   🌱 *Alisma plantago-aquatica*   🌿 *Sparganium americanum*

 *Sagittaria latifolia*   Characeae

Limestone   *Rorippa nasturtium-aquaticum*

In the Midwest ditches are sparser – though still very frequent – and most are beside roads. When hedges are present, as in S. Michigan, ditches beside these are shaded and usually without plants. In **Michigan** the main species are:

Most frequent	*Typha latifolia*	
Frequent	*Typha angustifolia*	
Less frequent	*Carex* spp.	*Phalaris arundinacea*
	Cladium mariscoides (northern)	*Scirpus atrovirens* agg. (northern)
	Equisetum fluviatile	
Drier	*Lythrum salicaria*	Grasses
Wetter	*Alisma plantago-aquatica*	*Scirpus validus*
	Lemna minor	Characeae
	Peltandra virginica (southern)	

Particularly after dredging:

Equisetum palustre   *Juncus nodosus*

In **Wisconsin** the main species are:

Most frequent	*Typha latifolia*	
Frequent	*Typha angustifolia*	
Less frequent	*Carex* spp.	*Scirpus atrovirens* agg.
	Cladium mariscoides	*Scirpus validus*
	Phalaris arundinacea	*Sparganium eurycarpum*
Wetter	*Lemna minor*	
Limestone	*Rorippa nasturtium-aquaticum*	

Much of **Minnesota** is swamps and ditches and indeed dykes are very common in these areas. The main species are:

Most frequent	*Typha angustifolia*	*Typha latifolia*
Frequent	*Carex* spp.	*Phragmites communis*

🌱 *Cladium mariscoides*　　🌿 *Polygonum lapathifolium*
　　　　　　　　　　　　　　　　　agg. (emptier channels)

🌱 *Glyceria canadensis*　　　🌱 *Scirpus atrovirens* agg.
　　(sides)

🌿 *Phalaris arundinacea*

Wetter　　🌱 *Alisma plantago-*　　🌱 *Sparganium eurycarpum*
　　　　　　　aquatica　　　　　　　(sides)

🌱 *Lemna minor*

Deeper dykes have some of the above species at the sides, but may
also have, in the centre:

🌱 *Elodea nuttallii*　　　🌱 *Sagittaria cuneata*

🌱 *Lemna minor*　　　　　🌱 *Sagittaria latifolia*

🌱 *Nuphar variegatum*

The vegetation of these dykes apparently depends mainly on summer
water level, and dredging (see Chapter 14).

In **Iowa**, *Typha* spp. again dominate.

In the east, ditches are sparser, and become increasingly so towards
the south. In **New Jersey**, **Delaware** and **Maryland**, the main
species are:

Most frequent　　🌱 *Typha angustifolia*　　🌱 *Typha latifolia*

Less frequent　　🌱 *Carex* spp.　　　　　🌱 *Scirpus cuperinus* agg.

　　　　　　　　🌿 *Phragmites communis*　🌱 *Sparganium americanum*

Drier　　　　　Land species

Wetter　　　　🌱 *Alisma plantago-*
　　　　　　　　aquatica　　　　　　*Potamogeton* spp.

　　　　　　　Ludwigia palustis

In **Virginia** and **North Carolina**, ditches are very sparse. The main
species are:

Frequent　　　🌱 *Scirpus cuperinus* agg.　🌱 *Typha latifolia*

Less frequent　　*Hibiscus moscheutos*　　*Lobelia cardinalis*
　　　　　　　🌱 *Juncus* spp.

Drier and wetter　　as above

Ditches form a uniform habitat, and the general habit of their vegetation is similar throughout. The variations are due mainly to the geographical distribution of the potential dominants (e.g. *Glyceria canadensis* and *Phalaris arundinacea* are northern species only) or to the summer water level.

CANALS

Only a few canals were recorded, and as these were steep-sided and with stony beds, plants were absent or very sparse (a situation comparable to that described for the Caledonian Canal in Chapter 14).

17

North American streams: vegetation types

The types of vegetation seen in each of the provinces and states visited (Fig. 15.1) are described. The primary controlling factor is flow regime (including depth and substrate), though geology, the geographical distribution of potentially luxuriant species and perhaps land use are also important influences.

Stream vegetation varies considerably over the area visited. Only the route marked on the map in Fig. 15.1 was surveyed, and there may be many other vegetation types in these states and provinces which are not described here.

QUEBEC

The lowland streams usually have the negligible summer flow and great seasonal fluctuation in discharge described in Chapter 15. Those rising in higher ground frequently have coarse bouldery substrates and little or no vegetation, while those with more silt and more vegetation usually rise in the more low-lying regions. The ground is mainly covered with glacial drift, but Resistant rocks outcrop in the hills. The lowlands are mainly arable, with some forest. Dark water, usually from forest peat, is common and some streams are turbid (see Chapter 15). Quieter, smaller channels require intermittent dredging; the vegetation probably recovers within 3 years (see Chapter 21). Stream vegetation depends in general more on flow and substrate regimes than on geological differences, though brooks on forest peat may have a distinctive flora.

In the lowlands, ditches may be the headwaters of brooks (Plate 12; see Chapter 16). The wider, deeper channels of the dykes may have, for example:

Alisma plantago-aquatica

Carex spp. (sides)

Eleocharis erythropoda agg.

Juncus effusus (sides)

Phalaris arundinacea (sides) (see Chapter 16)

Polygonum lapathifolium agg. (sides)

Sagittaria latifolia

Typha angustifolia

Typha latifolia

The principal channel species is *Alisma plantago-aquatica*, though *Typha* spp. may be luxuriant in the centre as well as at the sides. As the dyke passes into a brook (with slightly more ground slope), scour increases and these species decrease, while *Sagittaria latifolia*, which can tolerate some scour and considerable fluctuations in water level, increases. Because the summer water movement is so little, the vegetation of dykes and brooks is more similar than in Britain. In deeper water, emerged plants are more confined to the shallow sides and submergents may be present, such as:

Callitriche palustris   *Sparganium chlorocarpum*

Callitriche is always sparse. *Sparganium chlorocarpum* can grow in flooded or summer-dried channels, so is well adapted to fluctuating water levels.

Small summer-dry brooks may have, for example:

Equisetum palustre   *Lythrum salicaria*

As stream size and scour increase, vegetation becomes more confined to the sides, the total amount of vegetation decreases and *Sagittaria latifolia* is increasingly the most important species (Fig. 15.4). Once the channel is definitely that of a stream it is never choked with vegetation and may have only a very few plants, the amount depending on the scour and silting.

In the larger rivers vegetation is almost confined to silty places, and is principally composed of tough plants at the sides. These are deep-rooted and firmly anchored, are tolerant of drying, and die-back in winter, so are not very susceptible to summer drought or winter and spring scour (see Chapters 3 and 15). A few frail submergents may be present also, growing in almost still water in late summer.

Streams rising in higher hills have a swifter flow in summer, and so are without the phase of nearly still water. The substrate at the channel sides is coarser than in the other streams, and this and the spring scour prevent colonisation by emergents. Submergents such as *Potamogeton epihydrus* are occasionally present.

The species diversity in Quebec streams is very low according to North American standards, most sites having 0–2 species. The normal maximum number of species present is 4–6, and this number occurs only where there is considerable silt. Such a stream is illustrated in Plate 13 and resembles a British stream with *Nuphar* and *Sparganium*.

The important species include:

Most frequent	*Sagittaria latifolia*	
Frequent	*Lemna minor*	*Sparganium chlorocarpum*
	Potamogeton epihydrus	Short grasses
Less frequent	*Alisma plantago-aquatica*	*Phalaris arundinacea*
	Eleocharis erythropoda agg.	*Polygonum lapathifolium* agg.

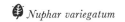 *Nuphar variegatum*   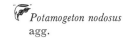 *Potamogeton nodosus* agg.

Potamogeton epihydrus is commoner in the north and west, and *Phalaris arundinacea*, *Potamogeton nodosus* agg. and *Sparganium chlorocarpum* in the south. Other submergents are rare, occuring in sites of, for example, more peat or less scour.

NOVA SCOTIA

The streams are on hills of Resistant rock and have swift nutrient-poor waters. The summers are cooler than in the other provinces and states described here. The vegetation is sparse and species-poor, resembling that of rivers in northern England and Scotland. Unlike the other parts of North America surveyed here, mosses are frequently found in the swifter streams. Slower streams have occasional tall monocotyledons at side, with, in shallow water, for example:

Nuphar variegatum   *Ranunculus* spp.

Nymphaea odorata   *Sagittaria* spp.

Potamogeton spp.   *Sparganium* spp.

This list is similar to those elsewhere in the chapter. *Ranunculus* is found on limestone (or gypsum deposits). In nutrient-poor streams, in bogs, connecting lakes, etc., there may be a sparse vegetation of:

    Eriocaulon septangulare   *Lobelia dortmanna*
    Isoetes spp.
(Information from [55].)

VERMONT

The streams seen are on grassy hills of Resistant rock and in the alluvial plain around Lake Champlain. The waters are usually clear, and the water level fluctuations are less than in Quebec.

In the hills, brooks may arise in shallow-sloping valleys, with soil, little scour, and tall monocotyledons in the channels, chiefly:

Sparganium americanum   *Phalaris arundinacea* (scarce)

Other brooks which rise on steeper slopes have more scour, coarser substrates and no river plants. Both types usually dry in summer. Scour increases downstream, where brook substrates are coarse even if water movement is little in summer.

Vegetation is abundant in silty places only. It is similar whether it occurs in the centre of a brook with perennial flow or at the sides of a river in the plain: neither have much scour. Fig. 17.1 shows the bands of luxuriant vegetation at the sides of a river, and also shows floating-leaved and semi-emerged plants growing well on dry mud in summer.

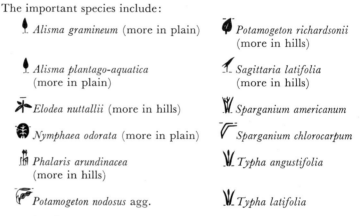

Lemna minor
Potamogeton epihydrus
Alisma plantago-aquatica
Sagittaria latifolia
Nymphaea odorata

Fig. 17.1. Silted river with plants on drying mud in July (Mud Creek, Vermont).

The important species include:

Alisma gramineum (more in plain) *Potamogeton richardsonii* (more in hills)

Alisma plantago-aquatica (more in plain) *Sagittaria latifolia* (more in hills)

Elodea nuttallii (more in hills) *Sparganium americanum*

Nymphaea odorata (more in plain) *Sparganium chlorocarpum*

Phalaris arundinacea (more in hills) *Typha angustifolia*

Potamogeton nodosus agg. *Typha latifolia*

As in Quebec, increased scour produces decreases in *Alisma* and increases in *Sagittaria*.

NEW YORK

The north-east of the State near Lake Champlain is a grassy plain on sandstone. Streams rising in this flat country may contain considerable vegetation even though the substrate is coarse.

Important species include:

Most frequent *Sparganium americanum*

Frequent *Elodea nuttallii* *Sagittaria* spp.

Lemna minor *Typha angustifolia*

Potamogeton nodosus agg. Grasses

Small summer-dry channels, including brooks arising in valley bottoms, often have tall monocotyledons such as:

Cladium mariscoides *Typha* spp.

Phalaris arundinacea Grasses

Larger streams, rising on higher land and thus having more scour, have less vegetation. The species are similar to those found in the small channels, though with rather more *Sparganium americanum*, and rather more submergents, such as:

   🗙*Elodea nuttallii*   🌿*Myriophyllum exalbescens*

ONTARIO

The Canadian Shield region, on Resistant rocks, is somewhat hilly, with woods, lakes and swamps, and frequent beaver dams. The rest of Ontario that was surveyed is fertile agricultural land with flat or rolling country, mainly on limestone or dolomite though often covered by glacial drift. The summer drop in water level is less than in Quebec, and the limestone streams are, of course, particularly stable since they are fed from springs. The stream vegetation is seldom choking but is species-rich, particularly on the Canadian Shield where human interference is least. Intermittent dredging is needed in silted channels in agricultural areas; their vegetation probably recovers in 2–3 years.

In the east there is no limestone influence, and the small brooks are usually summer-dry, generally containing land plants and perhaps:

   🌾*Phalaris arundinacea*   ⚘*Sagittaria latifolia*

Most of the streams here are medium-sized with at least some silt. There may be 6–9 species per site. As in Britain, any community is made up of a number of potentially frequent species which are characteristic of that stream type, with perhaps others which are rare in that community though often characteristic of other communities. Scour is less than in Quebec, and if the water is shallow enough, (e.g. 25–75 cm), as it usually is, plants are able to grow right across the channel. Tall emerged monocotyledons grow at the sides, with, inside them, the short large-leaved semi-emergents, and, in the centre, floating and submerged plants.

The most important species include:

Most frequent	*Sagittaria latifolia*	
Frequent	*Alisma plantago-aquatica*	*Sparganium americanum*
	Lemna minor	*Sparganium chlorocarpum*
	Nuphar variegatum	*Typha angustifolia*
	Potamogeton amplifolius	*Typha latifolia*
	Potamogeton nodosus agg.	Short grasses
	Sagittaria cuneata	Characeae

There is a wide range of other species which are sometimes present. Several are important in different stream types, for example:

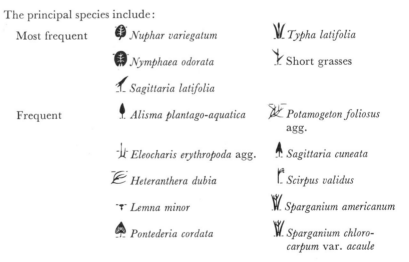 *Heteranthera dubia*   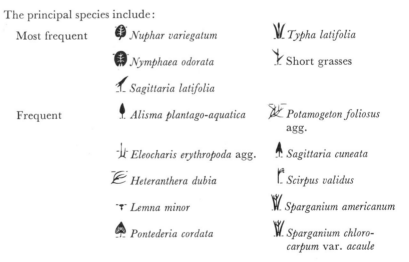 *Pontederia cordata*

On the Canadian Shield the small brooks are usually empty, whether they are summer-dry, or summer-flooded and stony. Most streams are medium-sized, of at least moderate depth, with silty or peaty substrates, little flow and peat-stained water. In this favourable habitat, 6–13 species per site is not uncommon, and the species diversity for the region as a whole is very high. A typical species-rich site is shown in Fig. 17.2.

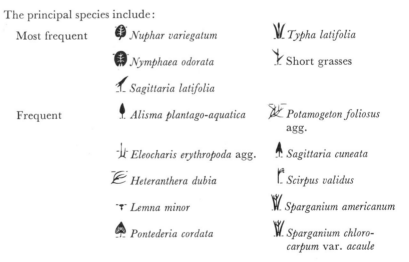

Typha latifolia

Scirpus validus

Sparganium americanum
Sagittaria cuneata
Nuphar variegatum
Alisma plantago-aquatica
Sagittaria latifolia
Nymphaea odorata
Pontederia cordata

Heteranthera dubia
Potamogeton amplifolius
Potamogeton foliosus

Fig. 17.2. Species-rich stream on Canadian Shield (Ontario). Note the distribution of emergents, both monocotyledons and dicotyledors, and that plants occur across the whole channel.

The principal species include:

Most frequent   *Nuphar variegatum*   *Typha latifolia*

Nymphaea odorata   Short grasses

Sagittaria latifolia

Frequent   *Alisma plantago-aquatica*   *Potamogeton foliosus* agg.

Eleocharis erythropoda agg.   *Sagittaria cuneata*

Heteranthera dubia   *Scirpus validus*

Lemna minor   *Sparganium americanum*

Pontederia cordata   *Sparganium chlorocarpum* var. *acaule*

In south-west Ontario there are minor variations in stream vegetation with geology and topography. Species richness is high in the north, sites often having 7–10 species and many having 4, but it decreases in the south. Because the bedrock is mainly limestone many streams have moderate or even fast flow in summer and the flow regime is less

290

variable than in the regions described above. Gravelly beds are more common and submerged species (typically *Potamogeton* spp. and *Heteranthera dubia*, rather than the British *Ranunculus* spp. and *Callitriche* spp.) may grow well over the channel. A few streams rise in tree swamps and are brown-stained even if they are also limestone springs. Most waters are clear, however, though locally they may be turbid enough to prevent submerged species growing more than *c.* 30 cm below the surface (see Chapter 15).

Some channel patterns are shown in Fig. 17.3. The limestone brook (*a*) is species-rich and has fringing herbs. The banks are relatively higher than in British chalk streams, indicating a flow regime more like that on British sandstone (see Chapter 12 and Fig. 10.2). A river with

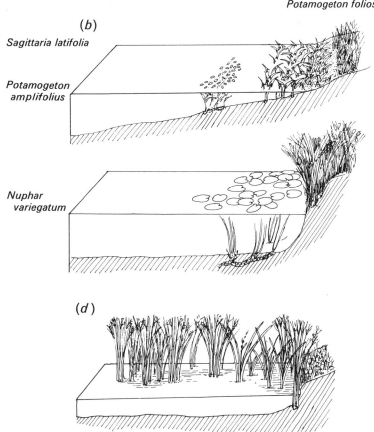

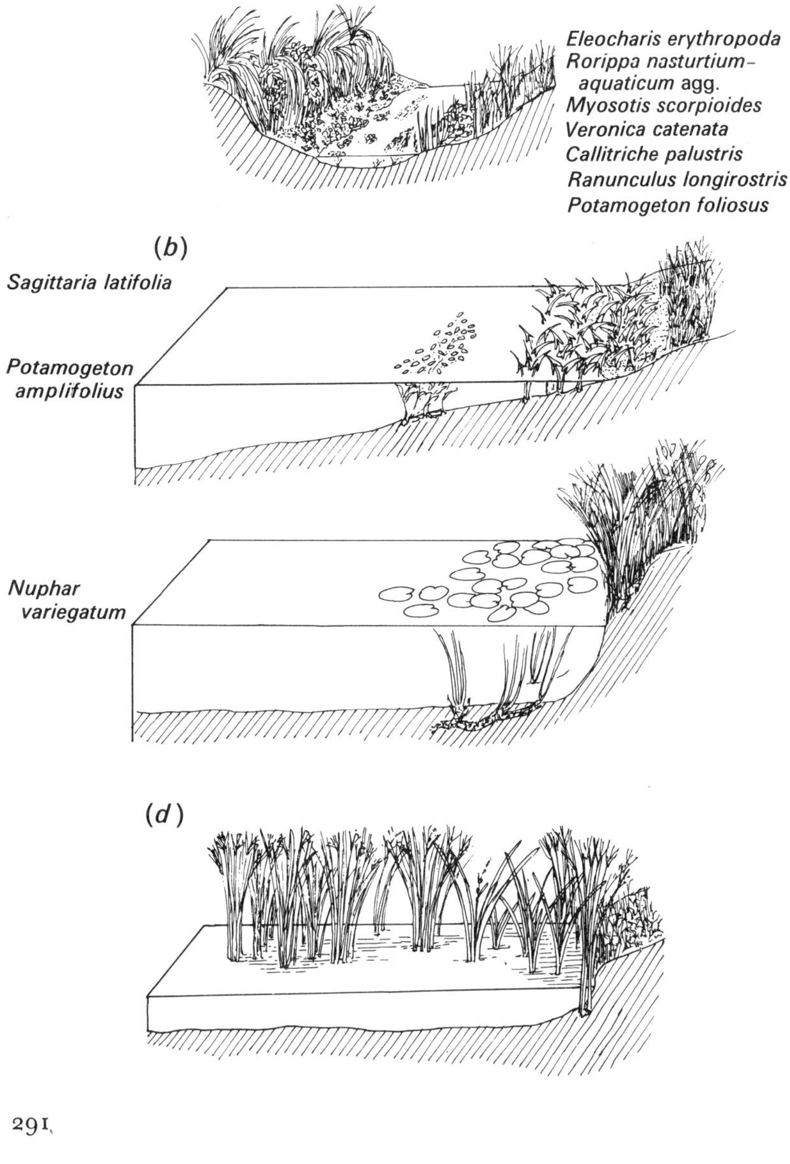

Fig. 17.3. Plant distributions in streams on limestone and clay (Ontario). (*a*) A limestone brook, with some fall in water level in summer but with perennial flow. It is species-rich with limestone plants (e.g. fringing herbs, *Callitriche palustris, Ranunculus longirostris*) and others (e.g. *Eleocharis erythropoda, Potamogeton foliosus*). (*b*) A stream with shallow silty sides, partly exposed in July, with a wide band of short emergents (*Sagittaria latifolia*) and sparser floating or submerged species confined to the sides (e.g. *Potamogeton amplifolius*). (*c*) A stream with steep banks without aquatic vegetation, though with plants on the channel bed (e.g. *Nuphar variegatum*). (*d*) A shallow silty stream with tall monocotyledons extending across the channel (e.g. *Scirpus validus, Sparganium americanum*). Contrast the plant patterns with Figs. 15.4 and 17.1.

291

wide shallow edges is shown in (*b*). Its vegetation consists of a wide band of short emergents and *Potamogeton* spp. nearer the centre. The steep slope and fairly shallow channel in (*c*) provide a good habitat for water-lilies. Dense emergents across a shallow silty river are shown in (*d*) (contrast this with Fig. 15.4, where there is more scour and summer drought).

Limestone streams resemble those of Wisconsin and Britain.

Fringing herbs	*Mentha piperita* agg.	*Rorippa nasturtium-aquaticum*
	Myosotis scorpioides	*Veronica catenata* agg.
Others	*Callitriche palustris*	*Sagittaria latifolia*
	Lemna minor	*Solanum dulcamara*
	Potamogeton foliosus	Short grasses
	Ranunculus longirostris	
Spring pools	*Hippuris vulgaris*	Characeae

Where there is rather less lime influence because of more drift on the surface, and in more agricultural areas with less marsh beside streams (see Chapter 15, Land use section), the species change:

Most frequent	*Potamogeton foliosus*	*Rorippa nasturtium-aquaticum*
Frequent	*Eleocharis erythropoda* agg.	*Typha angustifolia*
	Glyceria canadensis	*Typha latifolia*
	Lemna minor	Fringing herbs
	Phalaris arundinacea	Short grasses

There is still enough lime to prevent much occurrence of:

Alisma plantago-aquatica	*Scirpus validus*
Nuphar variegatum	*Sparganium americanum*
Nymphaea odorata	*Sparganium chlorocarpum*

When these brooks pass downstream over a non-lime glacial drift or bedrock, the lime species decrease and part-lime ones increase.

The principal species include:

| Most frequent | *Eleocharis erythropoda* agg. | Short grasses |
| | *Sagittaria latifolia* | |

Ontario

Frequent	Carex spp.		Ranunculus longi-rostris
	Glyceria canadensis		Scirpus validus
	Phalaris arundinacea		Sparganium chloro-carpum var. acaule
	Polygonum lapathifolium agg.		Typha angustifolia
	Potamogeton foliosus		Typha latifolia
	Potamogeton natans		Fringing herbs
	Potamogeton pectinatus		Characeae

Fringing herbs are still present, and most of the especially non-lime species (see above) are still unimportant. *Polygonum lapathifolium* agg. is fairly common as a result of the decrease of lime influence coupled with low scour.

This vegetation occurs both in the centre of medium-sized streams (e.g. 5–10 m wide) and the sides of rivers. There is less downstream variation than in Britain and once the channel is large enough to be classed as medium-sized then (provided the amount of lime influence remains constant) the community continues with little change.

When limestone influence is even less, turbidity increases, flow is slower and the substrates siltier. The change in species is considerable, though like species may still occur.

The important species include:

Most frequent	Sagittaria latifolia		Typha angustifolia
	Scirpus validus		Typha latifolia
	Sparganium americanum		Short grasses
Frequent	Alisma plantago-aquatica		Polygonum lapathi-folium agg.
	Eleocharis erythropoda agg.		Potamogeton ampli-folius
	Glyceria canadensis		Potamogeton foliosus
	Nuphar variegatum		Rorippa nasturtium-aquaticum

Large rivers flow over much glacial drift, and tend either to have sparse vegetation or to look like British clay rivers (with *Nuphar*, *Sagittaria*, *Scirpus* and *Sparganium*, etc.).

In drier parts without headwater springs or swamps, brooks rise as summer-dry gullies, often on sandy or gravelly slopes. At the sources, the beds may be grassed over or, if slightly damper, be covered by tall monocotyledons, e.g.:

Carex sp.	*Scirpus validus*
Cladium mariscoides	*Typha angustifolia*
Glyceria canadensis	*Typha latifolia*
Phalaris arundinacea	Other tall grasses

Passing downstream, scour increases and plants are absent or present only at the sides. Then when summer puddles are first found, short emergents (e.g. *Alisma plantago-aquatica*) or, if there is less scour, tall monocotyledons enter. *Polygonum lapathifolium* agg. may also enter above the zone of perennial flow. Once there is perennial flow, transition to the communities described above takes place.

MICHIGAN

The parts surveyed are lowland, mostly covered with thick glacial drift except for some Resistant rock outcropping in the Upper Peninsula. The south is mainly fertile arable land, and species diversity tends to be low except where swamps, etc. fringe the streams. The north and the Upper Peninsula are mainly wooded, with many tree swamps and peat-stained streams, and locally some infertile country. Forest tends to decrease the species diversity at any one site, though not that for the river as a whole (see Chapter 15). Even so, 7–10 species per site is not uncommon in the north and 6–7 in the more species-rich parts of the south. The water level is more stable than in most Ontario streams and is more like that of the limestone Ontario streams or British lowland streams. Submerged species can grow nearer to the edge as less of this edge dries out in summer. The side bands of emergents are narrow, though often species-rich. The flow types are similar to those of Britain, with much moderate flow in late summer. Snow-melt scour is less than in the regions described above.

A few southern rivers were surveyed for downstream patterning, as described in Chapter 11 for Britain. These streams rise in tree swamps and so have fairly deep, brown waters on peaty beds (Plate 14). There was no consistent downstream increase in stream width or depth before its confluence with a larger river. Current speed may increase downstream, and if it does then the substrate becomes coarser. With little downstream patterning in physical characteristics, there is also little patterning in vegetation. The tree swamp channels near the source of Augusta Creek contain, for example:

Carex spp.	*Phragmites communis*
Decodon verticillatus	*Potamogeton* spp.
Lemna minor	*Sparganium americanum*
Lemna polyrhiza	*Sparganium chlorocarpum*

Michigan

🍃 *Nuphar advena*   🌿 *Typha angustifolia*

🕷 *Nymphaea odorata* agg.   *Wolffia* sp.

🌲 *Peltandra virginica*

Farther downstream current increases and a tributary enters. This is from agricultural land and, as is typical of such streams (see Chapter 15), it is species-poor. As a result of these two influences the stream vegetation decreases and species more associated with flow enter, such as:

🌿 *Heteranthera dubia*   🌿 *Potamogeton foliosus*

🌿 *Potamogeton crispus*

A second stream, Gull Lake Outlet showed a similar though greater downstream change of vegetation. Here the increase in current is greater, flow type becoming intermittently fast, and the stream is species-poor.

Species of swifter flow include:

🌿 *Heteranther dubia*   *Potamogeton* spp.

🌿 *Myriophyllum exalbescens*   🌿 *Vallisneria americana*

🌿 *Potamogeton pectinatus*

In South Michigan two areas of species-rich streams were surveyed. In the first area most streams are slow and somewhat silted and there is an unusually wide range of frequent species, including some limestone plants (such as *Rorippa nasturtium-aquaticum* and *Solanum dulcamara*).

The principal species include:

Most frequent	🌿 *Lemna minor*	🌿 *Sparganium americanum*
	🌲 *Peltandra virginica*	🌿 Short grasses
	🌿 *Potamogeton nodosus* agg.	
	🌿 *Polygonum lapathifolium* agg.	
Frequent	🌿 *Alisma plantago-aquatica*	🌿 *Potamogeton pectinatus*
	🌿 *Carex ?stricta*	🌿 *Rorippa nasturtium-aquaticum* agg.
	🌿 *Decodon verticillatus* (swamp)	🌿 *Sagittaria latifolia*
	🌿 *Eleocharis erythropoda* agg.	🌿 *Scirpus validus*

Heteranthera dubia	*Solanum dulcamara*
Iris sp.	*Sparganium chloro-carpum*
Lemna polyrhiza (swamp)	*Typha angustifolia*
Nuphar advena	*Typha latifolia*
Nymphaea odorata agg.	*Wolffia* sp. (swamp)
Phalaris arundinacea	

The second area has streams of swifter flow and more luxuriant submerged plants. *Potamogeton* spp. are frequently abundant (e.g. Plate 15). Stream edges are well marked, with a fairly stable summer water level. The edge species are, however, the deep-rooted ones tolerating water level fluctuations, not the (limestone) fringing herbs.

The principal species include:

Frequent

Eleocharis erythropoda agg.	*Potamogeton foliosus*
Heteranthera dubia	*Potamogeton nodosus* agg.
Najas flexilis agg.	*Sagittaria* spp.
Nymphaea odorata agg.	*Scirpus validus*
Phalaris arundinacea	*Sparganium ameri-canum*
Polygonum lapathifolium agg.	*Utricularia vulgaris*
Pontederia cordata	Short grasses

Other parts of the south have sparser streams and less vegetation. The species present are those from the above lists that are the most ecologically tolerant, and are mainly emergents. Many streams are turbid, perhaps indicating excess agricultural silting.

Principal species include:

Lemna minor	*Sparganium americanum*
Phalaris arundinacea	Short grasses
Sagittaria latifolia	

Rorippa nasturtium-aquatium dominates in some small channels.

Near the sources, summer-dry brooks may be grassy or, if damper, may bear tall emergents such as:

Phalaris arundinacea	*Typha latifolia*
Typha angustifolia	

In central Michigan, woodland and peat-stained water increases, though streams remain somewhat species-poor. The vegetation alters slightly.

Principal species include:

Most frequent	*Lemna minor*	Short grasses
	Sparganium americanum	
Frequent	*Alisma plantago-aquatica*	*Potamogeton pectinatus*
	Eleocharis erythropoda agg.	*Sparganium chlorocarpum*
	Nymphaea odorata agg.	*Typha latifolia*
	Peltandra virginica	

Northern Michigan, like southern Michigan, is alkaline [50], though the land is more sandy and less fertile. Streams are often gravelly with moderate flow. The water is frequently brown as a result of the woodland which covers most of the countryside. The large-leaved, short edge species are unimportant here, which could be because the sandy ground produces a more unstable substrate.

Important species include:

Frequent	*Carex* spp.	*Scirpus validus*
	Eleocharis erythropoda agg.	*Sparganium americanum*
	Equisetum spp.	*Sparganium chlorocarpum*
	Juncus spp.	*Typha latifolia*
	Lemna minor	Short grasses
	Potamogeton pectinatus	Characeae
	Rorippa nasturtium-aquaticum	

The Upper Peninsula has slower, more silty streams, often with brown water. The proportion of important species that are submerged or part-submerged is over half. This is exceptionally high and shows a great contrast with, for example, Quebec.

Principal species include:

Most frequent	*Sparganium chlorocarpum*	
Frequent	*Callitriche palustris*	*Potamogeton illinoiensis* agg.

297

Ⲯ *Carex* spp.   Ⳁ *Sagittaria* cf. *cuneata*

-Ⳉ *Eleocharis erythropoda* agg.   Ⲯ *Sparganium ameri-*
                                       canum

Ⲙ *Nuphar variegatum*   ⳉ *Sparganium minimum*
                                       (east)

Potamogeton alpinus   Ⲯ *Typha latifolia*

Ⳡ *Potamogeton amplifolius*

Hippuris vulgaris occurs, but is not in limestone springs. This species also has two distinct habitats in Britain.

Michigan shows considerable geographical variation in species, for example:

Southern   Ⲯ *Decodon verticillatum*   ⳡ *Peltandra virginica*

             Ⲙ *Nuphar advena*

Northern - Ⳑ *Glyceria canadensis*   Ⳡ *Scirpus atrovirens*
                                                  agg.

             Ⲙ *Nuphar variegatum*

The east–west trend seen between Quebec and Michigan is also considerable and includes:

Eastern, more   Ⳁ *Sagittaria latifolia*

Western, more   Ⲯ *Callitriche palustris* (local)   Ⳁ *Sagittaria cuneata*

                 Ⲙ *Nymphaea*   ⳉ *Sparganium chloro-*
                                                           carpum

INDIANA

The small northern area surveyed is in the same arable country, and drainage basin as the adjacent parts of S. Michigan, with similar important species. Clear-water rivers are species-rich, and brooks are somewhat species-poor. It is reported that, farther south, agricultural silting prevents much plant growth in the streams.

WISCONSIN

Wisconsin is mainly lowland: in the south mostly pasture and arable, but with much woodland in the north. Glacial drift covers most of the land, and this is often sandy in the west (where also sandstone may outcrop at the surface) and clayey to the south-east. The bedrock in the south-east is usually dolomite, which outcrops locally. In this fertile arable area streams are often too turbid for plant growth. Vegetation in rivers large and swift enough to keep silt in suspension are damaged by agricultural silting [47] while streams large enough for carp (and not protected from these by impoundments) may suffer very serious damage from them (also see Chapter 15).

The general aspect of the streams, in the pattern of edge and channel plants, is like that of Britain.

Continuing the east to west trend, summer flow is greater here than it is farther east and the water level more stable, particularly on the dolomite. The maximum monthly discharge of the snow-melt floods is usually only 3–20 times the minimum summer flow [49], the lower ratios tending to occur with limestone springs or where there are frequent impoundments.

In the south east, where limestone influence and swamp-fringes are often present, species diversity is high, clear-water sites quite often having 8–12 species. Elsewhere, 4–7 species per site is the normal maximum, with most having 0–2 angiosperms, particularly in drier places.

Whole-river surveys have been done in the north on the Pine–Popple river system [56] and the Brule [57], and also in the south-east [58]. If variation in lime influence is excluded, there is remarkably little downstream change in vegetation, (as in Michigan and Ontario, see above). Loss of species below sewage works is recorded (see Chapter 22). Vegetation is more abundant on gravel than on silt, the latter being unstable because of the flow patterns or, in larger stream, carp.

Vegetation patterns in the Pine–Popple river system vary with:

beaver dams	flow
bed stability	shading
bed ice in winter	thick organic sediment (usually preventing plant growth)
depth	turbidity

The type of variation recorded in the Pine–Popple River includes:

Small channels:	*Potamogeton obtusifolius*	*Potamogeton epihydrus*
Large channels:	*Potamogeton nodosus*	
Lower reaches:	*Sparganium eurycarpum*	*Vallisneria americana*
Sheltered:	*Callitriche palustris*	*Lemna minor*
	Equisetum fluviatile	*Potamogeton berchtoldii*
Near impoundments:	*Ceratophyllum demersum*	*Nymphaea odorata* agg.
	Nuphar variegatum	

299

However, this type of variation is not of universal occurrence, even in Wisconsin. For example, *Sparganium eurycarpum* and *Vallisneria americana* can occur in upper reaches, and although *Nuphar variegatum* and *Nymphaea odorata* agg. avoid swift flow, they do not need impoundments. The contrast with Britain is striking (see Chapter 11) and any variation is usually due to inter-mittent, not downstream, changes in shelter, flow, etc. A hydroelectric power station, for instance, prevents plant growth in the zone where there are major fluctuations in discharge. The other streams studied showed a similar lack of downstream variation except for the alterations in limestone influence and the differences between small brooks and large rivers – though even here several species occur in both types (e.g. *Sparganium chlorocarpum*).

In the north, some streams have few plants, being stony and slow or having turbid or dark water. In swifter, siltier streams (i.e. those with more stable discharges), vegetation increases.

Important species include:

Frequent   *Elodea nuttallii*   *Sparganium chlorocarpum*

        Potamogeton spp.   Short grasses

        Sagittaria latifolia

Near the sources, the brooks are usually summer-dry. Somewhat lower, they may be stony and bare, or silty and with tall plants such as:

Phalaris arundinacea   *Typha latifolia*

Scirpus microcarpus   Grasses

Typha angustifolia

In the west, species-rich streams are like those of the south-east (below), but most are species-poor. Sandy hills run beside the Missi-ssippi River, and their effect on stream vegetation is described in Chapter 15 (p. 269). The sparse streams on the dry dissected plateau of sandstone and dolomite in the south-west are rather turbid, and usually without vegetation. In the dolomite valleys, however, the plants are those of species-poor limestone, e.g.

Polygonum lapathifolium agg.   *Rorippa nasturtium-aquaticum*

In the south-east the vegetation of limestone streams is like that of Ontario (see above) and England (see Chapter 12). The narrowest channels have 2–3 species, with 4–5 occurring a little downstream, as in Britain.

Principal species include:

Berula erecta   *Myosotis scorpioides*

Carex spp.   *Phalaris arundinacea*

Elodea ?canadensis   *Ranunculus longirostris*

Hippuris vulgaris (pools)   *Rorippa nasturtium-aquaticum*

Wisconsin

⋅⊤⋅ *Lemna minor*   ⋏ Short grasses

🜲 *Mentha piperita*   ⚘ Mosses

🜲 *Mimulus glabrata*

When the streams flow on to glacial drift the lime influence decreases, and non-lime species enter.

The first species to enter include:

⋇ *Elodea nuttallii*   ⋎ *Veronica catenata* agg.

⋎ *Potamogeton crispus*   🜲 *Zannichellia palustris*

𝅘 *Potamogeton natans*

Slightly farther downstream, there may be, for example:

🜲 *Potamogeton foliosus*   ⎾ *Scirpus validus*

⋏ *Sagittaria cuneata*   ⋎ *Sparganium chlorocarpum*

⋏ *Sagittaria latifolia*

Where lime and non-lime influences overlap, species of both groups can occur together, and there may even be 20 species at one site. The most species-rich sites often occur where the habitat is suitable for two or more plant communities, demonstrating yet again that species are seldom excluded from streams by competition.

Brooks without limestone springs tend to be summer-dry near the sources.

The first species to enter may be some of the following:

⋎ *Carex stricta*   ⋎ *Iris ?pseudacorus*

⊣⊢ *Eleocharis erythropoda* agg.   ⋏ *Sagittaria latifolia*

╁ *Equisetum fluviatile*   ⋎ Short grasses

The main species-rich rivers show little variation in species along their length once the channel has reached *c.* 0.5 deep, a depth which

Fig. 17.4; Species-rich brook of moderate flow (Wisconsin). The proportion of submerged species, in both number and in bulk of vegetation, is very high for North America.

Typha latifolia

Sparganium eurycarpum

Lemna minor
Potamogeton illinoiensis
Elodea nuttallii
Potamogeton zosteroformis
Potamogeton foliosus

301

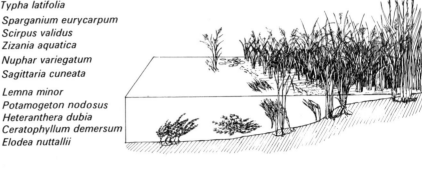

Typha latifolia
Sparganium eurycarpum
Scirpus validus
Zizania aquatica
Nuphar variegatum
Sagittaria cuneata

Lemna minor
Potamogeton nodosus
Heteranthera dubia
Ceratophyllum demersum
Elodea nuttallii

Fig. 17.5. Species-rich, slow-flowing stream with a wide swamp fringe (Mukwonago R., Wisconsin).

allows submerged and floating-leaved plants to grow well. (In deep rivers, of course, plants are confined to the sides.) Channel width varies from 3 m upwards. The beds are stable, with various substrate types, permitting the good establishment of several plant habits. The proportion of submerged species is consequently very high, both as regards luxuriance and diversity (e.g. Fig. 17.4). Fig. 17.5 shows a typical swamp fringe of a species-rich site (and an aquatic grass, *Zizania aquatica*, growing into deep water). The band of emerged plants is very narrow (Fig. 17.4) if the summer water level keeps high, but where the water level drops, leaving a wide band of damp mud in late summer, there is also a wide band of short emergents.

Principal species include:

Most frequent	*Ceratophyllum demersum*	*Potamogeton pectinatus*
	Elodea nuttallii	*Sagittaria latifolia*
	Lemna minor	*Sparganium eurycarpum*
Frequent	*Carex stricta*	*Potamogeton illinoiensis*
	Heteranthera dubia	*Potamogeton nodosus*
	Leersia oryzoides	*Sagittaria cuneata*
	Lemna polyrhiza	*Scirpus validus*
	Myriophyllum exalbescens	*Vallisneria americana*
	Nymphaea odorata agg.	Short grasses
	Phalaris arundinacea	

Some lake species are washed down from impoundments, and from temporary populations in the streams, e.g.

Myriophyllum verticillatum

Some streams have small springs, causing small seepage areas at the sides, particularly in the upper reaches. These are usually marked by patches of

 *Rorippa nasturtium-aquaticum*

If the seepage is set back a little from the river channel, it may also bear species such as:

Elodea sp.   *Solanum dulcamara* (sides)

Ranunculus sp.   *Zannichellia palustris*

The species-poor streams usually contain species from the above lists. As in other regions these more tolerant species are usually emergents, which are also morphologically tougher. In carp-infested waters toughness is an obvious advantage as it makes the plants less palatable to the fish.

MINNESOTA

The parts surveyed are lowland, mostly covered with calcareous glacial drift. In the north the land is mainly wooded, with swamps, marshes and some rough grassland. Here the watercourses are species-rich. The south is fertile agricultural land, on sandstone, with species-poor streams. Turbid water (as a result of fish) occurs in the far south. In the north-east the streams are species-poor, as in adjacent Wisconsin.

In most of the north, streams rise in swamps or tree swamps, and dykes as well as brooks are frequent. The smaller channels have peaty substrates, while the larger are silty with very slow flow. The waters are, of course, often brown and so there is less submerged vegetation than is usual in species-rich areas with a high summer water level. Channels with clear water not over 1 m deep may be choked with plants.

Important species include:

Most frequent	*Lemna minor*	*Potamogeton pectinatus*
	Phalaris arundinacea	
Frequent	*Carex* spp.	
	Ceratophyllum demersum	*Sagittaria cuneata*
	Elodea nuttallii	*Sagittaria rigida*
	Nuphar variegatum	*Scirpus validus*
	Nymphaea odorata agg.	*Sparganium chloro-carpum*
	Potamogeton natans	*Sparganium eurycarpum*

Emergents, such as *Zizania aquatica* (Fig. 17.5) can grow into the water.

One river (Rum River), was surveyed from its species-rich reaches in the swampy country into the arable sandstone land, where the river is almost without vegetation. The decrease in plants may be attributable to the river banks becoming steep and sandy, and so poor for anchorage, and the water becoming browner, thus confining plants to the shallow sides. There is also the change in land use.

In the south, streams are sparser and more turbid, with more water movement and somewhat coarser substrates. They are species-poor, with little submerged vegetation.

The more important species include:

⊤ *Lemna minor*   *Polygonum lapathifolium* agg.

Phalaris arundinacea   Short grasses

The Mississippi River in this area is very wide, with bands of river plants on muddy ground at the side.

The principal species include:

⊤ *Lemna minor*   *Sagittaria latifolia*

✳ *Lemna polyrhiza*   *Sparganium* sp.

Nelumbo lutea   *Vallisneria americana*

Nymphaea odorata agg. *Wolffia* sp.

IOWA

The small part of the north-east visited has dry hills with frequent summer-dry stony brooks without river plants, though some grass may be present if there are fewer stones. In the valleys most streams are species-poor. Some are turbid.

In streams which are not turbid the principal species include:

Glyceria canadensis   *Potamogeton crispus*

⊤ *Lemna minor*   *Sagittaria latifolia*

Phalaris arundinacea   Short grasses

NEW JERSEY

The Pine Barrens in the south are on infertile sand. Their streams are species-rich and most sites have 6–7 species. In the north the bedrock is more mixed and the land is fertile and arable. Most streams here are turbid with little vegetation, sites usually having only 1–2 river plants.

In the Pine Barrens, small channels in acid bogs may have:

Scirpus cyperinus   *Vaccinium macrocarpum*

Utricularia sp.   *Sphagnum* sp.

Tree swamps are common, with deep brown-water channels and slow flow. Small dicotyledons are frequent on the edges. Winter snow here is irregular, and the land is so flat that there is little force in snow-melt floods. The fringing herb habit can therefore occur, although the large-leaved short emergents and the tall monocotyledons are present also.

The principal species include:

Most frequent	*Sparganium americanum*	
Frequent	*Lemna minor*	*Polygonum hydropiper-oides*
	Ludwigia palustris	*Pontederia cordata*
	Nuphar advena agg.	*Scirpus cyperinus* agg.
	Nymphaea odorata agg.	*Utricularia* spp.

Southern species, seen in S. Michigan, occur here also, e.g.:

Decodon verticillatus   *Peltandra virginica*

In the species-poor usually turbid streams of the agricultural areas, the most frequent species are emergents.

The principal species include:

Ludwigia palustris	*Sparganium americanum*
Polygonum lapathifolium agg.	Short grasses
Sagittaria latifolia	

PENNSYLVANIA

Of all the states and provinces visited, this small part of Pennsylvania is the only one with consistently species-poor streams. Only one site recorded had much vegetation (with *Callitriche palustris*, *Rorippa nasturtium-aquaticum*, etc.; described in Chapter 15). Streams were recorded on various rock types: in the gorge of the Delaware River, on this river (bouldery, with scour), on the plain above, and in the hills to the south. Streams are somewhat sparse.

The only even slightly frequent species recorded include:

Polygonum lapathifolium agg.	Short grasses
Potamogeton richardsonii	

DELAWARE AND MARYLAND

The flat peninsula of Delaware and part of Maryland is on sands, gravels and clays, and is mainly arable. As well as the inland streams there are quiet coastal channels with brackish vegetation. South of the peninsula the land is somewhat hilly and streams are few. Coastal channels are like those of the peninsula. The inland peninsula streams are slow and silty, often somewhat turbid or brown, and species-poor.

The principal species include:

Ludwigia palustris   *Sagittaria latifolia*

Nuphar advena agg.   *Sparganium americanum* agg.

Peltandra virginica   Grasses

Pontederia cordata

On the mainland, vegetation is sparser.

Brooks often arise as summer-dry gullies, which may be empty, or, if somewhat damper, with, for example:

Polygonum spp.   *Typha latifolia*

Lower, the sides may have:

Ludwigia palustris   *Sparganium americanum* agg.

and, in deeper water, for example:

Pontederia cordata   *Sagittaria latifolia*

As streams approach the coast, species diversity and total vegetation often increase. This is partly due to an increase in the species listed above and partly because of the appearance of species more associated, in this region, with tidal influence (Fig. 17.6). Yet nearer to the coast the inland species are lost and species diversity decreases again. This is yet another illustration of habitat being more important than competition in influencing species-richness.

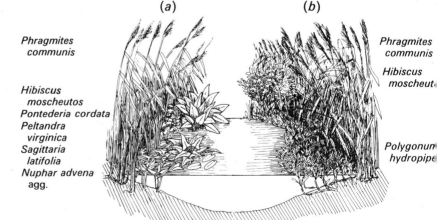

Fig. 17.6. Streams with both fresh water and tidal influence (Delaware). (*a*) A river with *Phragmites communis* marsh behind, and clumps of wide-leaved emergents. (*b*) A brook with *Phragmites communis* fringe and *Polygonum hydropiperoides* in shallow water.

The first species entering towards the coast include:

Hibiscus militaris (scarce)	*Lobelia cardinalis*
Hibiscus moscheutos	🌿 *Myriophyllum* sp.

Nearer the sea, *Phragmites communis* first increases at the edges and then moves further from the channel, being replaced at the edges by a fringe of *Spartina alterniflora*. By this state the other species are found only in marshes behind the *Phragmites communis* band where tidal influence is least. The water is very turbid and the channel bed silty, so the habitat is unsuitable for short channel species.

VIRGINIA AND NORTH CAROLINA

The streams surveyed are in the lowlands, mainly on sands and clays. Woodland and tree swamps are common. Streams rising in these have deep and silty channels, with brown water which may be kept turbid by frogs, turtles, etc. The water levels are relatively stable. Some rivers rise in the inland hills, and these are more likely to vary in water level. Streams are as sparse as in southern Maryland and fewer than in the other regions surveyed.

Coastal streams in Virginia have similar vegetation to those of Delaware and Maryland (see above).

Shade is heavy in the woods (Plate 1), and the breaks in the canopy beside the roads are fewer than in the more northern forests. Consequently the vegetation recorded may appear unduly sparse. However, channels are often exceptionally deep for their width (e.g. 1 m deep, 3 m wide) and thus have steep banks. Coupled with the dark-coloured water, this means a very poor habitat for plants, and plants often occur only in those channels which have gentle slopes at the sides. The streams are typically species-poor, most sites having 0–1 species and only a few having 4–6 species.

Because of the stable water levels there is often a luxuriant fringe of small-leaved herbs.

The most frequent species include:

Ludwigia palustris	*Polygonum lapathifolium* agg.
Polygonum hydropiperoides	*Sparganium americanum* agg.

Large-leaved short emergents and tall monocotyledons may also occur in this same habitat. For example:

Eleocharis spp.	*Sparganium americanum* ag.
Peltandra virginica	*Typha latifolia*
Pontederia cordata	Tall grasses
Scirpus validus	

In the channel, plants are usually absent. Those appearing in larger deeper channels include:

Hippuris vulgaris (large rivers)   *Nymphaea odorata* agg.

Lemna minor   *Potamogeton* cf. *epihydrus*

Nuphar advena agg.

Brooks which do not rise in tree swamps are summer-dry or summer-damp near the sources. As is usually the case, the small-leaved herbs (above) are found where there is more scour and the tall monocotyledons where scour is less.

Compared with the more northerly areas visited, there is some variation of vegetation with geographical location, for example, in the south:

Present   *Bacopa caroliniana*   *Peltandra virginica*

                       Decodon verticillatus

Absent or rare   *Alisma plantago-aquatica*   *Sagittaria* spp.

                       Cladium mariscoides   *Scirpus atrovirens* agg.

                       Glyceria canadensis   *Sparganium chloro-carpum*

18

Uses and benefits of river plants

Plants are part of the natural ecosystem of the river, and if the one is to be preserved, the other must be also. Briefly, the reasons why plants may be unwanted are if they hinder the passage of water for drainage purposes, or hinder boats, anglers, etc. The benefits of river plants are that they anchor in, and stabilise, channel banks; provide shelter, a substrate, and sometimes food, for invertebrate animals; and their clumps shelter fish, can protect some fish while they are spawning, and can provide invertebrates as food. Some plant species can be used as food for man or for birds, or as medicines for man. Plants also have considerable aesthetic value.

Recommendations are given for satisfactory amounts of vegetation in different stream types.

SOIL STABILISATION

Plant roots and rhizomes stabilise the substrate on the banks and on the channel bed. On the banks particularly they are most important for preventing erosion, and their loss often means major expenditure on bank maintenance. The tall monocotyledons or trees are the most effective group of plants for preventing erosion, followed probably by the shorter grasses. Silt banks, e.g. on silt alluvial plains, are perhaps the least stable of the common types of banks and so most care is needed to stabilise these.

During storm flows large bushy plants with poor anchorage are liable to be torn from the banks and the gaps left can start an erosion cycle. Where banks are deliberately cleared of plants, for example by the use of herbicides on dyke banks, then the risk of erosion may be serious. In narrow channels the stabilising effect of tall plants or bushes on the banks may be outweighed by the flood hazards they create, so the best protection from erosion then is trees and grass swards (see Chapter 14), which are able to stabilise the banks well while offering little resistance to flow. Though plants may also stabilise channel beds, their effect here is less than it is on the banks and they are not such an important factor in the maintenance of the watercourse. An exceptional situation is where boat traffic is dense over a silt bed. Here the silt is protected from disturbance while plants are present but if the plants are lost the disturbance prevents any recolonisation.

309

WATER PURIFICATION

When plants die they decompose. The dead organic matter is also a necessary part of the ecosystem, and dissolved substances are released from this into the soil and water.

River plants often oxygenate the water, and influence the concentration of dissolved carbon dioxide and the pH [7, 59]. The oxygen released by the plants is used by animals and by bacteria on decaying organic matter. If the oxygen demands of the water are great, as when organic effluents enter the stream, plants are most important for water purification. Some species, such as *Schoenoplectus lacustris*, kill coliform bacteria and aid water purification in this way. It is not known whether the effect is produced by itself or from small algae growing on it[60].

PLANTS AND ANIMALS

The small invertebrate animals found in streams usually live on the surfaces of plants and stones in the watercourse. Thus the more plants there are, the greater the surface area available and, generally, the more invertebrates there will be. Some invertebrates actually feed on the plants, though there are others that eat particles in the water, or other animals. Plants may also act as protection for a particular phase in the life cycle (e.g. eggs, larvae) of some species, where these would otherwise be eaten by other animals or (as in the case of mosquito eggs) washed away. Other species, such as dragonflies, use emerged plants to crawl up out of the water on when they change from their aquatic larval stage to the winged adult stage. In chalk streams it has been shown that the species of invertebrates present are not directly related to the species of plants, although their distributions are correlated because both are related to the flow type. In a chalk stream the plants – whatever their species – have more mayflies (Ephemeroptera) and blackflies (*Simulium*) in flow, while plants in sheltered places have more freshwater shrimps (*Gammarus*) [61]. However, in some still-water dykes some invertebrates may be associated with specific plants, and here animal conservation implies plant conservation.

Plants are a great benefit to some fisheries, increasing the food supply of the fish by increasing the invertebrates. In highly eutrophic water that already contains large quantities of detritus and algae, however, they are less necessary. They also provide shelter from storm flows, protection from predators for small fish, and for some species a sheltered environment for spawning. Some fish feed directly on the plants. Anglers tend to prefer fine-leaved species (e.g. *Myriophyllum* spp., *Ranunculus* spp.). This is partly through a misapprehension that these contain more food organisms (while, as explained above, it is in fact the flow type that controls the distribution of both plants and invertebrates) and partly because if plants are dense the fine-leaved species are less hindrance to angling than, for example, the large floating leaves of *Nuphar lutea*.

Birds often feed on the fruits and shoots of river plants (Table 18.1)

TABLE 18.1 *River plants eaten by birds*

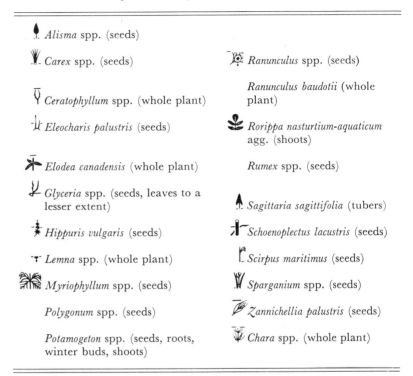

Alisma spp. (seeds)	
Carex spp. (seeds)	*Ranunculus* spp. (seeds)
Ceratophyllum spp. (whole plant)	*Ranunculus baudotii* (whole plant)
Eleocharis palustris (seeds)	*Rorippa nasturtium-aquaticum* agg. (shoots)
Elodea canadensis (whole plant)	*Rumex* spp. (seeds)
Glyceria spp. (seeds, leaves to a lesser extent)	*Sagittaria sagittifolia* (tubers)
Hippuris vulgaris (seeds)	*Schoenoplectus lacustris* (seeds)
Lemna spp. (whole plant)	*Scirpus maritimus* (seeds)
Myriophyllum spp. (seeds)	*Sparganium* spp. (seeds)
Polygonum spp. (seeds)	*Zannichellia palustris* (seeds)
Potamogeton spp. (seeds, roots, winter buds, shoots)	*Chara* spp. (whole plant)

Most data from [63].

and emergent plants provide shelter both for the birds themselves and for their nests (also see [62], [63] and the section on recommended vegetation below).

PLANTS USED BY MAN

Some plants can be used as food or medicine for man (Table 18.2). In Britain, however, watercress (*Rorippa nasturtium-aquaticum* agg.) is the only plant widely used now.

Because water plants take nutrients from the soil and water, it is possible to use them for lowering the nutrient status of the water body where eutrophication has been great. This is done simply by allowing the plants to take up the nutrients during growth and then cutting and removing them from the water. The method is effective if the plants concerned grow very quickly, so as to fill the water body before their season for rapid growth comes to an end, and they are then removed as

TABLE 18.2   *River plants eaten by man*

Species	Part(s) used	Use
Υ *Acorus calamus*	Underground parts	Medicinal
Ɣ. *Butomus umbellatus*	Underground parts	Medicinal
Ƴ *Glyceria fluitans*	Seeds	
Ɏ *Phragmites communis*	Underground parts	
⚘ *Rorippa nasturtium-aquaticum* agg.	Shoots	Food
⚲ *Sagittaria sagittifolia*	Underground parts	
Ψ *Typha* spp.	Pollen, underground parts	

These are British plants but their use is not necessarily British.

soon as the water is full. The plants are then allowed to grow again, both nutrients and other unwanted dissolved substances (such as compounds more toxic to animals than plants) having been removed from the habitat. This has proved successful in countries with hot summers, using submerged plants (e.g. [64], in ponds).

Submerged and floating plants are soft and soon decompose when removed from the water, so they can be used as compost and fertiliser for fields near the streams. Unfortunately rubbish (tins, bottles, polythene, etc.) is often too abundant for this to be feasible, as the rubbish cannot be removed cheaply from the plant mass and the weight of the plants makes it impractical to move cut plants far from the river. It is also possible to process the plants to varying extents in order to produce, for example, poultry food supplements, mulch or commercial fertiliser (e.g. [65], [66]), but these attempts have met with only limited success.

The tall emerged monocotyledons have many possible uses if they are available in sufficiently large quantities. In Britain the chief use of *Phragmites communis* is for thatching, while baskets and chair seats can be made from *Schoenoplectus lacustris*. Other crafts can produce many items from mats to bird cages [60] and all species can be used for animal bedding, packing material, insulation for horticulture, etc.

AESTHETIC VALUE
River plants add to the beauty of the landscape and are ornamental when transplanted to newly cleared areas or used as cut plants. The pleasure

of a canal holiday is enchanced by water-lilies and redswamp fringing the waterway. Unfortunately, dense boat traffic destroys vegetation in canals, rivers and lakes, though some protection can be given by baffles, etc. The plants also add to the pleasure of other activities in or beside the water, such as fishing, picnicking or walking, and may provide the habitat necessary for the birds wanted by bird-watchers. The variety and colour provided by river plants are usually much appreciated and it is unfortunate that the extent of this appreciation is usually not recognised until the plants are nearly lost, when it may be difficult and expensive to restore them.

VEGETATION RECOMMENDED FOR DIFFERENT STREAM TYPES

The main reason plants are sometimes considered undesirable in watercourses is the obstruction they can cause to water movement, boats, anglers and river gauges, sluices etc. The ways in which they can improve the watercourse are by stabilising banks and channels; providing food, substrates and protection for animals; oxygenating and purifying the water; adding to the aesthetic value; conserving rare species; and occasionally providing plants for practical use such as food, crafts, etc. The optimum vegetation is therefore that which produces as many of the desirable effects and as few of the undesirable effects as possible. The relative importance of each factor varies in different watercourses, and other influences may be important – for instance, swift rocky streams oxygenate water and so the function of plants in oxygenation is less important.

Mountain and upland streams

In the hills, the upper reaches are usually fairly swift, and natural oxygenation is good. Spate flows have sufficient force to sweep away thick vegetation before damaging floods occur and angling can take place unless plant cover is very thick. It is therefore quite satisfactory to have channels up to, say 75% full of plants – which means the normal natural maximum. If, however, flow is regulated by a reservoir, and spates are decreased in force or frequency, vegetation may become too dense. If rainfall, and consequently spates, are low for a decade or so (as in the 1960s), then again vegetation may come too thick for angling, river gauges, etc., though still satisfactory for flood prevention. In the flood plains downstream, the river banks are alluvial, and tall fringing monocotyledons are advantageous for stabilising the soil. If spate flows have little force then thick vegetation may intensify the effect of floods, but the floods are caused more by the large water discharges than by the thick vegetation, and so as long as the plants do not fill more than, say, half the channel, they probably do little harm.

As the flow decreases, oxygen in the water comes increasingly from plants, so in this respect plants are advantageous. Passing down from the hills the particle size of the substrate decreases, so that as the

313

boulders and stones disappear, animals requiring shelter rely more and more on plants. On the other hand, as flows get less there is some decrease in the need for shelter from flow, though the need for protection from other animals remains.

Recommendation.   (*a*) In the hills, encourage channel plants up to *c.* 75% full, unless clearer bands are wanted at the sides, for angling, or locally for gauges, etc. (*b*) In the flood plain, encourage tall fringing monocotyledons, and channel plants up to *c.* 50% cover (except locally for gauges, etc.).

Chalk streams

Their vegetation, particularly *Ranunculus*, can grow very dense and create serious flood hazards, though on the other hand the flow, though swifter than in other lowland streams, is slow enough to make it advisable to have plants for the oxygenation of organic effluents. From the angler's point of view *Ranunculus* can grow too dense and need to be controlled and short plants in the channel are the most desirable. The plants which occur naturally are those best suited to providing good food for fish, though after severe damage it may be advisable to import plants for recolonisation (see Chapter 1). In ordinary circumstances, if chalk streams have too few plants to support good fish populations, the remedy is to improve the physical characteristics and chemical quality of the river, and the plants will then invade and spread within a few years. In lower reaches the banks are on stable soil and subsoil, and though a fringe of tall plants is desirable, it is less essential than on banks more prone to erosion.

Recommendation.   (*a*) Upper brooks, where no floods occur and fishing is little, should be allowed to develop dense vegetation, for the conservation of this type of vegetation. (*b*) Middle reaches should contain *c.* 25% of plants, the most that is safe for flood control (see Chapter 19), and an amount adequate for fish, and water purification. Short plants are preferable to *Ranunculus*. Encourage trees, particularly alders (*Alnus glutinosa*) on the banks so as to decrease plant growth by shading and alter the species balance in favour of short plants. (*c*) Lower reaches should also contain *c.* 25% of plants, and have some tree shade. Encourage tall monocotyledons on the banks, particularly in flood plains.

Sandstone streams (lowland)

Depending on the stream, the recommendations vary between those for chalk and for clay, for the reasons given in those sections.

Clay streams

Clay covers much of lowland Britain, and so receives much of the country's organic and industrial effluents into its streams. Because the streams have slower flows than those on chalk, oxygenation from plants is important in ordinary conditions as well as for purifying effluents. However, thick plants in warm water can cause serious deoxygenation at night, when plants are taking up but not producing oxygen. Stream discharges are also less, and effluents often enter in brooks upstream; thus small, and even large streams may be composed mainly of effluents liable to cause deoxygenation. Plants are usually sparse in the upper reaches and confined to the shallower sides in lower reaches, so although *Nuphar lutea* may be a nuisance to anglers when abundant at the sides, the plants are seldom a flood hazard – except where slow silting flows lead to large stands of tall monocotyledons. Plants are needed to protect animals, and in some parts to increase the animal food supply, though again the slow silting stretches are an exception, for here there may be sufficient detritus for the food supply even without plants. Banks are often alluvial on lower reaches and so stabilisation by plant stands is advisable.

Recommendations. (*a*) Plant trees beside flatter brooks where tall monocotyledons can choke the channel, and beside those lower stretches where tall monocotyledons can again reach nuisance proportions. (*b*) Plant trees beside those small brooks where tall dicotyledons on the bank create a flood hazard (see Chapter 19). (*c*) Encourage narrow fringes of tall monocotyledons to stabilise the banks in lower reaches. (*d*) Encourage plants elsewhere (unless they seriously hinder fishing), for the physical characteristics of the stream will usually prevent them exceeding the 25% safety limit (see Chapter 19).

Canals

In the centre, plants, if present, must be submerged and short to allow the passage of boats. On the other hand, the rate at which boating on the canals is increasing suggests that boats will themselves soon do all the weed control necessary. Indeed, the destruction of canal plants by boat traffic is likely to lead to bank erosion, and the need for more piling, etc. There is negligible water movement, and some inflow of effluent, so plants are important for oxygenation (provided they are not too thick) and, as usual, for protection. As in the clay streams, the more eutrophic and detritus-rich channels have less need of plants to increase the food supply. Emerged or floating plants at the side can cause nuisance to anglers, but never constitute a flood hazard, as canals do not receive storm waters, and in general give much pleasure to those boating on, or walking by, the canals.

Recommendations. (*a*) Encourage tall fringing monocotyledons, even by specially protecting these against boats (if the cost of such protection is less than that of piling the sides). (*b*) Encourage as much other plant growth as is consistent with angling (and boat movement, in little-used canals). (*c*) Leave, or make when dredging, a shallow ledge at the side, which can bear dense vegetation in a partly protected habitat not needed for boats (see Fig. 10.2). (*d*) When boat density is high enough (see Chapter 21), stop weed control measures. Typically this is when plants no longer grow in the channel centre.

Dykes and drains

In these, the need for drainage water to move freely takes precedence over all other considerations, so few plants are permissible unless the channel lies outside the main drainage system. However, the banks are usually alluvial and unstable, and stabilisation by plants is particularly important. Tall plants are unsuitable growing in narrow channels because they also create obstructions to flow, so the best plants for stabilising the banks are trees, or, where these are not appropriate, short grasses. The considerations for fishing and for oxygenation of the water are the same as those for canals.

Recommendations.   (*a*) In narrow channels, encourage trees or short grasses on the banks. (*b*) In wide channels, encourage a narrow belt of tall monocotyledons, and trees or short grasses on the banks. (*c*) Keep channels on the main waterways free of plants in the centre. (*d*) Allow channels moving little water to develop thick vegetation for conservation, animal protection, etc.

19

Flood hazards created by river plants

Watercourses can be choked by vegetation, and then cause damage to low-lying surrounding land through ponding and flooding. The amount of vegetation which is safe from this point of view depends on the flow regime. Dykes and drains need to be kept nearly empty if water must move through them; lowland streams are safe if they are not more than a quarter full of plants; and in the hills, spates can normally remove any amount of vegetation, so channels can safely be choked with plants during low flows.

Plants can obstruct the flow of water and create flood hazards in various way: channels may be choked by emerged plants, or by floating and submerged ones; banks and edges may be covered by tall plants; cut river plants, land plants, branches, etc. may be washed downstream and become lodged to cause obstructions; or shoals may be deposited because plants are present. Plants that have become a nuisance can be removed or decreased by dredging, cutting, use of herbicides, shading by trees (see Chapter 21), having water too turbid or deep for river plants to grow in the channel, or deliberately increasing flash floods (e.g. from reservoirs) so as to wash away the plants and decrease substrate stability (see Chapters 3–5). It also appears that lowland flows have recently become more irregular, with less stable stream substrates. This is attributed partly to a change in rainfall [67], and perhaps partly to changes in farming practice. If continued it would decrease stream vegetation. It is important to know, though, exactly when plants do or do not constitute a hazard and need to be controlled. For example, many stretches of streams are prevented from developing choking vegetation because the flow, substrate or some other habitat factor prevents this, and yet often the plants along such stretches are routinely cut. Thus time and money could be saved by a greater botanical understanding of the river plants. Again, in many reaches, a distinction can be made between potentially explosive populations, which may create a hazard, and non-invasive sparse plants which are not hazardous. Channels which have been much altered by recent management have unstable plant populations, and the vegetation may be expected to change.

HAZARDS FROM PLANTS 'IN SITU'

Surplus water can occur as a result of either ponding or flooding. Ponding occurs when, although flow may be small, choking vegetation

317

Fig. 19.1

causes a rise in water level. This is commonly of the order of 30–50 cm, though 2 m has been reported (Fig. 19.1). It is harmful because land becomes waterlogged and under-drainage systems are destroyed, which mean farm production is reduced. Flooding results from run-off being unable to move freely along the channel after heavy rain. In contrast to ponding, the effect may be quite independent of the plants, may be increased by the plants, or it may be due only to the plants, as happens when the weight of water eventually forces its way through plants after land near by has been flooded for a while (Fig. 19.2). The damage caused by both ponding and flooding varies according to the type of land affected, being least if the land is unused, usually little if it is pastures and most if it is built over. Thus planning for flood prevention mean, taking land use as well as water levels into account. For example, in some built-up areas it may be an advantage to have water held up by plants if this means that storm water floods farming land upstream rather than the built-up area.

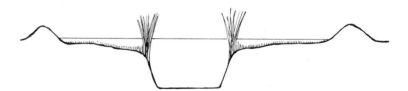

Fig. 19.2

The shape of the channel affects the height to which water may rise with safety. In a chalk river where there may be less than a 50-cm difference in level between the surrounding land and the water level before cutting plants, a pile-up of only 30 cm of water would create a flood hazard in a severe storm, whereas 30 cm is a negligible variation in level on the flood plain of a large hill-river with banks several metres high.

Before removing plants that are flood hazards the effect of the removal and the most desirable vegetation for the watercourse should be considered. Without management, vegetation is usually thickest between June and September, but cutting can itself lead to greater weed growth (see Chapter 21) and so the cutting of plants which are not flood hazards at the time of the cut may actually create a flood hazard later. Also, if one type of river plant is removed, another often takes its place, e.g. emergents are replaced by submergents which are then replaced by algae.

WATER MOVEMENT THROUGH VEGETATION

Any obstruction in a channel decrease its ability to transport water. Even sparse plants increase the roughness of a channel, and the greater the roughness the less the discharge which can pass through it (Fig. 19.3). If the plants are thick then although the water can flow freely

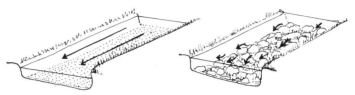

Fig. 19.3

above the plants it percolates only slowly through the plants themselves. This means that the effective size of the channel, upon which the discharge depends, is that of the part without plants (Fig. 19.4).

Fig. 19.4

If plants in a channel are to be cut, this can be done in various ways. They can be cut almost at ground level, or part way down, or narrow deep channels may be cut in the centre of the bed (Fig. 19.4). Because narrow deep channels carry more water than wide shallow ones of the same cross-sectional area of free water, this first type of cutting is preferable, other things being equal. When narrow deep swathes are cut by weed-cutting launches, a double swathe carries more than twice the water of a single swathe. Again, uncut channels with plants only at the sides carry more water than equivalent channels where the plants are spread evenly over the floor (Fig. 19.4).

The degree to which plants obstruct water movement depends on the type of plant, as well as on its position and its bulk. Resistance to flow increases the greater the perimeter of the plant parts in the water, and thus a fine-leaved species like *Ranunculus* is a greater obstruction to flow than the same weight of a large-leaved species like *Nuphar lutea* (Fig. 19.5). Hindrance to flow can also vary with season, being greatest in mid-summer when the plant parts are most rigid. Note that there is an important distinction, when considering resistance offered to flow, between the effect on the stream, as described here, and the effect on the individual plant, as described in Chapter 3. It was explained above that a certain weight of *Ranunculus* causes more ponding than the

Fig. 19.5

same weight of *Nuphar lutea*, but the streamlined shoots of *Ranunculus* are much less liable to damage from velocity pull than the bushy ones of *Nuphar lutea*.

OTHER FLOOD HAZARDS FROM PLANTS

Material being washed downstream can create obstructions, particularly where the channel is constricted, and may cause flooding. Some of this material is vegetation, such as filamentous algae, floating plants (e.g. *Lemna* spp.), cut river plants left in the channel, and crop debris, straw bales, branches and other plants washed in from the land. There is also rubbish of various kinds – even including caravans! It is the large and firm objects like trees and bales of straw – and not the river plants – that cause the worst flooding. Material in the channel can also cause blockages in sluices, etc., which can, particularly in flash floods, cause flooding by decreasing the amount of water that can pass through the sluice. Gauging stations can be blocked in this way.

Plants accumulate silt, and this can also cause a flood hazard if accumulated in large quantity. The amount depends on the flow type, the amount of sediment passing through the habitat and the type of plant. Emergents accumulate the most, particularly if their above-ground parts are still present in winter.

DYKES AND DRAINS

The gradient along dykes and drains is very small, values between 1:10000 and 1:40000 being common, so gravity is often inadequate to move enough water for drainage and pumps or flood gates are needed. The habitat can be considered as a long narrow pond, and because water moves so little under gravity, ponding is likely when there is extra water for some reason. (Because of the lack of slope, the difference between ponding and flooding is less than in other types of watercourse.)

For drainage purposes, all hindrances to water movement must be minimised, and this includes keeping plants sparse or absent. The only important exception is that dykes at the head of a drainage system, which receive little or no water from outside and merely pass their water to a lower dyke, can be choked with plants. Also, in the west and north where most plains are small, dykes are liable to some scour during storm flows from the hills, and if these can wash away plants then the dykes can, between storms, be more nearly choked with plants without causing flood hazards.

All types of river plants are much more of a flood hazard in dykes than they are in streams, but as the emergents accumulate most silt, and obstruct flow even when the water level has risen, they usually hinder water transport most. Tall plants on the bank create the greatest flood hazard in narrow dykes, because the banks occupy more of the cross-sectional area of the watercourse as the water level rises (Fig. 19.6;

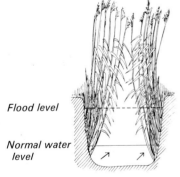

Flood level

Normal water level

Fig. 19.6

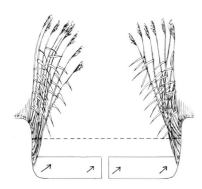

Fig. 19.7

Plates 16 and 17). They hinder flow and are dragged by flood water, and if pulled out into the channel impede water movement there. Short plants, in contrast, can serve the same function in preventing bank erosion without creating this hazard. In wider drains the banks occupy less of the cross-sectional area (Fig. 19.7) and larger channels are likely to be too deep for emergents anyway, except in a narrow band by the bank, so both tall (e.g. *Phragmites communis*) and short (e.g. *Alisma* spp.) emergents can grow at the sides without impeding water transport, provided they are non-invasive in the drain concerned. Sparse non-invasive plants can also be permitted in the channel. The main drains which collect water from a wide area, however, must be able to transport it quickly and so must be kept almost empty of plants.

A cover of free-floating plants (e.g. *Lemna* spp.) hardly decreases the capacity of the channel to move water and can be useful because its shading effect prevents submerged plants from growing below. In wider dykes and drains, though, particularly in exposed places or where the plant cover is thin, wind may blow the cover so that parts of the water surface are left clear and plants are able to grow below. Such a cover is only a nuisance where water has to pass through pumps, flood gates, or other constrictions which could be blocked by the plants.

CANALS

Canals receive little run-off and have stabilising channels which prevent much alteration in the water level. Because they do not have the function of moving water, they can safely contain more vegetation than drains. Floods are prevented by the general management procedures rather than by the absence of plants.

LOWLAND STREAMS

Lowland streams have a continuous flow of water under gravity, even though some stretches may be tide-locked or have water retained by gates, and because of the force and flow of the water, much more vegetation is permissible than in dykes. However, although storm flows may damage plants, enough vegetation is left behind to represent a flood hazard. More damage to the surrounding land can usually be expected from ponding than from flooding, and ponding is frequent in areas liable to choking plants if streams are not managed. Streams which can potentially have choking vegetation occur in a definable width–slope pattern (Fig. 6.5). If other factors are suitable, flatter streams up to *c.* 20–30 m wide and under *c.* 2 m deep may be choked, but outside this range management for flood control would seem unnecessary. The non-permissible amount of vegetation is that bulk which could, during a spell of warm weather, grow to a dangerous level. In general the vegetation is considered safe if the stream is 25% full (i.e. when looking at the stream it appears that 25% of the water

volume is occupied by plants). This level provides enough plant material for the conservation of the plants themselves, and the invertebrates,
fish and other animals dependent on them.

There are several patterns of plants which constitute a safe level of
vegetation (Fig. 19.8):

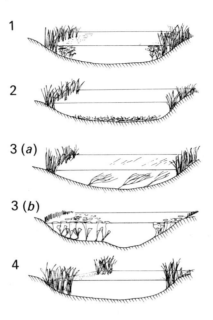

Fig. 19.8

1. Floating and submerged plants at the sides, as in many clay
streams.

2. A continuous layer of submerged plants, as in many chalk streams
in winter.

3(*a*), (*b*). Sparse plants growing throughout the water.

4. Emergents at the side.

The first is the arrangement most efficient for water transport. It is,
however, characteristic of clay streams, and chalk streams often have
the advantage over clay streams in terms of a steeper slope to the
channel and therefore a faster flow. If the surrounding land is such that
moderate ponding or flooding does no harm, choking plants are safe.
Tall herbs on banks of narrow brooks cover a good deal of the channel
cross-section when water level rises (Fig. 19.7), but it is usually sufficient
for the banks to be cut in late summer or autumn, before the winter
storm flows begin, and summer growth causes trouble only when the
bank is steep, storm flows are deep and shading is little.

Chalk streams usually have a good flow throughout their length.
Typically the substrate is stable, the water clear and its depth not
great, so submerged plants occur right across the channel (Fig. 10.2).
Growth is potentially very rapid, and backing up of water by the plants

is common, usually leading to rises of 30–50 cm. Tall emerged monocotyledons are seldom troublesome. Brooks usually have coarse substrates, little silting and fast flow, which confines luxuriant emergents
to local silted areas, and rivers often have steep non-silty banks, which
confine the tall emergents to narrow non-invasive bands (Fig. 10.2).
Emergents tolerating the coarse substrate (particularly *Schoenoplectus
lacustris*) may occur in the channel, but as long as the shoots are sparse
enough to be beaten down by the storm water (and do not exceed the
permitted bulk of vegetation) they do not create a flood hazard. Temporary flooding can be caused by *Rorippa nasturtium-aquaticum* agg., but
even the low force of storm flows in chalk streams can wash away
carpets of this species. In chalk streams, flood hazards are most likely to
be created by submerged plants.

Clay streams typically have less fall and slower flow, a more fluctuating flow regime and more water retention, more suspended and deposited sediment, and a less stable substrate. The emerged plants are more
troublesome than on chalk. Small brooks with little slope and soft beds
can be completely choked and downstream the emergents at the sides
accumulate silt and spread to form large stands. Dense stands are not
flattened by ordinary storm flows and may cover most of the channel
width, thus creating a flood hazard. Brooks with steeper slopes generally bear little or no vegetation, though fringing herbs are locally
luxuriant and may impede flow, particularly in winter. Usually, however, dense vegetation is found only in the silting downstream stretches.
Clay species are typically coarse-leaved, causing less resistance to flow
than the fine-leaved species typical of chalk streams. They also tend to
be shorter (1–2 m long, compared to 3–5 m for the usual length of
Ranunculus choking chalk streams), which means less choking occurs on
clay and so less management is usually required. Plants normally grow
right across the channel only if the water is *c*. 75 cm deep, the low light
and unstable substrate in the deeper areas confining plants to the sides,
so the floating and submerged plants often keep within the safe quarter-
full limit without any management. In at least some parts of Britain,
removal of *Nuphar lutea* is discouraged. With its large leaves and long
flexible stalks it is a plant that causes little resistance to flow, and its
removal, by dredging or herbicides, would probably result in its being
replaced by species that impeded water flow more. In general, flood
hazards in clay streams are most likely to be created by things washed
in from the land, though emergents are locally troublesome.

Sandstone streams tend to resemble chalk ones in plant type, but to
have fewer species and less bulk of vegetation (see Chapter 12).
Emergents are often intermediate in amount between those of chalk and
of clay.

Lowland streams on Resistant rocks are more likely to have flood
hazards created by oddments washed in from the land than by plants in
the streams.

UPLAND AND MOUNTAIN STREAMS

The force of water in these streams during severe spates can wash away any vegetation, and so flooding is usually caused by blockages of branches, fallen trees, crop debris and other rubbish being carried downstream rather than by rooted plants. In some upland streams, though, vegetation can become very dense between spates in the slower stretches, and ponding is a possible hazard. And even in the hills (usually upland rather than mountain) there may be some brooks which do not receive scour and become choked with emergents that cause local nuisances and may need management. In the flood plains below, flooding is frequent unless the channel is embanked, but this is not much affected by plant growth.

The width–slope patterns (Fig. 6.5) show that the wider flatter channels do not become choked, their substrates being unstable. Sandstone and limestone streams can, however, be full in rather flatter sites than can streams on Resistant rocks. Fine-leaved submergents are common in the channel, but tall emerged monocotyledons are usually restricted to the banks of slower stretches and the sides of some slower brooks.

20

Changes in flow patterns

Flow regimes can alter as the result of the construction or removal of weirs, sluices, reservoirs, etc., the regulation of rivers, the transfer of water from one river to another, or annual variations in rainfall. The effect of altering the discharge varies with the type of channel. Variations in silting alter nutrient status and the type of substrate available for roots to anchor in, while variations in flow type alter the velocity pull, battering and scour which influence the plants.

The effects of these changes on the vegetation can often be predicted.

STRUCTURES IMPEDING FLOW

A water mill is an ancient way of controlling the flow of a stream. The stream is usually divided into two: one channel, the mill stream to turn the mill wheel, the other for surplus water. In the mill stream the water is allowed to build up above the mill so that the maximum water force is available to turn the wheel when the gates are opened. Consequently flow is slower above, and faster below the mill, and there may also be a deep mill pool below, or, less often, above. Three habitats are thus created by the mill which differ in flow, substrate and trophic status from the original stream (see Chapters 2, 3, 4 and 9 for effect of these factors on plants) and consequently differ in their vegetation. For example, *Potamogeton pectinatus* was able to grow in a silted stretch of stream above a mill when the original stream – a near-clean chalk brook – was a mesotrophic habitat unsuitable for its growth. Frequently the mill stream is narrower than the main river, resembling a minor tributary, and yet its soil quality is the same as that of the main channel so it may bear vegetation unduly eutrophic for its size (e.g. thick algae in a brook 2 m wide). Other sidestreams may be created for ornament and they are influenced similarly, though they may also have foreign species planted in them for ornamental purposes.

Weirs, sluices and other flood gates are other impediments to flow. Their chief uses are for irrigation, drainage, flood control, navigation, ornament, fish pools and abstraction. Their main effect on the stream is that flow is slower above and faster below them. In hill streams species are frequently able to colonise the stretches of slower flow above small weirs in sites lying outside their width–slope pattern. For example:

Potamogeton alpinus   ⊤ *Sparganium emersum*

Potamogeton polygonifolius

325

Lowland streams have at least intermittent slow stretches, and the vegetation above the weir is the same as this. Some rivers, particularly lower stretches of clay streams, have structures at close enough intervals to retain deep slow water between them. The vegetation is that of deep slow silted rivers.

Flow may be varied from year to year when the gates controlling flow in a single channel are opened differently or when those controlling a stream with several channels (e.g. braided rivers, mill streams) alter the proportion of the water entering each channel. Such variations cause changes in the vegetation.

Examples of this are:

(*a*) *Braided stream.* The relative proportions of *Elodea canadensis* and *Ranunculus calcareus* in the minor channel varied with flow type and silting.

(*b*) *Mill stream.* The minor channel had, in different summers: (1) *Ceratophyllum demersum* dominant in a slow silted flow; (2) *Zannichellia palustris* abundant in water that was shallow with a moderate flow and sandy substrate (the silt having been washed away); (3) *Myriophyllum spicatum* and *Ranunculus fluitans* colonising the sand; and (4) in a flow that was again slow and silted these species becoming very unhealthy, particularly *Ranunculus fluitans*.

(*c*) *Single channel*, with *Myriophyllum spicatum*, *Potamogeton pectinatus* and *Ranunculus penicillatus*. In summers with shallow swift water *Ranunculus penicillatus* dominated, while *Myriophyllum spicatum* increased when the water was almost still.

(*d*) *Single channel*, with a carpet of *Berula erecta* in water less than 50 cm deep above a sluice. When the water depth was increased to *c.* 75 cm for over a year, *Berula erecta* grew much more slowly, and since erosion is easy in older patches (see Chapter 3), the carpet was almost destroyed through the lack of new shoots.

If minor changes in flow regime fall inside the tolerance limits of a plant then there is little overall effect on the vegetation, but if changes are more drastic, or if they move a site outside the tolerance limits of a species, then the vegetation alters markedly. The effect is greatest if the change is between a turbulent eroding flow and a slow silting one. A turbulent flow can be created both by increasing the discharge and by lowering the water level (which increases the influence of stones, etc. on the bed). In several chalk–clay streams it was found that in summers when there were shallower more turbulent waters *Ranunculus* and other chalky species increased and when there were deeper silting flows clay species increased.

Reservoirs are usually sited upstream and the water flowing from them is controlled. This means that both drought and storm flows are usually decreased and thus the substrate is stabilised also. This allows vegetation characteristic of a slower flow type to develop. In hill streams, plants increase. For example:

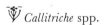

Callitriche spp.   Ranunculus spp.

Myriophyllum spp.   Sparganium emersum

And in a lowland stream a reservoir altered:
Ranunculus–Zannichellia palustris to *Enteromorpha–Sparganium emersum*

Reservoirs sometimes substantially alter the general quality of the stream water, and this may affect plants also. Dark-stained water decreases plant growth in deeper places (see Chapters 7 and 13). Alterations in temperature, dissolved gases and nutrients are unlikely to be sufficient to alter the vegetation. The extent of the effect of a reservoir on a river depends on the proportion of the total river discharge which comes from the reservoir. The effect is greatest and persists the farthest downstream if the other tributaries are few and small, but if there are many large tributaries joining the river below the reservoir then the regulatory effect of the reservoir will decrease as each enters. This may mean that plants can grow well in the stable substrate just below a reservoir, but then disappear as soon as a stream of comparable size joins the river. If much water is let out of a reservoir very rapidly, the effect is that of a damaging spate flow, and recovery could take some years (see Chapter 5).

The vegetation changes expected when flow is regulated can be predicted from the information in Chapters 2–6 and 12–13.

LOWERING OF THE WATER LEVEL

The ground water table in lowland Britain has fallen substantially during the past century as the result of man's activities, and this has affected the stream vegetation. Years of low rainfall have a similar effect. In the hills, man's use of the water is less and the rainfall is higher than it is in the lowlands, so the effect on the plants is also less than in the lowlands.

The lower discharges and spate flows which the lower water level causes in the hills lead to denser vegetation, both over a single summer (e.g. [68]) and over a decade or so (e.g. several northern rivers such as R. Eden, Cumbrian River Division, in the mid-1970s). The damage done to a single chalk stream by the lowering of the water level is charted in Fig. 20.1 for R. Darent (Thames Water Authority). Fig. 20.2 illustrates the situation diagrammatically; Plates 18 and 19 show the early stages of this effect in another chalk stream.

R. DARENT

(a) *1969*. Survey unfortunately scanty and done during storm flow, but *Ranunculus* was frequent.

(b) *1972*. Flow had decreased, the water was slower, and the substrate silty. *Ranunculus* had decreased (though at one site this was due to children paddling) and the few *Ranunculus* shoots seen were small, as is usual in slow flow. These small shoots cannot form a dense root weft, and are likely to become at least partly rooted in the silt above the gravel bed (see Chapter 3). *Myriophyllum spicatum* was frequent. This roots in silt more than in gravel and does not form a stabilising root weft, which, together with the bushiness of the shoots, means that plants are easily swept away by velocity pull (see Chapter 3).

(a)

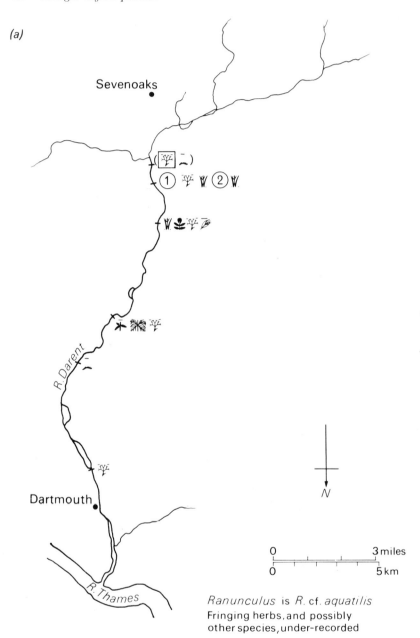

Fig. 20.1. Lowland chalk
river. R. Darent (Thames);
(a) 1969.

Ranunculus is *R.* cf. *aquatilis*
Fringing herbs, and possibly
other species, under-recorded

(The good growth of *Myriophyllum spicatum* in this chalk stream was attributable to the unusual amount of silt on the bed.) Thus by the end of 1972
the plants of the channel were potentially easy to wash away. In addition to
this, submerged trailing plants usually grow poorly or die in shallow water
(see Chapter 4). The damage, therefore, was due to a combination of flow,
silting and depth factor.

   (c) 1974.   The water was shallow enough to cause turbulence and much
of the silt in the centre had been removed. Most submerged plants were lost

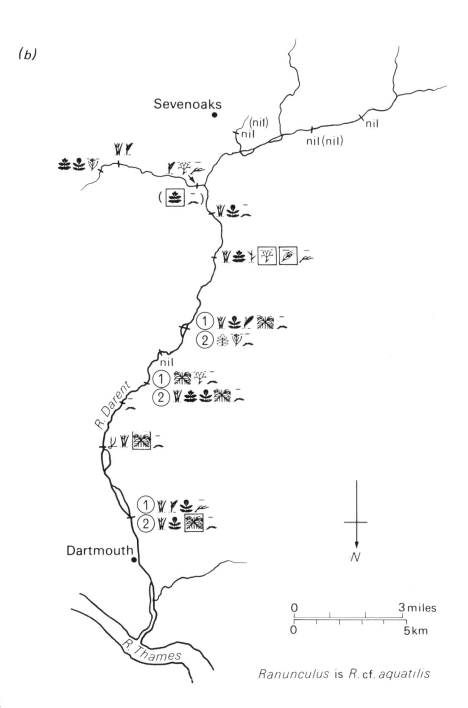

(b)

Fig. 20.1 *(cont.)*. Lowland
chalk river. R. Darent
(Thames); *(b)* 1972.

329

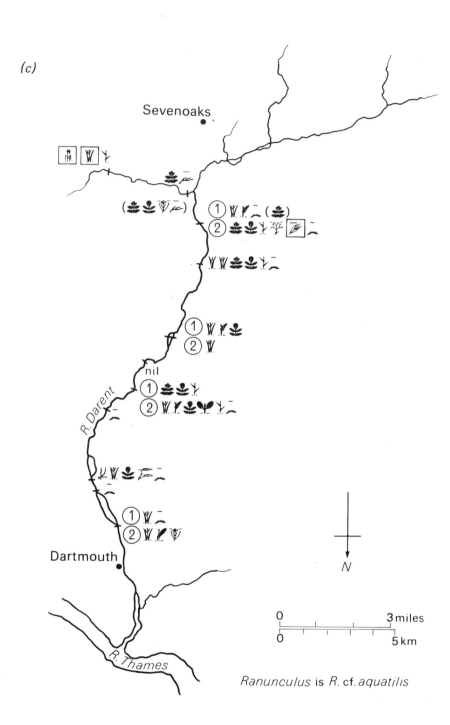

Fig. 20.1 *(cont.)*. Lowland
chalk river. R. Darent
(Thames); *(c)* 1974.

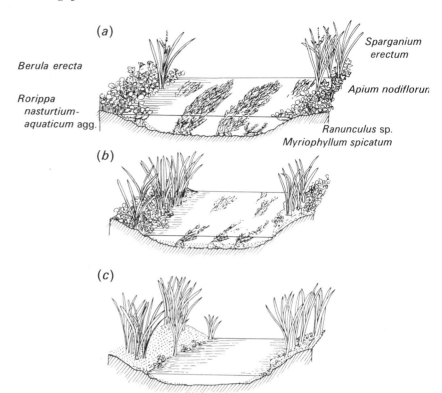

(a)

Berula erecta

Rorippa
nasturtium-
aquaticum agg.

Sparganium
erectum

Apium nodiflorun

Ranunculus sp.
Myriophyllum spicatum

(b)

(c)

Fig. 20.2. Changes in a chalk
stream due to a fall in water
level. (*a*) Original state. (*b*)
First stage after the fall in
water level. (*c*) Second and
stable stage after the fall in
water level.

with the silt, and the water was too shallow for submergents to grow (except
those of upper chalk brooks (Chapter 12), and these were absent anyway).
This same decrease in *Ranunculus* was seen in other streams between 1971 and
1974, and it could be attributed only to the change in water depth, as in
these other streams flow and substrate types did not alter. The gravel bed
was exposed over most of the channel. The fringing herbs had altered less
since 1972 than the plants in the channel, for although silting and storms
must have moved individual patches, the general habitat of shallow water at
the edge of a chalk stream had remained. *Sparganium erectum*, however, increased substantially and some bands were considerably larger than those
expected in a chalk stream. This species grows well on silt banks, and even
in the turbulent shallow flow of 1974 could both stabilise and accumulate silt.

In Hertfordshire, lowering of the water level has damaged fisheries and
plants in chalk streams, and R. Mimram is cited as a particularly bad case
[12]. Its vegetation resembles that of R. Darent in 1974, with fringing herbs
but few submerged plants in the shallow water. It is reasonable to conclude
the vegetation has been damaged in the same way as that of R. Darent.
Silted parts of R. Mimram bear dense stands of *Sparganium erectum*, attributable partly to the lowering of water level, as in R. Darent, though partly
also to the presence of some clay in the catchment.

Major changes in river plants due to a lack of water have been docu-
mented for chalk streams and been deduced for clay and sandstone
streams from the dryness of the upper tributaries (see Chapter 1). The

331

present status of river communities is unsatisfactory, and may deteriorate further. The cause of the fall in water level is irrelevant to its effects.
This makes predictions difficult, because while abstractions can be fore-
cast, droughts cannot, and when the discharge after abstraction should
be tolerable for the vegetation, droughts may further lower the water-
level and kill the plants. As described above, silting is an important
short-term cause of the damage from lowering of the water level.

There was an unusually severe drought in 1976, and the short-term
effects of this were probably similar to those of a long-term equivalent
lowering of water table by man's activities or a change in the climate.
From changes in streams surveyed in both 1976 and an earlier, normal,
year, Table 20.1 was constructed, showing the minimum safe water

TABLE 20.1 *Crucial water depths for different groups of river plants* (*1976*)

(*a*) *Streams*
The figures are the normal summer depths of the centre of the stream (as
described in Chapter 4).

	Damaging[a] (cm)	Critical[b] (cm)	Safe[c] (cm)
Fringing herbs	0	10	20
Tall emerged monocotyledons			
(excluding *Phalaris arundinacea*)	10	20	30
Water-supported plants[d]			
(*a*) Shallow-preference[e]	10	20	30
(*b*) Medium-preference	30	40	50
(*c*) Deep-preference[g]	40	50	60

(*b*) *Dykes, etc.*
(Fringing herbs are unimportant)
Tall monocotyledons tolerate summer drying
Water-supported plants have the same depth ranges as in streams

[a] Damaging means some loss in species presence, usually between two and
five out of ten occurrences.
[b] Critical means satisfactory for one summer, but leading to damage if
continued for long periods.
[c] Safe means recommended for the preservation of the species.
[d] The water-supported plants comprise submerged species and floating ones
rooted to the ground.
[e] Shallow-preference species include: *Callitriche* spp., (*Elodea canadensis* to
some extent), *Ranunculus aquatilis* (short-leaved), *Ranunculus peltatus* (short-
leaved).
[f] Medium-preference species include: *Ceratophyllum demersum, Elodea canadensis, Hottonia palustris, Myriophyllum spicatum, Myriophyllum verticillatum,
Ranunculus calcareus, Ranunculus penicillatus* and other medium-leaved species,
Schoenoplectus lacustris, Sparganium emersum, Zannichellia palustris (except in
chalk streams where shallower water is tolerated).
[g] Deep-preference species include: *Nuphar lutea, Ranunculus fluitans, Sagittaria
sagittifolia*

level for different groups of plants, the minimum level which is tolerable for most populations for fairly short periods, and the level which can be expected to cause substantial damage.

Table 20.1 can be used for predictions and planning. If a certain plant group is required in a stream (e.g. *Ranunculus* in chalk streams for fish), then the planned water level should not be less than the 'safe' level given in Table 20.1. The rare occasions when the water drops below this, as it did in summer 1976, are then likely to do little or no harm.

Although depth is the easiest factor to use in predictions, it is, of course, not the only one, as substrate and flow type are also important. These are summarised in the width–slope diagrams, and their use in predictions is discussed below. Briefly, if the water is too shallow for a species (as noted in Table 20.1) it will, eventually, die, and if originally absent it will not colonise the habitat. If, however, the water is deep enough, the species can only survive if the flow and substrate are suitable also.

WATER TRANSFERS

Water is becoming increasingly short in south and east England and the main industrial areas, and so will increasingly be moved there from the north and west, where rainfall is highest. This movement will usually be effected by transferring water from one river to another.

The effects of a water transfer depend on the amount of water transferred, the duration and season of the transfer, and the quality of the water. At one extreme, the stream discharge could be vastly increased on a permanent basis, which would completely change both the recipient and the donor stream below the point of transfer. At the other extreme, small amounts of water could be added to a stream solely during drought flows, and this would have a negligible effect on the plants. Large changes in discharge will influence the size, flow type and substrate of the river, and if these change sufficiently to alter the habitat for plants then new species will invade (providing the new habitat permits plant growth) and some time may elapse before the vegetation becomes stable. If the colonising species are present near-by and the new substrate is as stable as the original one (see Chapter 21), then a new stable vegetation can be expected in 1–3 years, but otherwise the process of stabilisation will take longer. A very slow increase in trophic status, due to silt of a higher nutrient content slowly accumulating in the recipient stream, may not influence the plants for a decade or more.

The transfer system is likely to start with a hill reservoir, the downstream effects of which have been described above. Then, below the point of transfer, the donor stream contains less water than before, and the recipient stream more. Regulation decreases spate force, and irregular rapid discharges from reservoirs increase spate force. Thus in the transfer system the river may alter its topographical flow type either towards a more or a less spatey flow regime. The new width–slope

333

pattern for vegetation can be predicted according to the width–slope patterns in Chapter 6 showing what range of plants can grow in different flow types. Subject to this overriding control, predictions for individual species may be made from the width–slope patterns for individual species also shown in Chapter 6 and from Table 20.1. If the transfer means that the water depth changes so that, for example, a plant able to grow in a certain width and slope in a depth less than 75 cm is subjected to water over 1 m deep, then this species will be lost. If, on the other hand, the change of depth is from 35 to 70 cm then this same species would be able to survive. Consequently, if the width, slope, present depth and expected depth of the river are known, then the survival of species which have a depth line on their width–slope pattern, or are in a river that is becoming shallower (Table 20.1) can be predicted. In general, donor streams which become very shallow may lose submerged vegetation (see above and Chapter 4). Predictions have been made for R. Tyne [46].

A discontinuous and irregular transfer may have other effects. The worst possible flow regime for plants is one alternating frequently (at less than 2-year intervals) between a turbulent eroding flow and a slow silting one (i.e. between the flows shown above and below the bridge in Plates 9 and 10), for as soon as, or even before, one community is established it is destroyed by the change in flow type. If the changes in flow stay within the width–slope–depth tolerance for the species (see above and Chapter 6), however, then discontinuous transfers, if made gradually, do no harm. Sudden increases in discharge have an effect equivalent to severe natural spates. The shape of the channel may be important in determining the effect on the plants of discontinuous transfers. A channel with steep sides and a flat bed will be, of course, deep during high discharges and shallow in low ones, but the area on which plants can grow remains much the same. A channel with gently sloping banks and shoals or banks at different levels across the bed, however, will have some areas wetted only during high discharges. The effect this has on the vegetation will depend on the duration of flooding of these areas, and on how much scour occurs during the flooding. For example, land plants can tolerate being submerged occasionally for days or even weeks, provided scour is low, and tall emerged monocotyledons (particularly *Phalaris arundinacea* though seldom *Phragmites communis*) can tolerate a water level fluctuating around ground level, again provided the substrate is not disturbed by scour. Fringing herbs are able to colonise an area more quickly but cannot tolerate scour, so can intermittently colonise parts with intermittent scour. Quick-invading submergents (e.g. *Callitriche* spp.) can colonise and spread during flooding lasting a few months, then die in the dry period and recolonise when water returns, while most slow-growing submergents can tolerate an occasional month or so of drought provided the substrate is silty so that the soil around the rhizomes remains damp.

The movement of water and at least some suspended sediment from the donor to the recipient stream during a transfer has the same effect on the plants as would the addition to the river of a tributary with that discharge (see Chapters 11–13). In general, because transfers will be made from the north and west to the south and east (see above), nutrient-poor water will be transferred to nutrient-rich rivers, and unless the discharge of the transfer is approaching that of the recipient stream the plants are not likely to be altered. If the donor is more nutrient-rich than the recipient stream, however, then the effects may be greater. Because trophic status increases downstream, then if a transfer is from the lower reaches of one river to the upper reaches of another, the trophic status of the recipient is likely to increase even if the two rivers are of equal trophic status (see Chapter 11). If the transfer leads to more silting, in either donor or recipient stream, a more eutrophic vegetation is likely to develop, and vice versa. Any alteration to the amount of silting will more than counteract any changes in the nutrient content of the water (or suspended sediment), and predictions of vegetation changes must take this into account. Transfers are sometimes likely to increase nutrient status if the silting is unaltered, and will usually do so if silting is increased. Plants will then be more luxuriant, or the species present will change, depending on the original trophic status and the amount of the increase. Water toxic to plants should not, of course, be transferred.

21

Maintenance and mechanical use of watercourses

Dredging, plant cutting, herbicides and shading are techniques used to maintain watercourses. Dredging is primarily done to keep a suitable depth and width in the channels, but of course it decreases vegetation too. The vegetation normally recovers in 2–3 years. Cutting is the traditional, and often the present way, of removing plants which constitute flood hazards. Streams with fast-growing Ranunculus *may need cutting up to five times each summer, but one or two cuts are more usual in streams with potentially choking vegetation. Herbicides are effective for weed control, but must be used with caution to prevent toxicity and bank erosion, etc. Shading by trees is a cheap and, in many circumstances, effective method of decreasing plants.*

This chapter describes the effects of channel maintenance on river plants, but not the practical aspects of maintenance. The equipment used for mechanical clearance of channels and banks is described in [69] and the approved herbicides, their methods of application and the expected results in [70] and [71]. Herbicides should be used only in accordance with the Ministry of Agriculture, Fisheries and Foods *Code of practice for the use of herbicides on weeds in watercourses and lakes* [72].

DREDGING

Watercourses are dredged mainly for channel maintenance, and only secondarily to decrease vegetation. Dredging may be done to form a new channel, to deepen an existing one, or to remove accumulated sediment (Plates 16 and 17). All or only part of the channel may be dredged.

In alluvial plains the subsoil is soft, so that after dredging a soft substrate is left which is stable only because there is no scour. In streams, if the hard bed is untouched by the dredging then plants recover quickly. If the hard bed is broken but the stream is silting, silt quickly accumulates and plants invade. If, however, the hard bed is broken, and this is left unstable, in a habitat without deposition of sediment, the vegetation returns very slowly: e.g. a chalk brook, dredged 50 cm deep, was not stable enough for plants even at the sides for 4 years. Recovery is also slow if the final water depth is too great for plants or when a coarse firm bed is left in a stretch with slow flow. In this last case neither the plants of swift flow nor those of soft soils can grow well and such places may remain bare until enough silt has accumulated to allow silt-rooting species to colonise.

336

Frequency of dredging required in different stream types:

(1) Mountain streams. Any accumulated silt is removed by spates so dredging is unnecessary.

(2) Upland streams. A few short stretches need dredging at fairly long intervals.

(3) Chalk streams. Those with little silting and swift flow need no dredging.

Where sedimentation is rather more (e.g. because some sandstone or clay occurs in the catchment), some stretches need occasional dredging. Most chalk streams have valuable fisheries, which means that the vegetation is important. Dredging is therefore kept shallow so as not to break the hard bed and disturb plant rhizomes, allowing vegetation to recover quickly.

(4) Sandstone streams; and (5) Chalk–clay streams. These have more sedimentation and are likely to need dredging at 10+ year intervals.

(6) Clay streams. These have much silting, and low-lying ones are typically dredged every 5–10 years.

(7) Dykes and drains. In those in the middle of alluvial plains there is little silting but no scour to move any silt that does accumulate. They are usually dredged every 5–12 years. In those at the edge of higher ground silting is greater, and dredging may be necessary as often as every 2–3 years.

Generally species return to a site within 2 years after dredging, and the amount of vegetation is back to normal within 3 years. Dredging usually restores the habitat to its earlier state, though occasionally a slow silting reach is changed to a swift gravel one, or vice versa. If this happens then naturally the potential species present will also change (e.g. *Ranunculus* will invade a newly formed gravel habitat). (Interestingly, *Nuphar lutea* has survived in 20–30 cm of swift water over gravel for over 3 years, though in a small form.) In many plants, particularly those of slow eutrophic habitats, the rooting level remains fairly constant while silt accumulates above (see Chapter 3). When the deposited silt is very deep (e.g. 1 m) the plants grow badly and dredging improves the vegetation.

Tall emerged monocotyledons recover relatively slowly after dredging because most have complex rhizome systems which they need to rebuild before they can dominate again. Other plants are likely to invade first, and then be shaded out by the regrowth of the tall species. For example, small silting clay brooks, regularly dredged, which have tall monocotyledons (e.g. *Sparganium erectum*) dominant for most of the time may have short species (e.g. *Potamogeton pectinatus*, *Sagittaria sagittifolia*) dominating for 2 years after dredging. The short species have parts buried in the undredged soil and can remain dormant with little above-ground vegetation during several years of heavy shading, and growing quickly when light is restored.

The speed of recovery of submerged and floating plants depends on what proportion of the propagules (fruits, stems, rhizomes, fragments, etc.) has been removed. Recovery can be within one growing season after shallow dredging, but if the plants are completely removed then recovery may take several years. However, in silting clay streams in Sussex dredging 15 cm or *c*. 125 cm deep has little effect on the rate of recovery of the vegetation, for silt quickly accumulates and plants soon grow well in this. Dredging is also usually done in short stretches, so propagules are able to invade freely from upstream. If a watercourse is isolated, invasion may be very slow. In the first few years after dredging, plants change least when little vegetation is present and most where tall emergents dominate.

Undamaged plant communities in shallow waters without scour are dominated by tall monocotyledons and are species-poor. In order to conserve the diverse and species-rich communities of dykes, etc. (see Chapter 14), dredging is a much better way of keeping the channels clear

TABLE 21.1   *Recovery after dredging*

(*a*)   *Great Ouse river division*

Approx. 30 sites with rather shallow dredging in 1971/2; recorded in 1972, 1973, 1974.

(1) Species growing better, or occurring only, in the first or second season:

Apium nodiflorum	*Myosotis scorpioides*
Callitriche spp. (from fragments)	*Potamogeton perfoliatus*
Elodea canadensis	*Rorippa nasturtium-aquaticum* agg.
Glyceria maxima	*Veronica beccabunga*
Lemna minor agg.	

(2) Species increasing to stable populations in the second or third season:

Agrostis stolonifera	*Potamogeton pectinatus*
Callitriche spp. (long-term plants)	*Sagittaria sagittifolia*
Mentha aquatica	*Sparganium emersum*
Nuphar lutea	*Sparganium erectum*
Phragmites communis	*Enteromorpha* sp.

(3) Sixteen sites had more vegetation in the second than the first season; 5 sites more in the third than the second.

Eleven sites had a similar amount of vegetation in the second and first years; 16 sites in the third and second.

338

(*b*)   *Lee Conservancy*

Approx. 20 sites of varying dredging depth; dredged in 1971/2; recorded in 1972, 1973, 1974.

(1) Species growing better, or occurring only in the first or second season:

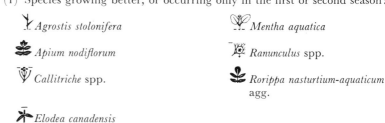

Agrostis stolonifera   *Mentha aquatica*

Apium nodiflorum   *Ranunculus* spp.

Callitriche spp.   *Rorippa nasturtium-aquaticum* agg.

Elodea canadensis

(2) Species increasing to stable populations in the second or third season.

Epilobium hirsutum   *Schoenoplectus lacustris*

Glyceria maxima   *Sparganium erectum*

Potamogeton natans   *Enteromorpha* sp.

(3) Five sites had more vegetation in the second than the first season; 3 sites had more in the third than the second.

Seven sites had a similar amount of vegetation in the second and first years; 17 sites had similar amounts in the third and second.

---

than spraying with herbicides, for dredging, unlike herbicides, provides a new substrate, does not introduce toxins, and does not harm invading propagules. The dredging should not, however, be so deep as to prevent submerged plants colonising: e.g. one dyke dredged to 1.5 m deep, with steep sides and water too turbid for light to reach the bed, bore no channel plants for at least 5 years. Deep drains bear good vegetation only if there are shallow ledges at the sides, and when large channels are dredged the final shape should include such ledges. (If dredging is used for weed control, rather than conservation, the reverse applies and the channels should have straight steep sides and be too deep for channel plants to grow.) In dykes on the Pevensey Levels and some clay streams in Sussex, there is a management cycle of a shallow dredge (15–30 cm), followed by herbicide in the second year, and spraying of potentially choking plants in the third year. For 2(–4) years thereafter no management is needed and this cycle is very satisfactory [73].

The changes in plants brought about by dredging in clay and part-clay streams of the Great Ouse and Lee River Divisions are shown in Table 21.1. In most sites the streams will never become choked with plants and vegetation is too sparse to constitute a flood hazard, so dredging is to remove sediment. Plants regenerating quickly from small fragments (e.g. *Callitriche* spp., *Elodea canadensis*) are likely to increase soon after dredging, and fringing herbs may also increase, though as silt builds up later anchorage becomes more difficult and they are washed away (see Chapter 3). Recovery is usually complete within 3 years. Sites choked with plants may not be choked again for *c.* 5 years after dredging or they may be full of plants in the first season. For example, one shallow dyke was choked in the first year by a fringing herb (*Apium nodiflorum*) and after that by a tall monocotyledon (*Glyceria maxima*). This pattern of choking fringing herbs in shallow channels soon after dredging (usually *Apium nodiflorum* or *Rorippa nasturtium-aquaticum* agg.) is found in many parts of Britain. Later, the plants are washed away by flow or are replaced by taller species.

Less detailed records from the Essex and Bristol Avon areas confirm that plants recover quickly after dredging, except that the development of choking vegetation may take several years.

The Pevensey Levels in Sussex are sprayed with herbicide as well as being dredged (see above). Tall monocotyledons are usually sparse, partly because the banks are steep and the dykes are deep, and partly because of the spraying. However they do, as is usual, dominate shallow dykes without recent management (see Chapter 14). In the first year after dredging there are slightly fewer angiosperms and rather more algae. In the second year *Hydrocharis morsus-ranae* and *Sagittaria sagittifolia* reappear, the former being characteristic of recently dredged sites. The regular management and turbid water (from the clay) mean that few dykes bear much vegetation.

In the Fens, species more frequent after dredging include:

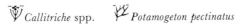

   Callitriche spp.   *Potamogeton pectinatus*

   Lemna minor agg.

Large deep-rooted plants of the channel re-appear later, for example:

   Nuphar lutea   *Sagittaria sagittifolia*

Disused canals are dredged when they are restored, and canals in use are occasionally dredged when silt has built up. In the first year after dredging the firm clay bottom is exposed and few plants are able to colonise. By the second year some silt has accumulated and quick-growing species (e.g. *Elodea canadensis*) invade. The vegetation recovers in 2–3 years [74]. Even after restoration of a largely-silted canal, species are able to appear quickly (e.g. eight species in 2 years). Most vegetation in canals that are in use is on shallow-water ledges at the sides, so to preserve their amenity value canals should be dredged in such a way as to leave this shallow ledge in place.

Less detailed records for other habitats and regions confirm the time-scale of recolonisation stated above.

CUTTING

Cutting is the traditional way, often still used, of removing plants which constitute flood hazards (but can itself cause flood hazards if the cut material is not removed but left to float downstream). A very important character of cutting is that its effects are predictable [69]. Spraying, on the other hand, is less predictable as its effects depend on, for example, the weather after the spray, and side effects may be unexpected. Dredging theoretically has predictable results if the type of dredging is known, but in practice its effects are variable. Removing plant growth by cutting means dredging is needed less often, and water movement is improved ([75]; see also Chapter 19). In flowing water, removing plants slows the development of shoals, or even erodes them, and in still water, removal prevents the continued accumulation of organic matter and humus.

Where tall plants are used as shelter by birds or mammals, cutting should not be done during critical stages of their cycles, e.g. nesting, and the cutting should be in short stretches so that the animals can always find cover nearby. Unfortunately, plants that are non-invasive in their habitat are often cut or sprayed unnecessarily (Plates 20, 21 and 22). This is harmful to the environment because cutting removes not only the plants but also the animals and algae associated with them. It likewise removes the shelter, protection, food and silt that the plants provided. It is also harmful if cut plants are not removed from the water because decaying plants require much oxygen for decomposition, and the resulting shortage of oxygen can kill fish and other animals. The general effects of cutting on the oxygen and carbon dioxide of river waters are described in [59].

Tall plants on the bank

These are cut to remove flood hazards in narrow channels, to keep paths clear for walking, to prevent bushes growing, or to improve the general look of the landscape (nettle beds are generally rather ugly). Tall bank plants die in winter but their shoots usually remain standing, so banks where they are a flood hazard should be cut at least in late summer or autumn before the winter storms. Paths may need cutting twice in summer if their plants are potentially tall and there is little trampling to help keep plants down. The denser the cutting, the greater the proportion of short plants.

Cutting dead shoots in winter has little effect on next summers' vegetation, but cutting in summer may have a long-term effect, depending on the species and its stage of growth.

Some examples are:

> *Urtica dioica* regrows quickly during the growing season and there is no long term effect of cutting.

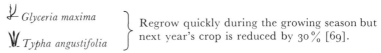

Glyceria maxima

Typha angustifolia } Regrow quickly during the growing season but next year's crop is reduced by 30% [69].

> *Phragmites communis* Has a strong internal rhythm of growing and dormant periods. Regrowth is quick after cutting during the period of rapid growth (in spring and early summer), and there are no long-term effects. Cutting in late summer means little regrowth and a substantially lower crop the following year [76].

Tall emergents in the channel

These decrease water flow and increase silting. If their shoots are sparse and easily flattened in storm flows then they are no more hindrance than short emergents (e.g. *Schoenoplectus lacustris* when sparse in chalk rivers), which have a similar effect but are less troublesome. When emergents are cut, soil is exposed and loose sediment is washed away. This may increase deposition somewhere else. Erosion–sedimentation patterns can, therefore, be altered by cutting. If the cut stubble of emerged plants is flooded, the rhizomes get inadequate oxygen and regrowth is decreased. This can be used as a technique to decrease growth [69]. In dykes and slow clay brooks, choking tall monocotyledons hinder flow at normal water levels, and do so even more after storms (see Chapter 19). Cutting prevents this, but must be done frequently, so a more satisfactory control measure is to dredge the main drains too deep for these plants. Dykes which carry little water can be cut in autumn, while those carrying more water need to be cut perhaps twice in summer, or be dredged approximately every 5 years with possibly one cut between dredges.

Submerged and floating plants

These accumulate less silt, partly because of their shape and partly because in the channel centres, where they grow, the current is swifter than at the sides. Deposited silt is eroded quickly after cutting in streams with swift flow (e.g. *Ranunculus* communities) and all plants rooted in this are lost also (e.g. *Elodea canadensis*, *Groenlandia densa*). Cutting is not needed in winter as these plants die-back either partially or completely, but may be required any time between late spring and early autumn when the plants grow thickly. Early-growing species in Southern England may need cutting in April (e.g. *Ranunculus calcareus*) while late-growing ones may not become thick before early July (e.g. *Sagittaria sagittifolia*). Most streams in fact are not cut at all. Some lowland streams are cut sporadically, when equipment happens to be available, or plant growth is unusually thick. The general pattern, though, in streams large enough for plants to be a hazard is:

0–1 cuts per year in clay streams,
0–3 cuts per year in soft sandstone streams,
1–5 cuts per year in soft limestone streams.

The warmer the weather the greater the plant growth, so a stream normally cut once a year may need cutting three times in hot summers. However, if there are frequent summer storms these prevent plants from becoming choking and so decrease the need for cutting. In long droughts the water may become too shallow for submerged plants to grow well (e.g. Chapters 4 and 20), and when a change of depth alters flow type, vegetation may be lost (e.g. the erosion of soft-soil plants by a turbulent flow) and cutting rendered unnecessary. Another possibility, reported from several parts of Britain, is that increased plant growth may occur after storms in May, and this is attributed to fertilisers washed out of the land.

In the best trout streams on chalk, cutting is for the benefit of the fishing, and only secondarily for flood prevention. The small-scale cutting patterns in summer are done according to local traditions. In autumn a general cut removes enough of the winter-green plants to allow the safe passage of the winter storm flows. In the rest of the country cutting is usually either right across the channel, or in the form of swathes in the centre made by weed-cutting launches.

Seasonal changes in the plant's growth cycle can be important. One reason why dyke vegetation can be particularly troublesome is that several potentially choking species with different periods of maximum development can replace each other during the summer (e.g. *Callitriche* spp., spring; *Elodea canadensis*, early summer; *Ceratophyllum demersum*, late summer), and after a cut more than one can spread very quickly again.

The main dominant of chalk streams is *Ranunculus*, (though *Rorippa nasturtium-aquaticum* agg. may be choking in brooks in later summer) and this genus' growth is greatest in early summer, after which it decreases. If *Ranunculus* is cut at this time, however, the new growth is quick and dense and choking vegetation is more likely to recur. This has been reported from various parts of England and has also been proved quantitatively [77]. Thus if the early summer growth does not present a flood hazard it should not be cut, or it may constitute a flood hazard later in the year.

A different type of growth cycle is shown by *Potamogeton pectinatus*. This too grows rapidly in early summer, but later in the year even cutting will not stimulate further growth. (This is a comparable cycle to *Phragmites communis* among the emergents.) Its clumps are smaller too, seldom exceeding 2 m long while those of *Ranunculus* are often 3–5 m. In circumstances particularly favourable for growth *Potamogeton pectinatus* may need cutting three times a year, but once is usual, even in choking stands.

Yet another growth type is shown by *Callitriche* spp. and *Elodea canadensis*. If undisturbed, *Callitriche* spp. reach their maximum size in spring, and *Elodea canadensis* in early summer. After cutting, luxuriant growth can occur any time between spring and autumn.

Cutting is the traditional form of management, and so the vegetation of rivers that are cut regularly consists of species with developmental patterns that can tolerate this. There is much similarity between the vegetation of frequently-cut chalk streams, and those spate-damaged hill streams with:

Fringing herbs   *Myriophyllum* spp.

Callitriche spp.   *Ranunculus* spp.

The vegetation of little-used canals is also similar to that of drains with little recent management.

HERBICIDES

Herbicides are a recent development, and treatment is still variable in that the herbicide used, its concentration, and when and where it is used may all differ from year to year in the same dyke system. Dalapon, dalapon–paraquat, maleic hydrazide and maleic hydrazide–2,4–D have been cleared in Britain for use on emergents; and diquat, dichlobenil, cyanatryn, chlorthiamid, terbutyrne and 2,4–D for use in water [71, 78].

Although emergents can be sprayed in any type of watercourse, aquatic herbicides may be used only where the water is not going to be used while still toxic for irrigation, drinking water for cattle or domestic water supplies. The means that aquatic herbicides cannot yet be used in streams or in alluvial plains where irrigation water is frequently abstracted from drains, but can be used in most dykes and drains.

Different herbicides have somewhat different effects, but in general delicate plants are killed easily while tough large ones are more resilient, though this varies with physiological, anatomical and phenological factors. More care and judgment are needed in the treatment of submerged than of emerged plants, since careless use of aquatic herbicides can lead to empty dykes. An empty dyke, or a dyke containing phytoplankton (which are either unaffected by the toxin, or developed afterwards in the absence of larger plants) is bad for bank stability, water turbidity, conservation, aesthetic value and (if the water has been left toxic) future use of water for irrigation or cattle. An effective combined use of herbicides and dredging has been described above (p. 339.)

Many of the comments made about cutting in the previous section apply equally well to spraying: for example, non-invasive plants need not be treated, and only short stretches should be treated at one time so that animals able to move can find alternative habitats. In contrast to cutting, where the cut parts of the plants remain alive in the water for several days or weeks and can be removed from the water at leisure, herbicides kill aquatic plants very quickly, and serious deoxygenation of the water can easily follow. Whenever possible, therefore, submerged and floating plants should be sprayed early in the year to prevent plants growing thickly, rather than later in the year, when they are already choking. Indiscriminate use of herbicides leads to a quite unnecessary loss of species and communities. It is, however, a cheaper way of controlling plants than cutting.

Emergents can be killed easily by herbicides, but unless the habitat is left barren and unstable the space left by them will be colonised by other plants. Before spraying, therefore, the preferred vegetation should be decided upon and any management directed towards this end. For instance, if, in a dyke used for water supply, replacement of emergents by submerged plants would be more troublesome than the original emergents, then the dyke should be deeply dredged so as to prevent any plant growth in the channel, rather than treating with herbicide. Tall bushy plants on the sides of narrow channels are the most dangerous type of bank vegetation because they both obstruct flow (see Chapter 19) and can be pulled away and start bank erosion (see Chapter 3). Tall monocotyledons are equally impeding, but cause less erosion. By using the growth retardant maleic hydrazide both these types of plants can gradually be replaced by a sward of short grasses, which binds the

344

soil but offers little hydraulic resistance to flow [70]. Some of the Lincolnshire Fenland dykes have been treated in this way. The emergents of the channel and the bottom of the bank can also be treated with maleic hydrazide, but this is not always successful. The major tall monocotyledons can be killed by dalapon or dalapon–paraquat. An overall kill is unsatisfactory for plant and animal conservation, but proper herbicide treatment, as in parts of the Romney Marsh, can lead to a sparse, non-dangerous stand with no loss in species. The same effect can be achieved by cutting but at greater expense.

Tall plants at the sides are a nuisance only if they cover much of the width of the channel. If they are confined to narrow bands, treatment is unnecessary for flood control, and harmful as regards bank stability and conservation. These bands may be above, at or below water level, the crucial point being whether, in the prevailing environment, they can spread. Invasive plants must be controlled, non-invasive ones should be left alone. In clay streams, tall emergents sometimes choke brooks and cover large parts of the channels downstream. Cutting is a good, though temporary remedy, whereas spraying for 2 years usually kills all of the plants and the problem does not then recur until the species have recolonised from elsewhere – which usually takes several years.

Systemic herbicides kill underground parts as well as shoots. This is an advantage for use on emergents growing on shoals in the channel, as the loose sediment is no longer stabilised and the shoals are easily eroded. The submerged and floating plants which can then invade are seldom troublesome in clay rivers, but often cause a flood hazard in dykes. In rivers, spraying emergents on shoals but leaving non-invasive narrow bands on the banks is satisfactory for both flood prevention and conservation.

TREES

Trees are the most old-fashioned, cheapest, and often most useful means of decreasing river plants and stabilising river banks. In Britain, trees have been planted for channel maintenance only as a pilot scheme, but more work has been done in Germany and the Netherlands [77, 79, 80]. The difficulty with tree planting is that while the trees are small they cannot fully maintain the channel, and their presence makes it difficult to use large maintenance equipment. However, if the saplings are as large as possible when planted, it will only be a few (4?) years before they cast sufficient shade to prevent choking river plants growing in the channel or on the banks. Trees and hedges are now often deliberately preserved in order to prevent plants blocking the channel (e.g. in the Sussex River and Water Division), and where recent schemes have removed shading, flood hazards have been created (e.g. in the Wye River Division). It is unfortunate that at present many channels have trees and bushes too far apart to be useful for weed control, and too close together to be convenient for use of large machines in channel maintenance.

In north-west Germany most dykes and streams are less than 4 m wide, and these can be shaded well from the banks. Alders (*Alnus glutinosa*), planted to provide a closed canopy with light shade, are the best form of management. This is satisfactory from the point of view of conservation because the plants can live in small breaks in the canopy and in very light shade (see Chapter 1) and for flood prevention because neither the plants on the bank nor those in the channel can grow thickly, while the tree trunks are sparse enough to cause little obstruction to flow. Alders are deeply rooted and are hardly ever blown over. Shorter-rooted trees may fall in strong winds. Trees live for many decades or even centuries, so replacement management is minimal. Coppicing (or pollarding) can be done at *c.* 15-year intervals if the trees are shading crops, but it is not necessary for stream maintenance in these circumstances. Sometimes more frequent lopping may be required.

The safe limit for channel plants in lowland streams is when they fill not more than a quarter of the channel (see Chapter 19). The reduction in light required to keep plants below this limit varies with the potential dominant of the stream (see Chapter 7). The light-requiring *Ranunculus* can be kept in control by continuous shade which allows 60–70% of full sunlight through, or dapple shade which allows 40–55% through. At the other extreme, the shade-tolerant *Sparganium emersum* needs a reduction in light to as low as 40–55% of full sunlight under a continuous canopy, and less than this in dapple shade in order to prevent excessive growth. A continuous light canopy of tree tops meets or exceeds these requirements. Bank plants may be more shade-tolerant than channel ones. In particular *Urtica dioica* can still grow tall when the channel is almost empty (though it is shorter and sparser than when it is in full light). In less eutrophic and more heavily shaded areas, bank plants are sparser. Except in the narrow steep-sided dykes, etc. (see Chapter 19), tall plants on the banks are only a nuisance if they are pulled away from the bank, and plants of some sort are desirable for soil stability. With the tree trunks to break the force of the water, tall plants are much less likely to be pulled from the bank.

The shading effects described above could, of course, be provided by any tree species capable of growing close to the water, though deciduous trees are preferable because these allow growth of river plants in early spring before the leaf canopy is complete. The great advantage of alders is in the second function of the trees: bank stability. Alders, unlike some other trees capable of growing in damp conditions have long deep roots going down below water level and perhaps even under the channel bottom (Plate 23). The trees can be planted in the 0.5 m above the water level. They will become established even on the steepest slopes, and so can be planted close to the water or on the bank above. Trees at the waterside take up less space and are less likely to shade crops, though trees farther away allow easy access to the channel. It is natural for some of their roots to lie outside the soil and this does not mean that the bank is falling down. Using trees for channel maintenance means that machines do not need to be used on the banks, which is an advantage because long-term use of such machines is liable to damage the banks. Plates 24 and 25 show brooks in the recommended stable low-maintenance state which is provided by alders. The tree and root patterns in Plate 22 have remained unchanged for 10 years.

Maintenance by alders is possible in channels which are:

(*a*) Channels with little sedimentation, therefore not requiring frequent dredging; or channels with more sedimentation, requiring regular dredging, where trees can be planted on one side of the stream only. This is successful if enough shade can be cast. Naturally only one bank is stabilised.

(*b*) Channels where all potentially choking plants can be shaded. That is, either when the tree canopy covers all the channel, or where physical factors prevent plants from growing in the middle and the side bands are shaded. Large chalk rivers are the only type of channel requiring maintenance that cannot be fully shaded.

Maintenance by trees is not possible where man-made banks are unsuitable for trees or in the rare situations where they are serious obstructions to flash floods.

GRAZING

Channels can also be managed by grazing (Plate 26). Sheep are the most effective animals for keeping down vegetation on the banks, as they graze close to the ground and are small enough to do little damage by trampling. Grazed areas develop into short grassland which stabilises the soil without impeding water flow. In many hill streams (where bank plants are in fact irrelevant to channel maintenance) tall plants are few because the land is grazed and grazing is very effective in flatter country where sheep are allowed on to the banks, e.g. in parts of Romney Marsh. In the channels themselves sheep grazing is effective only at the edges, or on dry firm beds, and cannot be used where the weight of the animals can damage the banks.

Some fish eat large quantities of water plants, and are useful for controlling these. Care is needed, however, because if the fish can grow well and breed (as did various carp species introduced into North America), they not only eliminate all plants, but, by stirring up the bottom silt, keep the water almost as liquid mud, which is very unsatisfactory. The grass carp (*Ctenophyaryngodon idella*) probably (1975) cannot breed in Britain. If this is proved true and the fish are introduced at densities allowing some, but not much plant growth, then this would be satisfactory from the points of view of flood prevention and animal conservation, while the fish themselves could be used for food. However, grass carp prefer to eat some plant species rather than others, so when introduced would be likely to kill the species they found most palatable and leave the rest. Thus the final composition of the main plant communities should be known, and considered satisfactory before recommending their introduction.

BOATS

Canals are principally used for navigation. The largest canals are the ship canals, and these are too deep and turbid to have much vegetation. The pleasure boat canals are used mainly by motor boats (Plate 27). These travel fast enough to stir up silt from the bed; their propellers cut plants, and, in thick vegetation, can become entangled in plants; and if boat traffic is dense, their wash is sufficient to sweep away bank plants. It is not easy to quantify the effect of boats on the plants. The only country-wide date available are the counts of boats in each stretch of canal in one day in August, recorded by the British Waterways Board. There are most in August, but single spot counts need not be representative of the annual boat density.

The different types of canal have different types of vegetation:

(*a*) Disused canals which are silted up. These are species-poor and commonly dominated by *Glyceria maxima*, with *Phragmites communis* and *Urtica dioica* on the banks.

(*b*) Disused canals over 1 m deep. These are species-rich both in the bed and at the sides (often 10+ species present per site). The channel is usually choked with submerged and floating plants.

(*c*) Canals which are little-used, up to 2 boats per mile on the spot counts. These are side arms of other canals, or are remote from the main centres of navigation, or are in process of being restored. They are species-rich in the bed and at the edges (often 8+ species per site) and the channel may be choked. It is in this group that plants may impede navigation and need controlling.

(*d*) Three groups of canals which have heavier boat traffic.

(1) Spot counts 3–11 boats per mile, averaging 7. These are unpolluted or little-polluted canals and they may have a rather short boating season. There is little or no bed vegetation, but in shallow water at the sides vegetation is still luxuriant and species-rich (often 5+ species per site).

(2) Spot counts 9–30 boats per mile, averaging 12. These canals may have a longer boating season and some receive industrial pollution. Bed vegetation is again negligible, but here the edge vegetation is species-poor and only intermittent.

(3) Spot counts 3–45 boats per mile, averaging 8. These are the canals which have heavy use, considerable pollution, or both. Plants are almost absent.

348

Boats and pollution thus both affect the plants in canals. Canals without plants on the edges must have their banks protected from scour in some other way and are usually piled.

Many rivers and drains are navigable for part or all of their length. The effects of boat traffic appear similar to those in canals, except that shorter or swifter stretches tend to have rowing craft only. Rowing boats and canoes damage plants less than motor boats, and river plants hamper their movement more.

TRAMPLING

Trampling is another common, though localised, form of mechanical disturbance. Cattle watering at a stream usually keep their watering place clear of plants and depending on the substrate, trampling may lead to a compacted firm substrate or a loose disturbed silt. Children (and indeed adults too) have a similar effect, though mostly at bridges on chalk streams and 'beauty spots' in the hills. These places have a combination of easy access, pleasant surroundings, bright sparkling water, a firm bed and, on chalk, an attractive edge of fringing herbs (until the plants are destroyed).

22

Pollution

The most important type of pollution for British river plants is the sewage and industrial effluent from town and city sewage works. Plants can be used to assess and monitor this pollution, as different species have different sensitivity thresholds. Potamogeton pectinatus *is the most tolerant species of polluted British rivers.*

Inorganic eutrophication (the raising of nutrient status), from fertilisers or sewage effluents of suitable quality, alters the vegetation to that of a clean stream of a higher nutrient status, and is easily diagnosed (when not masked by sewage pollution).

In most British rivers, heavy metal pollution from mines, industry or ore-bearing rocks, is not important to vegetation. Pollution from sediments is damaging in coal streams, where coal dust makes the water turbid and the substrate unstable, and town pollution is often present also. The vegetation may be severely damaged and Potamogeton pectinatus *is again the most characteristic species. Paper mill effluents, agricultural silting, etc., have lesser effects. Boats lead to substrate disturbance, turbidity and chemical pollution, and thus harm the vegetation. Herbicides and other agricultural chemicals may cause major damage (also see Chapter 21).*

Harmful pollution eliminates or decreases sensitive species, thus decreasing species diversity, and it increases any species actually favoured by the pollutant concerned.

The term 'pollution' is used here to describe the effects of those factors introduced into the watercourse by man which are not for the improvement of the watercourse, or for moving or supplying water. The effects of flow regulation and watercourse management are described in Chapters 20 and 21.

Pollution can consist of [81]:

(1) Dissolved substances, such as sewage and industrial effluents, fertiliser, herbicides, road washings and farm silage liquor. These alter the chemical status of the soil and water and may result in increased, decreased or unaltered plant growth.

(2) Sediments (sometimes with dissolved substances also) such as coal dust, paper mill effluents and silt from farmland. Disturbance caused by boats, cattle, paddling, etc. alters sedimentation and causes turbidity.

(3) Heat.

350

Under natural conditions many substances enter watercourses. For example, nutrients are washed out of soils; some chemicals are found locally in high concentrations (e.g. salt near Droitwich); and sediments are washed in from the land, fall in from banks, and are created by erosion of the channel bed and decomposition of organic matter. As rock type, topography and general ecology vary from place to place, the quantity and chemical composition of substances entering streams vary similarly (Chapter 8) Because of this large natural variation, human interference may create conditions which, although new to that site, occur naturally in other places. High chloride, for instance, is found in coastal channels receiving occasional sea flooding and in inland saltpan areas, as well as being a constituent of town effluents. This last is pollution, but the first two situations are not, even if the effects on the plants are similar in each case (other habitat factors being equal).

When human interference alters the incoming substances, the effects vary because natural streams vary. Adding a given concentration of a chemical may affect plants in a stream with a low natural level of that substance while having no effect on a stream containing more of the chemical. Thus a nutrient inflow which alters the trophic status of an oligotrophic stream may not change a eutrophic one. The dilution of the inflow, which depends on the discharge (on channel size and water movement), is also important, and effluents which seriously pollute small brooks may have no effect on large and swift rivers.

Plants (provided they are present in a stream) are a quicker and more convenient way of monitoring the state of a watercourse than are invertebrates or fish, for they remain stationary and are large enough to be identified from bridges. However, because the number of river plant species potentially present at a site is small, the method is less sensitive than invertebrate monitoring. Also, the factors deciding whether a watercourse is a suitable habitat may not be the same for plants and animals (for example, animals are less affected by soil than plants are) and plant monitoring does not necessarily assess water for its suitability for animal or human health.

SEVERITY OF POLLUTION

Any natural river chemical may harm plants if it is present in excess, and many substances are damaging even in low concentrations. Levels of pollution vary from river to river, and from day to day. For convenience they can be grouped into three categories, although these are arbitrary and all intermediate levels occur.

(1) *Streams with continuous very severe pollution*, and either with no plants at all, or with very sparse *Potamogeton pectinatus*, e.g. parts of: R. Douglas (Lancashire), R. Roch (Mersey and Weaver), Rhymney R. (Glamorgan), R. Tame (Trent). The water is visibly polluted, and plants cannot be used to assess pollution except by their absence from

sites which, if clean, would bear plants.

(2) *Streams with accidental very poisonous spillages*, usually lasting only a few hours. These spillages may be due, for example, to temporary faults in sewage works, to dumping of waste (e.g. road washings from over-turned tankers), or to toxic trade effluents reaching sewage works by mistake. The effects depend on the type of poison, its concentration, and the time it stays in any one part of the river. Damage varies from a few frail leaves being lost to a total kill of the entire vegetation. Short-term pollutions are unlikely to harm the soil or the underground plant parts, though, and the vegetation can regrow. The maximum effect of a temporary poison cannot be greater than that of thorough dredging, and is usually much less since it leaves below-ground parts and some tough above-ground parts alive. Vegetation recovers from dredging quickly, usually in not more than 3 years (see Chapter 21), so these acci-dental pollutions seldom have an important effect on vegetation, unless they recur frequently in the same stream. Their influence on fisheries is greater, both because a temporary loss of fish is economically undesir-able, and because fish populations, though not much harmed by dredging, recover only slowly after a kill.

Toxic pollutions may, however, have a long-term effect if the plants are already under stress from another cause. In one river studied, mild pollution had been causing some decline in vegetation for over a decade and when the plants were then killed by an accidental poisonous pollu-tion their recovery was very slow, being far from complete 3 years later.

(3) *Streams with continuous mild pollution*. Here the chemical composi-tion has been altered in such a way that, while plants still grow, the vegetation differs in species presence or abundance from that charac-teristic of a clean site in the same habitat. This is much the commonest type of pollution in Britain and over the years the vegetation has come to be in equilibrium with the pollution. Its effects are unfortunately often unrecognised, since observers merely note vegetation is present and do not realise that this vegetation is not that of a clean stream.

INORGANIC EUTROPHIC POLLUTION

Different species, and different groups of species, are characteristic of different nutrient regimes, and this is one cause of the different vegeta-tion in the upper and lower reaches of the same stream, and in streams on different rock types (see Chapter 8). Eutrophication is the increasing of the nutrients available to the plants and tends to change the nutrient regime to that of a clean stream of a higher trophic status. It can occur when fertilisers are added to the water. In a small chalk brook, fertilised effluent from a watercress bed led to:

Groenlandia densa   *Zannichellia palustris*

(see Chapter 12 for the unpolluted vegetation). Some sewage works effluents produce the same result, when this is not masked by the toxic effects described below, as when these same species were brought into a near-mesotrophic brook (Fig. 22.4; R. Witham, the site somewhat below the upstream effluent source).

The most eutrophic clean streams are the lower reaches of silted clay rivers. Eutrophication here cannot lead to the appearance of species characteristic of a higher nutrient status since the most eutrophic British species are already present. However, when such reaches are clean they normally also contain a few species typical of less eutrophic habitats (see Chapters 1, 8 and 12). The most eutrophic species, when occurring as a group, are:

Nuphar lutea   *Schoenoplectus lacustris*

Rorippa amphibia   *Sparganium emersum*

Sagittaria sagittifolia   *Enteromorpha* sp.

There are other species which, if present, do not alter this diagnosis of nutrient status:

Lemna minor agg.   *Sparganium erectum*

Potamogeton pectinatus

The presence of the following, however, indicates the habitat is less eutrophic, or, possibly, cleaner:

Alisma plantago-aquatica   *Potamogeton lucens*

Elodea canadensis   *Potamogeton perfoliatus*

Phalaris arundinacea   *Ranunculus* spp.

Polygonum amphibium   Fringing herbs

LIST OF SYMBOLS

Acorus calamus   *Myriophyllum spicatum*

Agrostis stolonifera and other small grasses   *Nuphar lutea*

Alisma plantago-aquatica   *Oenanthe crocata*

Apium nodiflorum   *Oenanthe fluviatilis*

Berula erecta   *Phalaris arundinacea*

Butomas umbellatus   *Phragmites communis*

Callitriche spp.   *Polygonum amphibium*

Carex acutiformis   *Potamogeton crispus*

Carex spp.   *Potamogeton lucens*

(*Catabrosa aquatica*, see *Agrostis*)

Ceratophyllum demersum

Elodea canadensis

Eleocharis acicularis

Eleocharis palustris

Eleogiton fluitans

Epilobium hirsutum

Equisetum palustre

Eriopharum angustifolium

(*Glyceria fluitans*, see *Agrostis*)

Glyceria maxima

Groenlandia densa

Hippuris vulgaris

Iris pseudacorus

Juncus articulatus

Juncus bulbosus

Juncus effusus (and other spp.)

Lemna minor agg.

Lemna polyrhiza

Lemna trisculca

Mentha aquatica

Menyanthes trifoliata

Mimulus guttatus

Myosotis scorpioides

Myriophyllum alterniflorum

Potamogeton pectinatus

Potamogeton perfoliatus

Ranunculus aquatilis

Ranunculus fluitans

Ranunculus omiophyllus

Ranunculus peltatus

Ranunculus spp.

Rorippa amphibia

Rorippa nasturtium-aquaticum agg.

Rumex hydrolapathum

Sagittaria sagittifolia

Schoenplectus lacustris

Scirpus spp.

Sparganium emersum

Sparganium erectum

Symphytum officinale

Typha latifolia

Utricularia vulgaris

Veronica anagallis-aquatica agg.

Veronica beccabunga

Zannichellia pallustris

Mosses

Benthic algae

Blanket weed (trailing algae)

Enteromorpha sp.

Other species

(a)

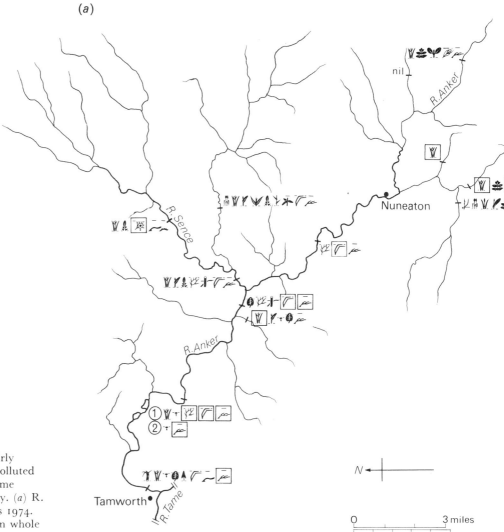

Fig. 22.1. Clean, fairly polluted and very polluted tributaries of R. Tame (Trent). Mainly clay. (a) R. Anker; most records 1974. For general notes on whole river diagrams see p. 177.

Species-richness usually indicates a habitat is both unpolluted and favourable for river plants in other factors also. In fairly clean clay rivers, sites on the lower reaches usually contain 5–8 species from all three of the about groups (e.g. R. Chelmer, Fig. 11.12). Exceptionally clean rivers may have 10–18 species per site (e.g. R. Blythe, Fig. 22.1). With pollution which is eutrophic but hardly toxic, such sites usually have 4–7 species from the first two groups (e.g. R. Soar, Trent). The Water Authority considers R. Soar as being somewhat polluted by nitrogen, etc. For plants, away from the main sewage works, it is eutrophicated more than poisoned and the slight loss in species

355

(*b*)

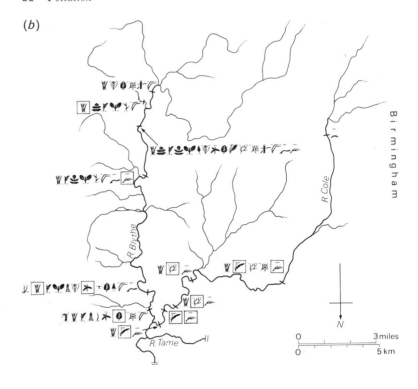

(*c*)

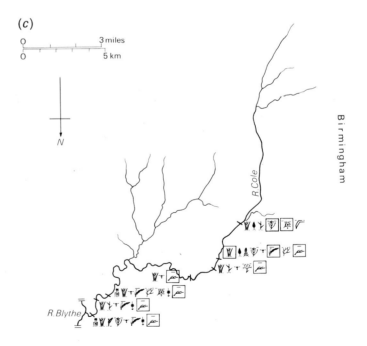

Fig. 22.1 (*cont.*). Clean, fairly polluted and very polluted tributaries of R. Tame (Trent). Mainly clay. (*b*) R. Blythe and R. Cole; most records 1974 (*c*) R.Cole 1976. For general notes on whole river diagrams see p. 177.

diversity is probably because of the (effective) loss of the third group of species given above, rather than to any damage to plants of the first two, more eutrophic, groups. In order to diagnose inorganic eutrophication there must be marked changes in the vegetation, such as the appearance of species which could not occur in a similar clean site, or the loss of a whole group of species (not of just one or two plants from that group).

Fertiliser run-off from agricultural land is another source of eutrophication, though at present the evidence for such run-off varies between regions. It appears, though, that some fertiliser (nitrates) is percolating down to the aquifers, contaminating the ground water and, in course of time, the streams. Fertilisers are potentially a worse hazard than effluents, for effluents come from point sources and so can, at least in theory, be cleaned before their water enters the rivers, while fertilisers affect the whole length of the river and so no cleaning-up is possible. Fertilisers have been added to agricultural land in large quantities for a short period only, and it may be a few more decades before their effects on river plants are known. It is not possible at present to detect differences in plants correlated with the amount of fertiliser added near the river (arable with mixed fertilisers, grassland with nitrogen, grassland without fertilisers), but minor effects could easily be masked by the difficulty of finding rivers comparable in other respects, and by the entry of sewage effluents (which have a much greater effect than do the fertilisers).

SEWAGE AND INDUSTRIAL EFFLUENTS

Effluents from towns and cities are the main sources of damage to British river plants from pollution. Their effects are:

(a) to cause the decrease and loss of the species most sensitive to such pollution;
(b) to decrease the species diversity at each site (this is the common and usual response to toxic pollution; [53], [82], [83];
(c) to increase any species actually favoured by this pollution (except where the pollution is too severe).

The widespread species of British river plants which are most tolerant to sewage and industrial pollution are described below.

1. *Very tolerant*

Potamogeton pectinatus. This is the only species which is actually brought into a habitat, and made more luxuriant there, by suitable levels of pollution (e.g. Plate 6). In clean habitats it occurs most often in the middle reaches of chalk–clay streams (where it can become luxuriant), and in the lower reaches of hill rivers with some hard sandstone in the catchment. *Potamogeton pectinatus* cannot, however, occur in all polluted streams, since some are too shallow or too swift, etc. for this species (see Chapters 2–4 and 6). The most polluted rivers to bear plants usually have sparse *Potamogeton pectinatus*; e.g. parts of R. Dearne (Yorkshire), R. Stour (Severn) and R. Tame (Trent).

Potamogeton pectinatus is also encouraged by sea salt (see Chapter 14), a coarse short form being commonly luxuriant in coastal dykes which have occasional sea flooding. It is also characteristic of saltpan areas [84] and may be abundant in salt- (and minor town-) polluted streams; e.g. R. Salwharpe (Severn). Elsewhere, the salt levels in ordinary sewage effluents do not appear to have an effect independent of that of the other elements in the pollution. Infrequently, luxuriant *Potamogeton pectinatus* can occur in a species-rich polluted site, suggesting more than one chemical factor may be involved.

2. *Tolerant*

Mimulus guttatus	*Sparganium emersum*
Potamogeton crispus	*Sparganium erectum*
Schoenoplectus lacustris	*Enteromorpha* sp.

These are all common in both clean and polluted stream, *Sparganium erectum* being the most widespread species in Britain. The other species occur in polluted places within their ordinary habitat range, but – unlike *Potamogeton pectinatus* – they are rarely, of ever, brought into other habitats merely because there is pollution. In South Wales coal streams *Mimulus guttatus* is often the first species to appear when pollution decreases.

3. *Fairly tolerant*

Agrostis stolonifera	*Nuphar lutea*
Butomus umbellatus	*Rorippa amphibia*
Glyceria maxima	The moss *Fontinalis antipyretica* [18]
Lemna minor agg.	

These all tolerate some pollution but, like the last group, are not brought in to habitats outside their normal range by pollution.

These groups are of the plants most commonly found in polluted areas. With some types of pollution, and in some types of stream, other species may occur with *Potamogeton pectinatus*. Most of the tolerant species are excluded from small brooks anyway (see Chapters 2–4 and 6), and in these semi-emergent species, such as *Apium nodiflorum*, can occur where there is considerable pollution. *Potamogeton pectinatus* is unusually tolerant to pollution in North America, as well as in Britain [53]. In North America, species are considered to be most sensitive to pollution when at the extremes of their geographic range [85, 86]. In such regions plants are under stress from an unsuitable climate, and so the additional stress of pollution is the more easily lethal. (Another example of interacting stresses is described on p. 64.)

The loss of species diversity in polluted streams is due to the small number of species able to tolerate pollution (Groups 1–3 above). The effect of the pollution is that where, in a clean site, there would be perhaps 5–10 species present covering a good deal of the bed, in a typical polluted site half or more of these species are absent, and some of the remainder may be growing less luxuriantly.

Fig. 11.12 (p. 193) of R. Chelmer shows two similar and adjacent tributaries: R. Can, which is clean, and R. Wid, which receives industrial and town pollution. Taking into account the fact that upper reaches of clay brooks are generally species-poor (see Chapter 12), it is clear that R. Can is relatively species-rich and R. Wid species-poor, with the main part of R. Wid having only pollution-tolerant species (see above). Other tributaries of R. Chelmer are also fairly species-rich and fairly clean. Its lower reaches are only mildly polluted and the vegetation consists of both the eutrophic and the less eutrophic species listed above. Diversity is high throughout, and *Potamogeton pectinatus* is absent from the sites recorded.

Fig. 22.1(*a*) shows three tributaries of R. Tame (Trent): – R. Blythe, R. Anker and R. Cole – which are comparable in all aspects except pollution. R. Tame has been mentioned above as one of the most polluted British rivers, bearing either no plants or sparse *Potamogeton pectinatus*. R. Blythe is clean enough to be used for water supply, and the main river is very species-rich. R. Anker is polluted from near the source and species-richness and pollution-sensitive species occur only in clean, rather small tributaries. The main river is species-poor, with few sensitive species and considerable *Potamogeton pectinatus*, though near the mouth the vegetation is recovering further from the source of pollution, and species diversity is increasing. R. Sence, the main tributary of R. Anker, is cleaner and more species-rich. R. Cole used to be seriously polluted. It was cleaned up in 1971/2 but in 1974 still had very few, mostly pollution-tolerant species. *Potamogeton pectinatus* was luxuriant, as was *Potamogeton crispus*. (It is rare to find luxuriant growth of *Potamogeton crispus* associated with pollution.) Pollution, of course, remains in the soil longer than in water, and because silt moves downstream, the upper reaches become clean quicker than the lower ones. The lowest site here remained unchanged from the clean-up until 1975, when 2 less tolerant species appeared, showing recovery was in progress. The *Ranunculus* at one middle site in 1974 was an early sign of recovery. The change in 1976 (Fig. 22.1*c*) is dramatic. Species diversity is greatly increased, and the pollution-tolerant *Potamogeton pectinatus* and *Potamogeton crispus*, though still present, are hardly luxuriant. The river, except for the upper site recorded, is not yet clean for plants, but in 1976 could no longer be classed as seriously polluted.

R. Strine (Severn) showed a similar period between clean-up and recovery, though here the pattern was complicated by drought and dredging. In 1972 (around the time of the clean-up) *Potamogeton pectinatus* was the most important species, and the tolerant *Sparganium emersum* the only other abundant one. Lowering of the water level probably eliminated *Potamogeton pectinatus* from the upper reaches. By 1975 *Potamogeton pectinatus* was no longer luxuriant, and was nearly absent in stretches where the polluted soil had been removed by dredging. *Sparganium emersum* was still growing well and it seemed likely that a clean vegetation pattern would develop.

Fig. 22.2 shows the effect of Stoke-on-Trent and Newcastle-under-Lyme on R. Trent. The river has exceptionally low species diversity, with *Potamogeton pectinatus* luxuriant and other species effectively absent. When this polluted stream is diluted by the entry of the clean R. Penk and R. Sowe, species diversity improves, though the plants present are pollution-tolerant.

Figs. 22.3 and 22.4 are of rivers with many effluent sources, the largest of which are marked. Most affect the vegetation, though considerable recovery is shown between sources. R. Avon (Severn), shown in Fig. 22.3, takes sewage from the large towns of Rugby and Coventry. Rugby sewage brings dominant *Potamogeton pectinatus* into R. Avon,

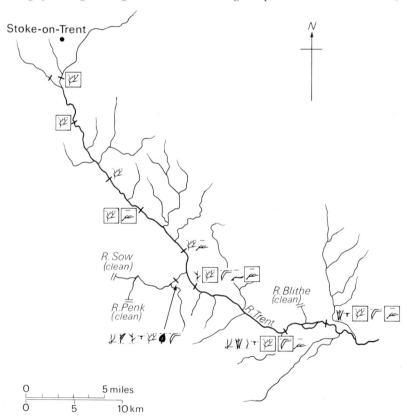

Fig. 22.2. Very polluted R. Trent downstream of Stoke-on-Trent. Mainly clay; 1974.

and greatly decreases species diversity (Plate 6). There is some recovery downstream, though diversity remains fairly low, and sensitive plants remain rare or absent. Coventry sewage has a similar effect in R. Sowe, except that downstream recovery is less, so that when R. Sowe enters R. Avon, R. Avon becomes more polluted, more species-poor and has more *Potamogeton pectinatus*. Even though smaller amounts of sewage enter farther down, there is a recovery in that species diversity increases somewhat, and *Potamogeton pectinatus* decreases. However, comparing the lower R. Avon with the lower R. Chelmer (Fig. 11.12),

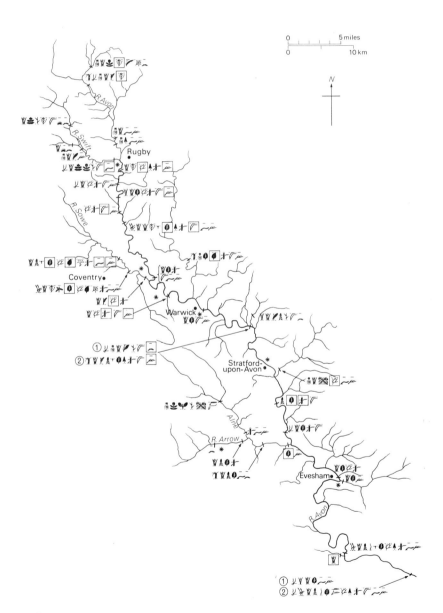

Fig. 22.3. Intermittently polluted R. Avon (Severn). Mainly clay; most records 1975. Main effluent sources starred.

Fig. 22.4. Intermittently polluted R. Witham (Lincolnshire). Mostly clay, but upper R. Witham on soft limestone; most records 1975 and 1974. Main effluent sources starred.

♣ Land species in brook

the cleaner river is more species-rich and has species which are pollution-sensitive and typical of less eutrophic areas. (The contrast with R. Blythe in Fig. 22.1, is even greater.) R. Soar, described above, is intermediate between R. Avon and R. Chelmer. Two tributaries of R. Avon, R. Alne and R. Arrow, are inadequately recorded on Fig. 22.3, but the clean R. Alne bears dense vegetation while the polluted R. Arrow contains much less [87]. Below the confluence of these two rivers, the sites recorded bear only tolerant species, and diversity is not high.

The upper R. Witham (Lincolnshire) in Fig. 22.4 shows loss of species diversity below most of the major effluent inflows. Most pollution-tolerant species do not grow in small brooks, and the most polluted brook sites here contain only algae. The sewage from Grantham has the greatest effect on the vegetation, with species diversity low for a considerable distance and *Potamogeton pectinatus* luxuriant. The effluents entering farther downstream on R. Witham and R. Brant have similar but shorter effects. The upper effluent on R. Brant has little effect on the sites recorded. A solely eutrophic effluent in a clay brook is not easy to detect because of the naturally high trophic status of the site. In contrast, the more mesotrophic habitat of upper R. Witham shows eutro-

362

phication (see above). There is less *Potamogeton pectinatus* here than in R. Avon: R. Avon receives sewage from larger towns.

The examples described so far have been lowland, usually clay rivers. The same effects, however, occur on other rock types, provided the flow in hill streams is not too strong for plant growth (see Chapters 2, 5, 6 and 13). R. Don and R. Dearne (Yorkshire) on Coal Measures and Resistant rocks have species-rich sites only on clean tributaries without fierce spates. The main rivers are polluted, species-poor, and with *Potamogeton pectinatus* as the most frequent species. R. Ribble (Lancashire, 1973) on mountainous Resistant rock has a clean tributary, R. Hodder, which has little but pollution-sensitive vegetation; a seriously polluted one, R. Calder, with widespread *Potamogeton pectinatus* and little else (in fact the flow in R. Calder appears too fierce for *Potamogeton pectinatus* in its ordinary habitat range); and R. Ribble itself, which is clean above but becomes slightly polluted and eutrophicated (upstream of the confluence with R. Calder), with a species-rich flora including *Potamogeton pectinatus*.

Pollution cannot, however, be assessed solely by species lists. R. Leadon (Fig. 11.6) receives harmful organic pollution from Ledbury, though recovery is shown farther downstream and the vegetation of Fig. 11.6 fits this pattern. However, this river has a very fluctuating water level, much silting and considerable shading, all of which decrease vegetation, and even some fairly large tributaries are species-poor. The main river may indeed have its vegetation damaged by pollution, but this cannot be proved by the species found there because of the other unfavourable factors involved.

Species diversity may vary with large fluctuations in weather. For instance, fringing herbs increase when there are few storms, as do submerged plants in hill streams. This must be considered when interpreting vegetation records.

HEAVY METAL POLLUTION

Many metals, such as iron, zinc and lead, may enter rivers in trade effluents, in streams draining from mines, and in general run-off from ore-bearing rocks. Plants seem to be little affected by the usual metal concentrations in most British streams, and a country-wide survey (see Chapter 8) showed no detectable correlation of plant patterns with the variability shown by any or all of ten heavy metals (beryllium, cadmium, chromium, cobalt, copper, iron, lead, manganese, nickel, tin and zinc in acid solution). This is so even in sites generally considered to be polluted. For example:

R. Dee (Dee and Clyd), trade effluent from chemical factory
R. Hayle (Cornwall), copper, etc. effluent from mines
R. Lydney (Wye), iron effluent from mine
Canal, Coventry, metal trade effluent

Of course, effects of trade effluents from towns could well be masked by the other effects of sewage effluents (see above), and as plant monitoring is not yet fully developed some plant variation may occur. With extreme metal pollution, as in tiny brooks in mining areas, damage may occur. The very severely polluted rivers, e.g. R. Tame (Trent) contain so many toxic pollutants that it is not now possible to say which of these pollutants harm plants the most.

POLLUTIONS IN WHICH SUSPENDED OR DEPOSITED SEDIMENTS ARE IMPORTANT

Coal mine pollution produces turbidity and deposited sediment. Turbidity decreases light penetration, confining plants to shallower water, decreasing both the total amount of vegetation present and (selectively) the most light-sensitive species (see Chapter 7). Sedimentation causes unstable substrates and extra sediment moving during faster flows (see Chapter 3). In the lowlands, plants can grow well with extra coal dust sediment in brooks where there is little storm flow. In the hills, however, the swifter normal flow and the spate flows make the habitat very unstable, and therefore the plants sparse. Coal rivers usually have sewage and industrial pollution also, and *Potamogeton pectinatus* is the most tolerant species (in Greater Manchester, however, *Sparganium emersum* is the more common). This plant can anchor in and partly stabilise fine sediment, and as it grows afresh each season from winter buds it is not affected by the wash-out in winter which would be very damaging to plants with winter rhizomes. (Winter buds are scattered by wash-out but enough remain in suitable sites for regrowth.) The next most tolerant species in coal streams are emergents at the edges, and these are naturally less affected by the coal dust. *Ranunculus*, which is light-sensitive and has winter rhizomes, can recolonise and increase rapidly in suitable streams if coal pollution is removed [13].

Lead mining can cause much deposition of sludge, and any smothering sediment causes at least temporary damage (see [88] and Chapter 3).

Paper mill effluents are lighter and do not accumulate on the channel bed. Light penetration is decreased of course, but when particles are pale, plants can grow well below the depth of water visible when looking down on the stream. In R. Don (Dee and Don), the following grow satisfactorily:

Potamogeton × *sparganifolius*   *Sparganium emersum*
Ranunculus cf. *aquatilis*

Sewage and industrial effluents of poor quality may be turbid, but their chemical effects (see above) are the more important to river plants.

Boats stir up silt on channel beds causing turbidity and unstable substrates, and dredging off the loose sediment may improve vegetation.

Boat propellors also cause mechanical damage, and the chemical pollution from rubbish, sewage, Elsan fluids, etc. can be considerable. Canals are sometimes polluted by industrial effluents as well as by boats. The most pollution-tolerant canal species are:

꒛ *Glyceria maxima*   ⍊ *Sparganium emersum*

⊤ *Lemna minor* agg.   ⩊ *Sparganium erectum*

⨯ *Potamogeton pectinatus*

Silt is always washed out from the land, but the amount is much increased by deep ploughing, and by the conditions causing rapid run-off and flash floods [67]. This extra silt (see Chapter 3):

(1) increases nutrient status;
(2) favours deep-rooted (silt) species, and discouraged shallow-rooted (gravel) ones;
(3) decreases substrate stability by increasing the amount of loose sediment;
(4) increases turbidity;
(5) increases abrasion (but not significantly).

The effects are most noticeable in clay catchments, where silt is most abundant. The increase in sediment is less than the amount occurring in coal streams, however, and so the plants are affected less. (Also see Chapter 20, lowering of water level.)

Sediments washed out from clay, sand or gravel works increase the amount of sediment of these in streams. The effects on vegetation are similar to those from agricultural silting, except that sand and gravel hardly affect nutrient status and affect rooting differently to silt (see Chapter 3). Such sediments enhance the natural sedimentation–erosion patterns (e.g. R. Don, Dee and Don), except that in swift streams with little deposition the extra sediment may not affect plants (e.g. R. Wheeler, Dee and Clwyd).

Cattle trampling, swimming, etc., also cause disturbance, and with sufficient disturbance plants are damaged and, later, killed. The tall monocotyledons are the most resistant, as they are large and receive least trampling.

HERBICIDES AND PESTICIDES

Herbicides sprayed onto crops may drift onto, or be accidentally sprayed on, watercourses and cause damage to plants (e.g. herbicides containing 2,4-D harm emerged dicotyledons). Intentional spraying of watercourse emergents is described in Chapter 21.

Chemicals entering the water may come from:

(1) Aquatic herbicides applied to dykes, drains and, to a lesser extent, canals. In a few years' time, they may also be applied to

flowing water (see Chapter 21).

(2) Land herbicides washed out from the land [89] or falling in from above.

(3) Herbicides and pesticides dumped into watercourses as the easiest method of disposing of them, or being washed in from accidental spillages.

The effects on the plants are greatest in dykes and drains, both because herbicides are deliberately applied in these and because they have very little water movement, so the chemicals remain in the habitat instead of being quickly washed out to sea. Unfortunately for plant growth, herbicides do drift with water movement within the dyke system, and so affect areas far from those that are actually sprayed. The dyke species most tolerant to management and pollution (not necessarily only herbicides; see Chapter 21) are:

Agrostis stolonifera   *Lemna minor* agg.

Callitriche spp.   *Enteromorpha* sp.

With the exception of *Callitriche* spp. these are among the plants most tolerant to sewage pollution. Most of the pollution-tolerant stream species, however, do not grow well in dykes anyway (see Chapters 3, 12, 13 and 14).

The decrease of species diversity in North America where arable land reaches right up the channel (Chapter 17) could well be due to herbicides, but there is at present no proof.

HEAT

Vallisneria spiralis is not native to Britain, but it occurs where heated effluents enter rivers in some lowland regions [84].

RECENT CHANGES IN POLLUTION

Some examples of recovery after rivers have been cleaned up have been described above. In R. Tees (Yorkshire), in *c.* 30 years [44] there has been a decrease in *Elodea canadensis*, an increase in *Ranunculus fluitans*, and the appearance of *Enteromorpha* sp. in freshwater parts. In the habitat there has been a decrease in coal dust and an increase in organic pollution in the tributary R. Skerne, which increased eutrophication and pollution in the lower R. Tees.

Unfortunately, we have reliable evidence concerning vegetation in the nineteenth and early twentieth century for only a few watercourses, and most of the surveys are of streams which were, and remain, fairly clean (e.g. [5, 6], records of rare *Potamogeton* spp.). It is, however, generally agreed that the vegetation of Fenland dykes has seriously declined in both quality and quantity since around 1950, coinciding with the increased use of herbicides. *Chara* was reportedly widespread and abundant in dykes, but is now almost absent from both dykes and streams. In North America, where pollution is less, *Chara* is more common (see Chapter 17). In general, British rivers with the worst industrial pollution are being cleaned, but the potential dangers from agricultural pollution are considerable.

Glossary

Some of these terms are explained more fully in the text. Some are terms in general use, which are, in this book, used in a more limited sense, and are defined here.

Abrasion   Rubbing off or away.

Abstraction   Drawing off water (e.g. from below the ground, from soft limestone or sandstone).

Accretion   Deposition of material on the bed of a stream either generally over the bed, or in the form of shoals (see Sedimentation).

Acid   Poor in nutrients, of low pH (see Alkaline, Basic, Dystrophic, Oligotrophic, Mesotrophic and Eutrophic).

Agg. (aggregate)   Used after a plant name to denote an aggregate of species which are difficult to identify separately, e.g. *Rorippa nasturtium-aquaticum* agg. for *Rorippa nasturtium-aquaticum*, *Rorippa microphylla*, and the hybrid between them.

Algae   Small green plants, not composed of stems, roots and leaves. Strictly, chlorophyll-containing thallophytes, which usually grow immersed in water (fresh or marine). *Chara* and *Enteromorpha* are unusually large, and are the only algae named in this book.

Alkaline   With relatively high amounts of bases, such as lime; with pH above 7 (see Acid, Basic, Dystrophic, Oligotrophic, Mesotrophic and Eutrophic).

Alluvial plain   Flat tract of country composed of alluvium.

Alluvium   Deposits of silt, sand, etc. left by water flowing over land which is not permanently submerged; especially those left in river valleys and deltas.

Anatomical   Of the study of plant structure, especially at the microscopic level (see Biological, Morphology, Physiological).

Angiosperm   Higher plants; seed plants with their seeds enclosed in an ovary; including all plants described in this book except for algae, mosses, *Equisetum* and *Azolla* (see Macrophyte, Dicotyledon, Monocotyledon).

Anion   Electronegative substance such as nitrate, phosphate, sulphate and chloride (see Cation).

Aquatic   (1) Living or growing in or near water. (2) An aquatic plant or animal.

Aquifer   Rock which yields water.

Arable   Land fit for ploughing and tillage; not grassland, woodland, or built-up land; bearing crops.

Arithmetic scale   Scale on which the terms differ by a constant difference, as in 1, 3, 5, 7, 9 (see Geometric scale).

Bank   Sloping margin of a watercourse, in this book normally used for that part above normal water level.

Basic   Rich in lime or similar alkaline mineral, and probably in other nutrients also (see Acid, Alkaline, Dystrophic, Oligotrophic, Mesotrophic and Eutrophic).

Bed (of river)   Bottom or floor of watercourse.

369

Bed (of watercress etc.) Flat open channel or space in which plants are cultivated; also the plants which grow in it.

Biological Of the structure and functioning of animals (see Anatomical, Morphology, Physiological).

Biomass Total weight of living things present at any one time (see Production, Productivity).

Biotic Of plants or animals (see Organic, Chemical and Physical factors, and Mechanical management).

Bog Wet spongy ground, consisting chiefly of decayed moss and other plants; nutrient-poor and acid (see Fen, Peat, Swamp).

Boulder Large particle of stone which is too large to be picked up, or to be picked up easily, by hand (see Rock, Stone, Gravel, Sand, Silt, Mud).

Brackish Of water, intermediate between fresh and sea.

Braided Of a stream divided into two or more (more or less) parallel channels.

Brook Small stream (see Stream, River, Creek).

Calcareous Chalky; composed of, or containing, lime or limestone (see Alkaline, Basic).

Canal Artificial watercourse uniting rivers, lakes or seas for the purpose of inland navigation or (in North America) for irrigation.

Carbon Essential element in organic compounds and hence life.

Carbon dioxide (CO_2) A gas; the main source of carbon for photosynthesis (see Gases).

Carpet Covering of plants resembling a carpet in spread and (more or less) in uniformity (see Clump).

Catchment Natural drainage area or basin, wherein rainfall is caught and channelled to a single exit point (see Watershed).

Cation Electropositive substance, such as calcium, sodium, copper or manganese (see Anion).

Chalk Soft, white, comparatively pure limestone, consisting of calcium carbonate (see Clay, Sandstone, Oolite, Resistant rock, Lime, Limestone, Calcareous).

Channel Bed and sides of a watercourse.

Chemical factors Those of a chemical nature; variations in concentration of, for example, nitrate, calcium or pollution (see Biotic and Physical factors, and Physiological).

Chemical management (of watercourses) Herbicides, pesticides and their application (see Biotic and Mechanical factors).

Clay Soft rock, or stiff earth, consisting mainly of aluminium silicate and derived mostly from the decomposition of felspathic rocks (see Chalk, Limestone, Resistant rock, Sandstone).

Clump Tuft, cluster; in separate patches, not forming a smooth carpet (see Carpet).

Coliform bacteria Bacteria potentially harmful to man; in the effluent from sewage works, and of faecal origin.

Community plant The plants present in a site and their social ordering (see Population, Stand).

Competition Striving (rivalry) of two or more for the same objective.

Consolidated Made solid, firm or compact.

Correlate To have or establish a mutual relation of association between different structures, characteristics, etc.

Cover Extend over, occupy.

Creek (1) In Britain, a narrow inlet in the coastline of the sea, tidal estuary, etc. (2) In North America, a branch of a main river, a tributary river, a small stream.

Critical velocity Water speed at which particles of a specified size will move on the channel bed.

Culvert Conduit or drain of brick or timber, etc. to convey water beneath, for example, a road.

Current Portion of a body of water moving in a definite direction; that which runs or flows (see Discharge, Flow, Velocity).

Cusec Measure of discharge: a cubic foot of water per second (see Discharge).

Cuticle Outer skin of leaves, stems, etc. composed of a fatty substance not easily permeable by water. Often little or negligible on submerged plant parts.

Debris Remains of plants, etc. broken down or destroyed (see Humus, Detritus, Peat).

Deoxygenation Deprivation of free oxygen, leading to death of animals and plants requiring this for survival.

Detritus Debris and other broken-down material (e.g. from river beds), usually broken down more than that referred to as debris (see Humus, Peat).

Dicotyledon One of the two main groups of angiosperms, having two seed-leaves. Typically with leaves relatively wide and shoots bushy (see Angiosperm, Monocotyledon).

Discharge Total volume of water per unit time flowing through the channel (see Flow).

Disperse To spread about, to distribute from a source.

Dissolved gases See Gases.

Ditch (1) A long narrow hollow dug to receive or conduct water. Usually 0·5–2 m wide. Found throughout the country, particularly on poorly drained soils, but in this book termed 'dyke' when occurring in wetlands. (2) In North America used synonymously with ditch, dyke, drain and sewer in Britain.

Dolomite Rock consisting of both calcium carbonate, as limestone, and magnesium carbonate (see Limestone).

Dominant Of a plant species occupying the most space in, and controlling the character of, a plant community (see Luxuriant, Performance).

Drain Drainage channel, the larger channels of the Fenland drainage system, receiving water from dykes, usually by gravity, and passing it to the rivers, etc., usually through pumping systems. Usually 6–20 m wide.

Drainage order Analyses of the pattern of tributaries of a river. See p. 74.

Drift, glacial Material deposited during the glaciations of the Quaternary era (as opposed to the solid rock of more ancient origin).

Dyke An artificial watercourse for draining marshy land and moving surface water. Usually 2–4 m wide, found beside roads and between fields, draining into larger drains, sewers, rivers or the sea. Found in East Anglia and Fenland, and, by extension, elsewhere.

Dystrophic Of negligible nutrient content; acid and usually composed of, or stained with, bog peat. By extension, used to describe species characteristic of such a habitat. (See Oligotrophic, Mesotrophic, Eutrophic and Peat.)

Ecology Study of plants and animals in their homes; mutual relations between plants and animals and their environment (see Ecosystem Environment, Habitat).

Ecosystem The land and water, the plants and animals in these, the climate, and the functioning of all these together (see Ecology, Environment, Habitat).

Effluent Outflow from sewage works, factories, farms, etc.

Emerged (of plant parts)   Above water (see Submerged, Floating, Free-floating).

Emergent   A plant mainly or entirely above water (see Submergent).

Energy   Capacity to do work; radiant energy (light energy) is used in photosynthesis (see Photosynthesis).

Environment   Conditions or influences, under which a plant (or animal) lives or develops; its surroundings (see Ecology, Ecosystem, Habitat).

Ephemeral   Short-lived, transitory (see Perennial).

Epidermis   Outer layer of tissue (of cells) on plant parts.

Eriphyte   Plant living on another plant; usually of algae living on larger plants.

Erode   Wear away.

Erosion   Scour; the removal of material from the channel of a stream (see Abrasion).

Established   Securely or permanently settled.

Eutrophic   Of high nutrient regime, with ample or even excess nutrients for plant growth present. By extension, used to describe species characteristic of very nutrient-rich habitats. (See Dystrophic, Oligotrophic, Mesotrophic.)

Eutrophication   Raising of nutrient status.

Fen   Low land, now or formerly covered with shallow water, or intermittently so covered. Any peat developed is alkaline (contrast bog peat) because of the high base status of the waters, derived from the land around. Fen silt is alluvial and nutrient-rich. Particularly the Fenland of E. England, and by extension, other regions. (See Bog, Peat, Swamp.)

Fertiliser   That which fertilises land; especially describing nutrients added as powders or sprays.

Flash flood   Storm flow in which discharge rises very rapidly, due to a combination of heavy rainfall and quick run-off from the catchment (see Flood, Spate, Storm flow).

Floating   Lying flat on and in the plane of the water surface, or occasionally just below it (see Emerged, Free floating, Submerged).

Flood   (1) An overflowing of water over land. (2) A storm flow (see Flash flood, Storm flow).

Flood gate   Contrivance for stopping or regulating the passage of water (see Lock, Sluice, Weir).

Flood hazard   That which, by obstructing water movement, may or will cause flooding.

Flood meadow   Grassland intermittently flooded by a stream (see Alluvial plain, Flood plain, Water meadow).

Flood plain   Flat land beside the lower reaches of a river which, at least in the past, was flooded during high discharges (see Alluvial plain).

Flora   List or descriptive catalogue of the plants of an area, etc. (see Vegetation).

Flow   Water movement; quantity of water moving (see Discharge).

Free floating (of a plant)   Floating and not anchored to the substrate (see Emerged, Submerged).

Fringing herbs   Group of semi-emergent, somewhat bushy short dicotyledons, commonly fringing the edge of certain stream types, and occurring more sparsely in a wider range of types.

Gases   Used here for the two gases, oxygen and carbon dioxide, which are required for plant life, and which reach emerged and floating plant parts through the air, and submerged ones through the water in which they dissolve (and, where relevant, by transport from aerial parts).

Gauging station   Structure to measure water discharge, normally with a small weir.

Geology   Geological features (rock types, etc.) of a district.

Geometric scale   Scale on which the ratio between the successive quantities is constant, as in 1, 3, 9, 27, 81 (see Arithmetic scale).

Glacial drift   Material deposited during the glaciations of the Quarternary era (as opposed to the solid rock of more ancient origin).

Gradient   (1) Of slope: amount of inclination to the horizontal. (2) Of temperature, etc.: rising or descending by regular degrees.

Gravel   Strictly, particles over 2 mm wide. Here used for particles large enough to be easily picked up separately, and under *c.* 2 cm wide (see Boulder, Stone, Sand, Silt, Mud).

Ground water level   Plane below which the rock or soil is saturated with water (see Water table).

Gypsum   Calcium sulphate (the material from which plaster of Paris is made).

Habit   Characteristic mode of growth and appearance of a plant or animal (see Morphology, Ecology).

Habitat   Kind of locality in which a plant or animal naturally lives and grows, as, the fast-flow habitat (see Ecosystem, Environment, Site).

Hard rock   Here used to describe those very hard types of rock which do not erode easily and so can potentially, and often actually, form hilly and mountainous country. They include Carboniferous limestone, Coal Measures, Resistant rocks and Old Red Sandstone, etc. (see Soft rock).

Headwater   Stream near the source of a river.

Heavy metal   Metallic element of high specific gravity, such as cadmium, copper, lead.

Herb   Plant of which the aerial stem does not become woody or persistent.

Herbicide   Chemical which kills plants; used for weed control (see Pesticide).

Highland   High or elevated land. Used here in a general sense for hilly ground, or, with the capital H, for the Highlands of N. Scotland (see Lowland, Hill, Upland, Mountain).

Hill   Natural elevation of the earth's surface. Used here in a general sense (see Highland, Upland, Mountain).

Histogram   Chart in which the relative frequency of different items is indicated by the height of a bar (strictly, the area of the bar, but here bars are of equal width in each chart).

Hummock   Protuberance arising above the general level of a surface.

Humus   Vegetable mould; dark-brown or black substance resulting from the slow decomposition of organic matter (see Debris, Detritus, Peat).

Hybrid   Offspring of two plants of different species.

Hydraulic   Pertaining to water as conveyed through channels.

Impoundment   In North America, a pond caused by a dam across a stream, used for water supply, water power etc.

Inorganic   Not formed from plant or animal parts (except when completely broken down); mineral (see Mineral, Organic).

Invertebrate animals   Animals without backbones, comprising insects, spiders, snails, worms, etc.

Laminar flow   Flow where water particles appear to move in smooth paths, and one layer of water slides over another. Viscous forces are strong compared to inertial forces (see Turbulent flow).

Lime   Calcium carbonate (see Chalk, Oolite, Calcareous).

Limestone   Rock consisting chiefly of carbonate of lime (calcium) (see Chalk, Oolite, Clay, Sandstone, Resistant rock).

Lock (of canal, etc.)   Portion of the channel shut off above and below by gates and provided with sluices to let the water out and in; used to raise or lower boats from one level to another (see Canals, Sluice).

Lode   Navigable watercourse in the Fenland, usually developed from a stream. Commonly 3–10 m wide (see Drain, Dyke, Stream).

Lowland   Low-lying land. Used here in a specialised sense, defined on p. 225 (see Upland, Mountain).

Luxuriant   Growing profusely. Used of stands which cover a considerable proportion of the ground and/or occupy a considerable proportion of the water (see Dominant, Performance).

Macrophyte   Large plant; the plants discussed in this book, the higher plants (angiosperms), horsetails, water fern, mosses and the large algae *Chara* and *Enteromopha*.

Mechanical management   Maintaining a watercourse by dredging, cutting, etc. (see Biotic, Chemical).

Mesotrophic   Of moderate nutrient regime. By extension, used to describe species characteristic of these habitats (see Dystrophic, Oligotrophic, Eutrophic).

Mesozoic   Geological name for the period, and hence describing the rocks formed, between *c.* 180 million and *c.* 50 million years ago (see Pre-Cambrian, Palaeozoic, Tertiary, Quarternary).

Metabolism   The process by which nutritive material is built up into living matter, or by which the complex substances of protoplasm are broken down to perform special functions (see Photosynthesis, Respiration).

Microhabitat   Subdivision of a habitat, in which one or more environmental influences differ somewhat to those of other parts of the same habitat, as, stony and silty patches on a stream bed (see Habitat).

Mill (water mill)   Building fitted with machinery for grinding corn, in which the power is provided by a wheel on a stream. Most British mill wheels are disused.

Mill pool   Pool or pond often found in a mill stream somewhat downstream of the mill wheel; sometimes upstream to give a head of water.

Mill stream   Stream turning a mill wheel. Normally constructed as a branch from the main stream, with structures for water regulation. Most British mill wheels are disused.

Mineral   Natural substance of neither animal nor vegetable origin; inorganic (see Organic).

Monocotyledon   One of the two main groups of angiosperms (seed plants), having one seed-leaf. Typically with narrow leaves (some wide), with parallel veins (see Angiosperms, Dicotyledon).

Moraine   Glacial drift deposited around the side or end of a glacier.

Morphology   Study of the form of plants and animals, and the structures etc. which influence that form (see Anatomical, Habit, Physiological).

Mountain   Large hill. A mountain stream is here defined in a specialised sense, see p. 225 (see Lowland, Upland).

Mud   Strictly, particles below 0·002 mm diameter. Here included within 'silt' (see Boulder, Stone, Gravel, Sand, Silt, Mud).

Mulch   Partly decayed vegetation, perhaps mixed with earth, used as a manure, etc. (see Humus, Detritus).

Niche, ecological   Place or position suited to a particular plant species or community (see Habitat).

Nutrient   Serving as nourishment; normally used of inorganic substances necessary for plant growth, such as calcium, phosphate, etc. (see Dystrophic, Oligotrophic, Mesotrophic, Eutrophic).

Oligotrophic   Low in nutrients. By extension, used to describe species characteristic of this habitat (see Dystrophic, Mesotrophic, Eutrophic).

Oolite   Pale soft limestone, composed of small rounded granules, with a lower proportion of calcium carbonate than chalk (see Limestone, Clay, Sandstone, Resistant rock).

Organic Of, pertaining to, or composing plants or animals (see Biotic, Inorganic).

Outcrop The cropping out or exposure of a rock type at the surface (see Rock).

Oxygen (O₂) A gas used in respiration by plants and animals (see Gases).

Oxygenate To supply with oxygen (see Deoxygenate).

Palaeozoic Geological name for the period, and hence describing the rocks formed, between *c*. 500 million and *c*. 180 million years ago (see Pre-Cambrian, Mesozoic, Tertiary, Quarternary).

Particle Portion of matter.

Peat Plant material stored and partly decomposed under water. Found in fens (alkaline peat), bogs (acid peat) and swamps (also see Debris, Detritus, Humus).

Perennial flow Of a stream flowing throughout the year.

Perennial plant Plant which remains alive through a number of years.

Performance, good or **high** Growing profusely. Used of stands which cover a considerable proportion of the ground and/or occupy a considerable proportion of the water (see Dominant, Luxuriant).

Perimeter Outer boundary of an area or surface.

Pesticide Chemical which kills pests, usually used for those killing small animals which are dangerous for crop production or human health (see Herbicide).

pH Measure of acidity in terms of hydrogen ion concentration [H+], ranging from 0 or less at the acid end through 7 (neutral) to 14 or more at the alkaline end. Most British rivers have a pH between 6.8 and 8.5.

Phenology Study of the periodic phenomena of plants, such as leafing or flowering.

Photosynthesis Conversion in plants of carbon dioxide and water into carbohydrates in the presence of light (see Respiration).

Physical factors Those factors of a material nature (e.g. flow, substrate textures, water turbidity), as opposed to chemical and biotic factors.

Physiological Of the functions of plants or animals (see Anatomical, Biotic, Chemical factors, Physical factors).

Phytoplankton Floating microscopic plants, including desmids and diatoms.

Piled banks Those with piles (pointed stakes) driven into the bed of a river or canal for the support of a wall, bridge, steep-sided bank, etc.

Plant community See Community.

Plateau Elevated tract of comparatively flat or level land. When dissected, gullies or valleys are cut into this table-land.

Pollutant Substance covering pollution.

Pollution The alteration of the chemical status of a watercourse by human interference. By extension, any alteration by man which is not for the benefit or proper maintenance of the river and its ecosystem (see p. 350 and Effluent).

Ponding Rising of water level, even though flow may be small, because a channel is choked with plants.

Population Group of plants at a site (see Community, Stand).

Porous Having minute interstices through which water, air, etc. may pass.

Pre-Cambrian Geological name for the period, and hence describing rocks formed, earlier than *c*. 500 million years ago (see Palaeozoic, Mesozoic, Tertiary and Quaternary).

Production Weight of new organic matter made by plants (or animals) over a period (see Biomass).

Productivity Rate at which new organic matter is formed (see Biomass).

Propagule   Plant part used for propagation, such as fruit, seed, corm, bulb, rhizome fragment, winter bud, etc.

Quarternary   Geological name for the period, and hence describing the materials deposited, during the last 1 million years. Glacial deposits are termed 'drift' (see Pre-Cambrian, Palaeozoic, Mesozoic and Tertiary).

Reach   Portion of a river etc. which can be seen in one view. Hence 'lower reaches' for the lower or downstream end of a river and 'upper reaches' for the part nearer the sources (see Stretch).

Reen   See Rhyne.

Reservoir   Receptacle specially constructed to contain and store a large supply of water for ordinary uses (see Impoundment).

Resistant rock   Very hard rock resistant to both erosion and solution; including andesite, basalt, gneiss, granite, Millstone Grit, schist, shale and slate (see Limestone, Clay, Sandstone).

Respiration   Conversion of carbohydrates to carbon dioxide and water, using oxygen and releasing energy for growth etc. (see Photosynthesis).

Rhizome   Perennial, horizontal, root-like stem, usually underground but sometimes floating (see Stolon, Propagule).

Rhyne   A system of dykes and drains. S. Wales, S.W. England (see Ditch, Dyke, Drain, Sewer).

River   A large stream of water flowing in a channel towards the sea, a lake, or another stream (see Stream, Brook, Creek).

Rock (bedrock)   Material composing the hard surface of the earth (see Clay, Limestone, Sandstone, Resistant rock).

Rock (of particle size)   (1) Bedrock exposed in the channel, or large particles of the size of boulders. (2) In North America, particles of the size of stones. Not used in this sense in this book.

Root weft   Layer of closely interwoven roots.

Saltation   Particle movement by leaping or bounding along the bed of a river.

Sand   Strictly, cohesionless material between 0.06 and 2 mm wide. Here used for particles large enough to be easily seen as separate particles when held in the hand, but too small to be easily picked up separately (see Boulder, Stone, Gravel, Silt, Mud).

Sandstone   Rock composed of consolidated sand (see Clay, Limestone, Resistant rock).

Sediment   Particles which fall by gravity in water; mud, silt, sand, gravel, stones and boulders.

Sedimentation   Deposition of material on the bed of a stream, either generally over the bed or in the form of shoals (see Accretion, Silting).

Semi-   Half, partly, to some extent.

Sewer   In S. England, an artificial watercourse for draining marshy land and moving surface water into a river or the sea. Often 2–6 m wide. (see Ditch, Dyke, Drain, Rhyne.

Shoal   Place where the water is of little depth; a submerged bank or bar.

Shoot   Stem together with its leaves.

Silt   Strictly, cohesionless material between 0.002 and 0.06 mm diameter. Here used to include mud, and those particles too fine to be seen individually from a little distance (see Boulder, Stone, Gravel, Sand and Mud).

Silting   Depositing silt (see Accretion, Sedimentation).

Site   Place, position, situation; place occupied by some specified plant or object (see Habitat).

Sluice   Structure for impounding the water of a river, canal etc., provided with an adjustable gate or gates by which the volume of water is regulated or controlled (see Flood gate, Weir, Lock).

Soft rock   Here used to describe the softer types of rock which erode easily and always form lowland landscapes. They include chalk, oolite, clay, Tertiary sandstones, etc. (see Hard rock).

Soil   Earth, substrate.

Solid rock   All rock types except Glacial Drift and recent alluvial deposits (see Rock).

Spate   Large discharge or storm flow caused by heavy rains, etc. in hill streams where the water force is great (see Storm flow, Flood).

Species   Group of plants (or animals) having certain common and permanent characteristics distinguishing it from other groups.

Specific gravity   Relative heaviness of any kind of matter (the ratio of the weight of a given volume to that of an equal volume of some substance, usually water, taken as a standard).

Spring   Flow of water rising or issuing naturally out of the ground.

Stable   Able to maintain its place or position. A stable river is one which over a long period of time does not progressively scour or accrete, though it may both scour and accrete at various times.

Stand   Group of plants at a site, normally of a single species (see Community, Population).

Stolon   Annual, horizontal, leafy stem, borne above or below ground (see Rhizome).

Stone   Strictly, a particle over 2 cm wide. Here used for particles larger than gravel which can be easily picked up by hand (see Boulder, Gravel, Sand, Silt, Mud).

Storm flow   The large water discharge following heavy rain (see Spate, Flood).

Stream   Course of water flowing continuously along a bed on the earth, forming a river or brook.

Streamlined   The shape of a moving body, e.g. a fish, which gives the smallest amount of resistance as it passes through its medium, e.g. water; the similar shape of some stationary objects in moving water.

Stretch (of river)   A continuous length or distance; an expanse of water (see Reach).

Submerged (of plant parts)   Within the water (see Emerged, Floating, Free-floating).

Submergent   A plant within the water (see Emergent).

Subsoil   Stratum of soil lying immediately under the surface soil.

Substrate   Material near the surface of the bed of the watercourse; the rooting medium; the soil.

Suspended   Of particles diffused through the water.

Suspension   The condition of being suspended, as particles in water.

Swamp   (1) In North America and, by derivation, Britain. A piece of wet spongy ground, a marsh, fen or bog. (2) In North America, a river in such a swamp.

Swathe   Space covered by a sweep of a mower's scythe or the scythes of a weed-cutting launch.

Tall monocotyledons   Group of tall emergent aquatics with long narrow leaves, forming dense stands which shade out all shorter plants.

Terrestrial   Of or pertaining to the land as distinct from the water.

Tertiary   Geological name for the period, and hence describing the rocks formed, between *c.* 180 million and *c.* 1 million years ago (see Pre-Cambrian, Palaeozoic, Mesozoic, Quaternary).

Texture (of a soil)   The character of a substrate as to its being fine, coarse, etc.; the structure of its constituents.

Topography   Features of a region or locality collectively.

Toxic   Poisonous.

Transpiration   Evaporation of water from plant leaves etc. into the air.

Tributary   A stream or river which flows into another.

Trophic   Of or pertaining to nutrition (see Nutrient, Dystrophic, Oligotrophic, Mesotrophic, Eutrophic).

Turbid   Thick or opaque with suspended matter; not clear; cloudy.

Turbulence   Agitation, disturbance, commotion of the water; movement within a limited space rather than along a single direction. Strictly, turbulent flow occurs when the viscous forces are weak compared to the inertial forces (see Laminar flow).

Unstable   Apt to change or alter.

Upland   Hilly country. An upland stream is here defined in a specialised sense, see p. 225 (see also Lowland, Mountain, Hill).

Vegetation   Plants in general; the plant life at a site (see Flora).

Vegetative   Of all parts of a plant other than its flowers and fruits and their stems.

Velocity   Speed of motion, normally along the length of the stream (see Current, Flow, Discharge).

Watercourse   A stream of water, a river or brook; an artificial channel for the movement of water. The general term for water channels, including all the other types defined here.

Water meadow   Grassland beside chalk streams, kept intermittently wet under a particular form of management. Now mainly disused.

Watershed   (British: for North American, see Catchment.) Tract of ground between two drainage areas.

Water table   Plane below which the rock or soil is saturated with water (see Ground water level).

Weathering   Disintegration resulting from the action of the atmosphere (including rainfall, and river action) on substances exposed to this.

Weft   See Root weft.

Weir   Barrier or dam to restrain water (see Flood gate, Sluice).

Winter-bourne   Brook on soft limestone which has water flowing in a well-defined channel in winter, but dries in summer.

Bibliography

This bibliography is in two parts: first, a list of references cited in the text, by number in the order in which they are cited, and second, a list of additional references, cited under the chapter to which they are most relevant.

The two papers by Butcher [5, 6] summarise many of the phenomena and factors which this book describes, but are not cited as such throughout.

1. Arber, A. (1920). *Water plants*. Cambridge University Press, London.
2. Sculthorpe, C. D. (1967). *The biology of aquatic vascular plants*. Edward Arnold, London.
3. Tansley, A. G. (1949). *The British Islands and their vegetation*. Cambridge University Press, London.
4. Spence, D. H. N. (1964). The macrophytic vegetation of freshwater lochs, swamps and associated fens. In *The vegetation of Scotland*, ed. J. H. Burnett, pp. 306–425. Oliver & Boyd, Edinburgh.
5. Butcher, R. W. (1927). A preliminary account of the vegetation of the River Itchen. *J. Ecol.* **15,** 55–65.
6. Butcher, R. W. (1933). Studies on the ecology of rivers. I. On the distribution of macrophytic vegetation in the rivers of Britain. *J. Ecol.* **21,** 58–91.
7. Westlake, D. F. (1968). The biology of aquatic weeds in relation to their management. In *Proc. 9th Brit. Weed Control Conf.*, pp. 372–9.
8. *The University Atlas* (1946). George Philip & Son, London.
9. *Rainfall atlas of the British Isles* (1926). Royal Meterological Society, London.
10. Haslam, S. M., Sinker, C. S. & Wolseley, P. A. (1975). *British water plants*. Field Studies Council, London.
11. Clapham, A. R., Tutin, T. G. & Warburg, E. F. (1962). *Flora of the British Isles*. Cambridge University Press, London.
12. Parkin, L. (1973). Last throw on the Mimram. In *Association of River Authorities Year Book and Directory 1973*, pp. 199–205. National Water Council, London.
13. Whitton, B. A. & Buckmaster, R. C. (1970). Macrophytes of the River Wear. *Naturalist* **914,** 97–116.
14. Holmes, N. T. H. & Whitton, B. A. (1975). Macrophytes of the River Tweed. *Trans. Bot. Soc. Edinb.* **42,** 369–81.
15. Holmes, N. T. H. & Whitton, B. A. (1975). Submerged bryophytes and angiosperms of the River Tweed and its tributaries. *Trans. Bot. Soc. Edinb.* **42,** 383–95.
16. Haslam, S. M. (1971). Physical factors and some river weeds. In *Proc. Eur. Weed Res. Coun., 3rd int. Symp. Aquatic Weeds 1971*, pp. 29–39.
17. Westlake, D. F. (1967). Some effects of low velocity currents on the metabolism of aquatic macrophytes. *J. exp. Bot.* **18,** 187–205.
18. Dr N. T. H. Holmes, personal communication.
19. Mr T. Brabben & Mr K. R. C. Powell, personal communication.
20. Hynes, H. B. N. (1970). *The ecology of running waters*. Liverpool University Press.

21. Andrews, F. W. (1945). Water plants in the Gezira canals. A study of quatic plants and their control in the canals of the Gezira cotton area (Anglo-Egyptian Sudan). *Ann. appl. Biol.* **32,** 1–14.

22. Haslam, S. M. (1973). Some aspects of the life history and autecology of *Phragmites communis* Trin. *Pol. Arch. Hydrobiol.* **20,** 79–100.

23. Jones, H. R. & Peters, J. C. (1976). Physical and biological typing of unpolluted rivers. In *EIFAC Symposium, Helsinki.*

24. Jones, H. R. & Peters, J. C. (1977). In *Water Research Centre Technical Report.* (In press.)

25. Dr B. A. Whitton, personal communication.

26. Westlake, D. F. (1977). In *Freshwater Ecosystems*, ed. E. D. Le Cren. International Biological Programme Synthesis Volumes. Cambridge University Press, London. (In press.)

27. Mr D. F. Westlake, personal communication.

28. Spence, D. H. N. & Chrystal, J. (1970*a*). Photosynthesis and zonation of freshwater macrophytes. I. Depth distribution and shade tolerance. *New Phytol.* **69,** 205–15.

29. Spence, D. H. N. & Chrystal, J. (1970*b*). Photosynthesis and zonation of freshwater macrophytes. II. Adaptability of species of deep and shallow water. *New Phytol.* **69,** 217–27.

30. Westlake, D. F. (1973). Aquatic macrophytes in rivers. *Pol. Arch. Hydrobiol.* **20,** 31–40.

31. Bristow, J. M. & Whitcombe, M. (1971). The role of roots in the nutrition of aquatic vascular plants. *Amer. J. Bot.* **58,** 8–13.

32. Bristow, J. M. (1975). The structure and function of roots in aquatic vascular plants. In *The development and function of roots*, ed. Torrey & Clarkson, pp. 221–36. Academic Press, New York & London.

33. Denny, P. (1972). Sites of nutrient absorption in aquatic macrophytes. *J. Ecol.* **60,** 819–29.

34. Nicholson, S. A., Levey, R. A. & Clute, P. R. (1975). Macrophyte-sediment relationships in relationships in Chautanqua Lake. *Verh. int. Verein. Limnol.* **19,** 2758–64.

35. Casey, H. & Westlake, D. F. (1974). Growth and nutrient relationships of macrophytes in Sydling Water, a small unpolluted chalkstream. In *Proc. Eur. Weed Res. Coun. 4th int. Symp. Aquatic Weeds 1974*, pp. 69–76.

36. Westlake, D. F. (1975). Macrophytes. In *River ecology*, ed. M. Owens & B. A. Whitton, pp. 106–28. Blackwell, Oxford.

37. Goulder, R. & Boatman, D. J. (1971). Evidence that nitrogen supply influences the distribution of a freshwater macrophyte, *Ceratophyllum demersum*. *J. Ecol.* **59,** 783–91.

38. Seddon, B. (1972). Aquatic macrophytes as limnological indicators. *Freshwater Biol.* **2,** 107–30.

39. Casey, H. & Downing, A. (1976). Levels of inorganic nutrients in *Ranunculus penicillatus* var. *calcareus* in relation to water chemistry. *Aquat. Bot.* **2,** 75–9.

40. Westlake, D. F. (1975). Primary production of freshwater macrophytes. In *Photosynthesis and productivity in different environments*, ed. J. P. Cooper, pp. 189–206. Cambridge University Press, London.

41. Westlake, D. F., Casey, H., Dawson, F. A., Ladle, M., Mann, R. H. K. & Marker, F. H. (1972). The chalk-stream ecosystem. In *Productivity problems of freshwaters*, ed. Z. Kajak & A. Hillbricht-Ilkowska, pp. 615–37. IBP–UNESCO, Kazimierz Dolny, Poland.

42. Berrie, A. D. (1972). Productivity of the River Thames at Reading. *Symp. zool. Soc. Lond.* **29,** 69–86.

43. Mann, K. H., Britton, R. H., Kowalczewski A., Lack, T. J., Mathews, C. P. & McDonald, I. (1972). Productivity and energy float at all trophic levels in the River Thames, England. In *Productivity problems of freshwaters*, ed. Z. Kajak & A. Hillbricht-Iklowska, pp. 579–96. IBP–UNESCO, Kazimierz Dolny, Poland.

44. Rudescu, L., Niculescu, C. & Chivu, I. P. (1965). *Monografia stufului den delta Dunarii*. Editura Academiei Republicii Socialiste Romania, Bucharest.

45. Whitton, B. A. & Dalpra, M. (1968). Floristic changes in the River Tees. *Hydrobiol.* **32,** 545–50.

46. Holmes, N. T. H., Lloyd, E. J. H., Potts, M. & Whitton, B. A. (1972). Plants of the River Tyne and future water transfer scheme. *Vasculum* **57,** 56–78.

47. Professor S. G. Smith, personal communication.

48. Kohler, von A. Vollrath, H. & Beisl, E. (1971). Zur Verbreitung, Vergesellshaftung und Ökologie der Gefass-makrophyten im Fliesewassersystem Moosach (Münchener Ebener). *Arch. Hydrobiol.* **69,** 333–65.

49. *Water resources data for Wisconsin 1973*. United States Department of the Interior, Geological Survey.

50. Dr B. Hellquist, personal communication.

51. Dr H. M. Dale, personal communication.

52. Mitchell, R. S. (unpublished data). Effects of Radford army ammunition plant wastes on aquatic vegetation in the New River, Virginia.

53. Kulberg, R. K. (1974). Distribution of aquatic macrophytes related to paper mill effluents in a southern Michigan stream. *Amer. Midl. Nat.* **91,** 271–81.

54. Moyle, J. B. (1945). Some chemical factors influencing the distribution of aquatic plants in Minnesota. *Amer. Midl. Nat.* **34,** 402–20.

55. Dr M. J. Harvey, personal communication.

56. Smith, S. G. (unpublished date). Aquatic macrophytes of the Pine and Popple river system, Florence and Forest Counties, Wisconsin.

57. Thomson, J. W., Jr (1945). A survey of the larger aquatic plants and bank flora of the Brule River. *Trans. Wisc. Acad. Sci. Arts Lett.* **36,** 57–76.

58. Smith, S. G. (1973). *Ecological studies of the surface waters of the Whitewater Creek watershed, Walworth, Rock and Jefferson Counties, Wisconsin.* University of Wisconsin Water Resources Centre, Madison.

59. Westlake, D. F. (1960). Water weed and water management. *Inst. Publ. Health Eng. J.* **59,** 148–60.

60. Haslam, S. M., Klotzli, F., Sukopp, H. & Szczepanski, A. (1979). Management. In *Ecology of wetlands*, ed. J. Kvet, A. Szczepanski & D. F. Westlake. Cambridge University Press, London. (In Press.)

61. Dr D. M. Whitcomb, personal communication.

62. Haslam, S. M. (1973). The management of British wetlands. I. Economic and amenity use. *J. environ. Manage.* **1,** 303–20.

63. Haslam, S. M. (1973). The management of British wetlands. II. Conservation. *J. Environ. Manage.* **1,** 345–61.

64. McNabb, C. D., Jr & Tierney, D. P. (1972). *Growth and mineral accumulation of submerged vascular hydrophytes in pleioeutrophic environs*. Institute of Water Resources, Michigan State University, Technical Report 26.

65. Lange, S. R. (1965). The control of aquatic plants by commercial harvesting, processing and marketing. In *Proc. 18th Sth Weed Control Conf.*, pp. 536–42.

66. Little, E. C. S. (1968). *Handbook of utilisation of aquatic plants. A compilation of the world's publications.* FAO, Rome.

67. Green, F. H. W. (1975). The effect of climatic and other environmental changes in water quality in rural areas. In *Science, technology and environmental management*, ed. R. D. Hey & T. D. Davies, pp. 123–36. Saxon House Lexington.

68. Ayreshire River Purification Board. *Fifteenth annual report, to 31 December 1968*.

69. Robson, T. O. (1974). Mechanical control. In *Aquatic vegetation and its use and control*, ed. D. S. Mitchell. UNESCO, Paris.

70. Robson, T. O. (1972). Recommendations for the control of aquatic weeds. In *Weed control handbook*, 7th edn, ed. J. Fryer & R. Makepeace. Blackwell, Oxford.

71. Robson, T. O. (1973). *The control of aquatic weeds*. Ministry of Agriculture, Fisheries and Food, Bulletin 194.

72. Ministry of Agriculture, Fisheries and Food (1975). *Code of Practice for the use of herbicides on weeds in watercourses and lakes*.

73. Sussex River and Water Division, personal communication.

74. Mr V. E. Tomlins, personal communication.

75. Robinson, G. W. (1971). Practical aspects of chemical control of weeds in land drainage channels in England and Wales. In *Proc. Eur. Weed Res. Coun., 3rd int. Symp. Aquatic Weeds 1971*, pp. 297–302.

76. Haslam, S. M. (1968). The biology of reed (*Phragmites communis*) in relation to its control. In *Proc. 9th Brit. Weed Control Conf. 1968*, pp. 392–7.

77. Dr F. H. Dawson, personal communication.

78. Mr T. O. Robson, personal communication.

79. Lohmeyer, W. & Krause, A. (1975). Über die Auswirkungen des Gehölbewuchses an kleinen Wasserläufen des Munsterlandes auf die Vegetation im Wasser und an den Böschungen im Hinblick auf die Unterhaltung der Gewasser. *Schweiz. Reihe Veg.* **9,** 105 pp.

80. Hermens, L. C. M. (1975). Levend groen. II. Groene beken in Limburg. *Tijdschrift* **86,** 473–81.

81. Haslam, S. M. (1975). River vegetation and pollution. In *Science, technology and environmental management*, ed. R. D. Hey & T. D. Davies, pp. 123–86. Saxon House/Lexington.

82. Lachavenne, J.-B. & Wattenhofer, R. (1975). *Contribution a l'étude des macrophytes du Leman. Commission Internationale pour la protection des Eaux du Leman contre la pollution*. Conservatoire Botanique du Geneva.

83. Volker, R. & Smith, S. G. (1965). Changes in the aquatic vascular flora of Lake East Okobogi in historic times. *Iowa Acad. Sci.* **72,** 65–72.

84. Hynes, H. B. N. (1960). *The biology of polluted waters*. Liverpool University Press.

85. Stuckey, R. L. (1971). Changes of vascular aquatic flowering plants during 70 years in Put-in-Bay harbor, Lake Erie, Ohio. *Ohio J. Sci.* **71,** 321–42.

86. Stuckey, R. L. & Wentz, W. A. (1969). Effect of industrial pollution on the aquatic and shore angiosperm flora in the Ottawa River, Allen and Putnam Counties, Ohio. *Ohio J. Sci.* **69,** 226–42.

87. Mr A. Churchyard, personal communication.

88. Proctor, H. G. (1971). Aquatic macrophytes in the Wheel of the Tees. *Vasculum* **56,** 59–66.

89. Bovey, R. W., Burnett, E., Richardson, C., Merkle, M. G., Baur J. R. & Knisel, W. G. (1974). Occurrence of 2, 4, 5–T and picloram in surface runoff water in the blacklands of Texas. *J. environ. Qual.* **6,** 1–4.

ADDITIONAL REFERENCES

Chapter 1

Chancellor, R. J. (1962). *The identification of common water weeds.* Bull. minist. Agric. Fish. Food, London, no. 183.

Cook, C. D. K. (1961). *Sparganium* in Britain. *Watsonia* **5,** 1–10.

Cook, C. D. K. (1966). A monographic study of *Ranunculus* subgenus *Batrachium* (DC) A. Gray. *Mitt. Bot. München* **6,** 47–237.

Fryer, A. & Bennet, A. (1900). *The* Potamogetons *of the British Isles.* L. Reeve, London.

Howard, H. W. & Lyon, A. G. (1950). The identification and distribution of the British Watercress species. *Watsonia* **1,** 228–33.

Hutchinson, E. L. (1975). *Treatise on Limnology. III. Limnological Botany.* Wiley-Interscience, New York, London.

Jones, H. (1955). Studies on the ecology of the river Rheidol. I. Plant colonisation and permanent quadrat records in the main stream of the lower Rheidol. *J. Ecol.* **43,** 462–76.

Keble Martin, W. (1965). *The concise British flora in colour.* Ebony Press/ Michael Joseph, London.

Macan, T. T. (1974). Running water. *Mitt. int. Verein. Limnol.* **20,** 301–21.

Macan, T. T. & Worthington, E. B. (1951). *Life in lakes and rivers.* Collins, London.

Oglesby, R. T., Carlson, C. A. & McCann, J. A. (1972). *River ecology and man.* Academic Press, New York & London.

Owens, M. & Whitton, B. A. (eds.) (1975). *River ecology.* Blackwells, Oxford.

Chapters 2–4

van Donselaar, J. (1961). On the vegetation of former river beds in The Netherlands. *Wentia* **5,** 1–85.

Heslop-Harrison, Y. (1955). *Nuphar* Sm. *Biol. Fl. Brit. Isles. J. Ecol.* **43,** 342–64.

Howard, H. W. & Lyon, A. G. (1952). *Nasturtium* R. Br. *Nasturtium officinale* R. Br. (*Rorippa nasturtium-aquaticum* (L.) Hayet). *Biol. Fl. Brit. Isles. J. Ecol.* **40,** 228–45.

Jones, E. N. (1931). The morphology and biology of *Ceratophyllum demersum.* *Stud. nat. Hist. Iowa Univ.* **13,** 11–33.

Laing, H. E. (1941). Effect of concentration of oxygen and pressure of water upon growth of rhizomes of semi-submerged water plants. *Bot. Gaz.* **102,** 712–24.

Meyer, B. S., Bell, F. H., Thompson, L. C. & Clay, E. I. (1943). Effect of depth of immersion on apparent photosynthesis in submerged vascular aquatics. *Ecology* **24,** 393–9.

Peltier, W. H. & Welch, E. B. (1969). Factors affecting growth of rooted aquatics in a river. *Weed Sci.* **17,** 412–16.

van der Voo, E. E. & Westhoff, V. (1961). An autecological study of some limnophytes and halophytes in the area of the large rivers. *Wentia* **5,** 163–258.

Yeo, R. R. (1965). Life history of Sago Pondweed. *Weeds 1965* **13,** 314–21.

Chapter 7

Meyer, B. S. & Heritage, A. C. (1941). Effect of turbidity and depth of immersion on apparent photosynthesis in *Ceratophyllum demersum.* *Ecology* **22,** 17–22.

Bibliography

Spence, D. H. N. (1972). Light on freshwater macrophytes. *Trans. Bot. Soc. Edinb.* **41,** 491–505.

Westlake, D. F. (1964). Light extinction, standing crop and photosynthesis within weedbeds. *Verh. int. Verein. Limnol.* **15,** 415–25.

Westlake, D. F. (1966). The light climate for plants in rivers. *Symp. Brit. ecol. Soc.* **6,** 99–119.

Chapter 8

Allenby, K. G. (1966). The manganese and calcium contents of some aquatic plants and the water in which they grow. *Hydrobiol.* **27,** 498–500.

Allenby, K. G. (1968). Some analyses of aquatic plants and water. *Hydrobiol.* **32,** 486–90.

Almestrand, A. (1951). Studies on the vegetation and hydrochemistry of Scanian Lakes. II. Ion determinations in lake waters. *Botan. Notis. Suppl.* **2,** 142–74.

Anderson, R. R., Brown, R. G. & Rappleye, R. D. (1966). The mineral content of *Myriophyllum spicatum*. L. in relation to its aquatic environment. *Ecology* **47,** 844–6.

Auerbach, S., Prufer, P. & Weise, G. (1972). CO_2–Stoffwechsel submerser Wasserpflanzen vom Typ *Fontinalis antipyretica* L. bei intermittierender Beleuchtung. *Biol. Zentralbl.* **91,** 463–75.

Bombówna, M. (1971). Sklad chemiczy wody potoków Polskich Tatr Wysokick ze szczególnym uwzgledneiniem Suchej Wody. (The chemical composition of the water of streams of the Polish High Tatra Mts, particularly with regard to the Stream Sucha Woda.) *Acta Hydrobiol.* **13,** 379–91.

Brink, N. & Widell, N. (1967). Eutrophication in a small stream in Central Sweden. *Schweiz. Z. Hydrol.* **29,** 336–60.

Bristow, J. M. (1909). The effects of carbon dioxide on the growth and development of amphibious plants. *Can. J. Bot.* **47,** 1803–7.

Butcher, R. W., Pentelow, F. T. K. & Woodley, J. W. A. (1930). Variations in composition of river waters. *Int. Revue ges. Hydrobiol. Hydrogr.* **24,** 47–80.

Carr, J. (1969). The primary productivity and physiology of *Ceratophyllum demersum*. 2. Microprimary productivity, pH and the P/R ratio. *Aust. J. mar. freshwater Res.* **20,** 127–42.

Casey, H. & Newton, P. V. R. (1973). The chemical composition and flow of the River Frome and its main tributaries. *Freshwater Biol.* **3,** 337–53.

Cowles, R. P. & Schwitalla, A. M. (1923). The hydrogen-ion concentration of a creek, its waterfall, swamp and ponds. *Ecology* **4,** 402–15.

Crisp, D. J. (1970). Input and output of minerals from a small Watercress bed fed by chalk water. *J. appl. Ecol.* **7,** 117–40.

Dijkshoorn, W., Sujitno, J. S. A. & Ismunadji, M. (1974). Potassium uptake by rice plants and interaction with other cations. *Plant Soil* **40,** 525–34.

Edwards, R. W. & Owens, M. (1962). The effects of plants on river conditions. IV. The oxygen balance of a chalk stream. *J. Ecol.* **50,** 207–20.

Edwards, R. W. & Owens, M. (1965). The oxygen balance of streams. *Symp. Brit. ecol. Soc.* **5,** 149–72.

Haslam, H. W. (1975). Geochemical survey of stream waters and stream sediments from the Cheviot area. *Rep. Inst. Geol. Sci.* 75/6.

Hutchinson, G. E. (1970). The chemical ecology of three species of *Myriophyllum* (Angiospermae, Haloragaceae). *Limnol. Oceanogr.* **15,** 1–5.

Köhler, A., Wonneberger, R. & Zeltner, G. (1973). Die Bedeutung chemischer und pflanzlicher 'Verschmutzungsindikatoren' im Fliessgewassersystem Moosach (Münchener Ebene). (Chemical data and aquatic vascular plants as indicators for pollution in the Moosach river system near Munich). *Arch. Hydrobiol.* **72**, 533–49.

Laurie, E. M. O. (1942). The dissolved oxygen of an upland pond and its inflowing stream, at Ystumtiun, North Cardiganshire, Wales. *J. Ecol.* **30**, 357–81.

Moss, B. (1976). The effects of fertilization and fish on community structure and biomass of aquatic macrophytes and epiphytic algal populations: an ecosystem experiment. *J. Ecol.* **64**, 313–42.

Owens, M. & Edwards, R. W. (1963). Some oxygen studies in the River Lark. *Proc. Soc. Water Treat. Exam.* **12**, 126–44.

Seddon, B. (1967). The lacustrine environment in relation to macrophytic vegetation. In *Quarternary palaeoecology*, ed. E. J. Cushing & H. E. Wright, pp. 205–15. Yale University Press, New Haven & London.

Stake, E. (1967). Higher vegetation and nitrogen in a rivulet. *Schweiz. Z. Hydrol.* **29**, 107–24.

Stake, E. (1968). Higher vegetation in a small stream in Central Sweden. *Schweiz. Z. Hydrol.* **30**, 353–73.

Thomas, G. W. & Crutchfield, J. D. (1974). Nitrate-nitrogen and phosphorus contents of streams draining small agricultural watersheds in Kentucky. *J. environ. Qual.* **3**, 46–9.

Westlake, D. F. (1961). Aquatic macrophytes and the oxygen balance of running water. *Int. Ver. theor. angew. Limnol. Verh.* **15**, 499–504.

Westlake, D. F. (1966). A model for quantitative studies of photosynthesis by higher plants in streams. *Int. J. Air Water Pollut.* **10**, 883–96.

Wium-Andersen, S. (1971). Photosynthetic uptake of free CO_2 by the roots of *Lobelia dortmanna*. *Physiol. Plant.* **25**, 245–8.

Chapter 9

Carr, J. L. (1969). The primary productivity and physiology of *Ceratophyllum demersum*. I. Gross macroprimary productivity. *Aust. J. mar. freshwater Res.* **20**, 115–26.

Edwards, R. W. & Owens, M. (1960). The effects of plants on river conditions. I. Summer crops and estimates of net productivity of macrophytes in a chalk stream. *J. Ecol.* **48**, 151–80.

Mathews, C. P. & Kowalczewski, A. (1960). The disappearance of leaf litter and its contribution to production in the river Thames. *J. Ecol.* **57**, 543–52.

Owens, M. & Edwards, R. W. (1961). The effects of plants on river conditions. II. Further crop studies and estimates of net productivity of macrophytes in a chalk stream. *J. Ecol.* **49**, 119–26.

Owens, M. & Edwards, R. W. (1962). The effects of plants on river conditions. III. Crop studies estimates of net productivity of macrophytes in four streams in southern England. *J. Ecol.* **50**, 157–62.

Westlake, D. F. (1965). Some basic data for investigations of the productivity of aquatic macrophytes. *Mem. Ist. ital. Idrobiol.* **18**, 229–48.

Westlake, D. F. (1900). The weight of water-weed in the river Frome. In *Association of River Authorities Year Book*, pp. 1–12.

Chapter 10

Ladle, M. & Casey, H. (1971). Growth and nutrient relationships of *Ranunculus penicillatus* var. *calcareus* in a small chalk stream. In *Proc. Eur. Weed Res. Count. 3rd int. Symp. Aquatic Weeds*, pp. 53–64.

Chapters 11–13

Farrar, C. F. (1921). *Ouse's silent tide*. Sydney Press, Bedford.

Bibliography

Gehu, J. M. (1961). Les groupements végétaux du Bassin de la Sambre Française. I. **10**, 69–148.

Jones, H. (1955). Studies on the ecology of the river Rheidol. I. Plant colonisation and permanent quadrat records in the main stream of the lower Rheidol. *J. Ecol.* **43**, 462–76.

Jones, H. (1956). Studies on the ecology of the river Rheidol. II. An ox-bow of the lower Rheidol. *J. Ecol.* **44**, 12–27.

Kohler, von A., Vollrath, H. & Beisl, E. (1971). Zur Verbreitung, Vergesellschaftung und Okologie der Gegass-makrophyten im Fliesswassersystem Moosach (Munchener Ebene). (The distribution, the phytosociological composition and the ecology of the vascular macrophytes in the Moosach river system near Munich.) *Arch. Hydrobiol.* **69**, 333–65.

Malmer, N. (1960). Some ecologic studies on lakes and brooks in the south Swedish uplands. *Botan. Notis.* **113**, 87–116.

Pekkari, S. (1965). The north east corner: notes on aquatic vegetation. *Acta Phytogeog. Suec.* **50**, 209–14.

Quennerstedt, N. (1965). The major rivers of Northern Sweden. *Acta Phytogeogr. Suec.* **50**, 198–204.

Schneider, von S., Dewes, E., Krause, A., Kroesch, U., Lorenz, D. & Miosaga, G. (1974). *Gewasseruberwachung durch Fernerkundung die mittlere Saar*, pp. 49–70. Selbstverlag der Bundesforschungsanstalt für Landeskunde und Raumordnung, Bonn–Bad Godesberg.

Weber-Oldecop, D. W. (1970). Wasserpflanzengesellschaften im östlichen Niedersachsen. I. *Int. Revue ges. Hydrobiol.* **55**, 913–67.

Weber-Oldecop, D. W. (1971). Wasserpflanzengesellschaften im östlichen Niedersachsen. II. *Int. Revue ges. Hydrobiol.* **56**, 79–122.

Chapters 15–17

Aiken, S. & Gillett, J. M. (1974). The distribution of aquatic plants in selected lakes in Gatineau Park, Quebec. *Can. Field Nat.* **88**, 437–48.

Countryman, W. D. (1968). *Alisma gramineum* in Vermont. *Rhodora* **70**, 577–9.

Dore, W. G. & Gillett, J. M. (1955). *Botanical survey of the St Lawrence Seaway area in Ontario.* Botany and Plant Pathology Division, Science Survey, Canada. Department of Agriculture, Ottawa, Ontario.

Fassett, N. C. (1957). *A manual of aquatic plants.* (With Revision Appendix by Eugene C. Ogden.) University of Wisconsin Press, Madison.

Haynes, R. R. (1974). A revision of North American *Potamogeton* subsection *Pusilli* (Potamogetonaceae). *Rhodora* **76**, 564–649.

Lindsay, A. A., Petty, R. O., Sterling, D. K. & Van Asdall, W. V. (1961). Vegetation and environment along the Wabash and Tippecanoe rivers. *Ecol. Monogr.* **31**, 105–56.

Chapter 18

Edwards, R. W. (1968). Plants as oxygenators in rivers. *Water Res.* **2**, 243–48.

Gaevskaya, N. S. (1966; transl. 1969). *The role of higher aquatic plants in the nutrition of the animals of freshwater basins.* Nauka, Moscow. (National Lending Library for Science and Technology, Boston Spa, Yorks.)

Haslam, S. M. (1972). *The Reed.* Norfolk Reed Growers Association, Norwich.

Krull, J. N. (1970). Aquatic plant–macroinvertebrate associations and water fowl. *J. Wildlife Manage.* **34**, 707–18.

Lindstrom, H. V. & Sandstrom, W. M. (1939). Nutritive value and chemical composition of certain freshwater plants of Minnesota. III. The nature of the carbohydrates of species of *Elodea, Myriophyllum, Ceratophyllum, Ruppia* and *Ranunculus. Min. Agric. Exp. Stn. Techn. Bull.* **136,** 43–7.

Sansom, C. H. (1947). *Cultivation of watercress.* Bull. Minist. Agric. Fish., no. 136.

Sawyer, F. (1952). *Keeper of the stream.* A. & C. Black, London.

Westlake, D. F. (1959). The effects of organisms on pollution. *Proc. limnol. Soc. Lond.* **170,** 171–2.

Westlake, D. F. (1960). Water weed and water management. *Inst. Publ. Health Eng. J.* **59,** 148–60.

Chapter 20

Queenerstedt, N. (1958). Effect of water level fluctuation on lake vegetation. *Verh. int. Verein. Limnol.* **13,** 901–6.

Chapter 21

Barrett, P. R. F. & Robson, T. O. (1974). Further studies on the seasonal changes in the susceptibility of some emergent plants to dalapon. In *Proc. 12th Brit. Weed Control Conf.,* pp. 249–53.

Brooker, M. P. & Edwards, R. W. (1973). Effects of the herbicide paraquat on the ecology of a reservoir. I. Botanical and chemical aspects. *Freshwater Biol.* **3,** 157–75.

Chancellor, A. P. (1958). *The control of aquatic weeds and algae.* HMSO, London.

Grizzell, R. A. & Neely, W. W. (1962). Biological controls for waterweeds. In *Trans. 27th N. Amer. Wildl. Nat. Resour. Conf.,* pp. 107–13.

Ogg, A. G., Bruns. V. F. & Kelly, A. D. (1969). Response of Sago Pondweed to periodic removal of top growth. *Weed Sci.* **17,** 139–41.

Robbins, W. W., Crofts, A. S. & Raynor, R. N. (1942). *Weed control.* McGraw-Hill, New York & London.

Robinson, G. W. & Leeming, J. B. (1969). The experimental treatment of some waters in Kent with diuron to control aquatic weed growth. In *Association of River Authorities Year Book 1969,* pp. 58–63.

Robson, T. O. (1973). Recent trends in weed control in freshwater. *OEPP/ EPPO* **3,** 5–17.

Robson, T. O. & Fearon, J. H. (1976). *Aquatic herbicides.* British Crop Protection Council, London.

Stott, B. & Robson, T. O. (1970). Efficiency of Grass Carp (*Ctenopharyngodon idella* Val) in controlling submerged water weeds. *Nature, Lond.* **226,** 870

Thompson, T. W. (1972). Control of Water Milfoil in Wisconsin. In *Meeting Weed Sci. Soc. America, St Louis,* pp. 61–2 (abstr.).

Chapter 22

Agami, M. (1973). Effects of water pollution on the flora and vegetation of the Alexander and Yarkon rivers. Abstract of MSc Thesis, Tel-Aviv University.

Alabaster, J. S., Garland, J. H. N., Hart, I. C. & Solbé, J. F. de L. G. (1972). An approach to the problem of pollution and fisheries. *Symp. zool. Soc. Lond.* **29,** 87–114.

Anon. (1960). The control of aquatic and marginal weeds. In *Abstr. Meet. Weed Soc. Amer., 1960,* pp. 62–8.

Carpenter, K. E. (1926). The lead mine as an active agent in river pollution. *Ann. appl. Biol.* **13,** 395–401.

Cope, O. B. (1966). Contamination of the freshwater system by pesticides. In *Pesticides in the environment and their effects on wildlife*, ed. N. W. Moore. *J. Appl. Ecol. 3 (Suppl.)* 33–44.

Department of the Environment (1971). *Report of a river pollution survey of England and Wales*, vol. 1. HMSO, London.

Devon River Authority (1968). *Annual report for the year ended 31 March 1968.*

Edwards, A. M. C. (1975). Long term changes in the water quality of agricultural catchments. In *Science, technology and environmental management*, ed. R. D. Hey & T. D. Davies, pp. 117–22. Saxon House & Lexington, Farnborough & Lexington.

Edwards, D. (1969). Some effects of siltation upon aquatic macrophyte vegetation in rivers. *Hydrobiol.* **34**, 29–36.

Ericksen Jones, J. R. (1940). A study of the zinc-polluted river Ystwyth in North Cardiganshire, Wales. *Ann. appl. Biol.* **27**, 368–78.

Forsberg, C. (1964). The vegetation changes in Lake Takern. *Svensk bot. Tidskr.* **58**, 44–55.

Gorham, E. & Gordon, A. (1963). Some effects of smelter pollution upon aquatic vegetation near Sudbury, Ontario. *Can. J. Bot.* **41**, 371–8.

Grahn, O., Hultberg, H. & Landner, L. (1974). Oligotrophication – a self-accelerating process in lakes subjected to excessive supply of acid substances *Ambio* **3**, 93–4.

Haller, W. T., Sutton, D. L. & Barlowe, W. C. (1974). Effects of salinity on growth of several aquatic macrophytes. *Ecology* **55**, 891–4.

Hayes, S. A. (1967). The herbicidal activity of various organic acids on the growth of *Elodea canadensis*. In *Proc. 20th Sth Weed Control Conf.*, pp. 294–7.

Houghton, G. V. (1964). The River Stour (Essex and Suffolk) – hardness, chloride and nitrate content. *Proc. Soc. Water Treat. Exam.* **13**, 145–52.

Houghton, G. V. (1972). Long-term increases in planktonic growth in the Essex Stour. *Proc. Soc. Water Treat. Exam.* **21**, 299–308.

Hynes, H. B. N. (1965). A survey of water pollution problems. *Symp. Brit. ecol. Soc.* **5**, 49–63.

Hynes, H. B. N. & Roberts, F. W. (1962). The biological effects of synthetic detergents in the River Lee, Hertfordshire. *Ann. appl. Biol.* **50**, 779–90.

Klausner, S. D., Zwerman, P. J. & Ellis, D. F. (1974). Surface runoff losses of soluble nitrogen and phosphorus under two systems of soil management. *J. environ. Qual.* **3**, 42–6.

Krause, A. (1972). Einfluss der Eutrophierung auf die Makrophytenvegetation der Oberflächengewässer. *Ber. Landwirtsch.* **50**, 140–6.

Lewis, K. (1973). The effect of suspended coal particles on the life forms of the aquatic moss *Eurhynchium riparioides* (Hedw.). I. The gametophyte plant. *Freshwater Biol.* **3**, 251–7.

Litar, M. & Agami, M. 1976. Relationship between water pollution and the glaa of two coastal rivers of Israel. *Aquat. Bot.* **2**, 23–41.

Morgan, N. C. (1970). Changes in the fauna and flora of a nutrient-enriched lake. *Hydrobiol.* **35** 545–53.

Newton, L. (1944). Pollution of the rivers of West Wales by lead and zinc mine effluent. *Ann. appl. Biol.* **31**, 1–72.

Nicholls, K. H. (1974). Nutrients in subsurface and runoff waters of the Holland Marsh, Ontario. *J. environ. Qual.* **3**, 31–5.

Nuttall, P. M. (1972). The effects of sand deposition upon the macroinvertebrate fauna of the river Camel, Cornwall. *Freshwater Biol.* **2**, 181–6.

Olsen, S. (1964). Vegetations aendringer i Lyngby Sø Bidragtilanalyse af kulturpåvirkninger på vand- og sumpplantevegetationem. *Bot. Tidsskr.* **59**, 273–300.

Owens, M., Garland, J. H. N., Hart, I. C. & Wood, G. (1972). Nutrient budgets in rivers. *Symp. zool. Soc. Lond.* **29**, 21–40.

388

Owens, M. & Wood, G. (1968). Some aspects of the eutrophication of water. *Water Res.* **2,** 151–9.

Pasternak, K. (1973). Rozprzestrzonienie metali cìezkich w wodach plynach w rejionie wystepowania naturalnych zlóz oraz przemyslu cynku i olowiu. (The spreading of heavy metals in flowing water in the region of occurrence of natural deposits and of the zinc and lead industry.) *Acta Hydrobiol.* **15,** 145–66.

Römkens, M. J. M. & Nelson, D. W. (1974). Phosphorus relationships in runoff from fertilised soils. *J. environ. Qual.* **3,** 10–31.

Thornton, I., Watling, H. & Darracott, A. (1975). Geochemical studies in several rivers and estuaries used for oyster rearing. *Sci. Total Environ.* **4,** 325–45.

Usk River Authority (1968). *Third Annual Report for the year ended 31 March 1968.*

Westlake, D. F. (1959). The effects of biological communities on conditions in polluted streams. *Symp. Inst. Biol.* **8,** 25–31.

Index